The Safe Handling of Chemicals in Industry

Volume 1

The Safe Handling of Chemicals in Industry

P. A. CARSON M.Sc., Ph.D., A.M.C.T., C. Chem., F.R.S.C., M.I.O.S.H
Safety Liason Manager, Unilever Research, Port Sunlight
C. J. MUMFORD B.Sc., Ph.D., D.Sc., C. Eng., M.I. Chem. E.
Senior Lecturer in Chemical Engineering, University of Aston

Volume 1

Copublished in the United States with
John Wiley & Sons, Inc., New York

Longman Scientific & Technical
Longman Group UK Limited
Longman House, Burnt Mill, Harlow,
Essex CM20 2JE, England
and Associated Companies throughout the world

*Copublished in the United States with
John Wiley & Sons, Inc., 605 Third Avenue, New York, NY 10158*

© P. A. Carson and C. J. Mumford 1988

All rights reserved; no part of this publication may be reproduced, stored in a retrieval system, or transmitted in any form or by any means, electronic, mechanical, photocopying, recording, or otherwise without either the prior written permission of the Publishers or a licence permitting restricted copying in the United Kingdom issued by the Copyright Licensing Agency Ltd, 33–34 Alfred Place, London, WC1E 7DP.

First published 1988

British Library Cataloguing in Publication Data
Carson, P. A.
  Safe handling of chemicals in industry
  1. Hazardous substances – Safety measures
  I. Title   II. Mumford, C. J.
  363.1'79   TP149

  ISBN 0-582-00304-0

Library of Congress Cataloging-in-Publication Data
Carson, P. A., 1944–
  Safe handling of chemicals in industry / P. A. Carson, C. J. Mumford.
     p.   cm.
  Includes index.
  ISBN 0-470-20886-4 (Wiley)
  1. Hazardous substances – Safety measures.   I. Mumford, C. J.
II. Title.
T55.3.H3C37   1988
604.7 – dc 19

Set in Linotron 202 10/12 pt Times

Printed and bound in Great Britain
at the Bath Press, Avon

Contents

Preface		vi
Acknowledgements		viii

Volume I
1	Introduction	1
2	General occupational safety	9
3	Physico-chemical principles and safety	75
4	Flammable materials handling	122
5	Toxicology	185
6	Control measures for handling toxic substances	269
7	Ambient air analysis for hazardous substances	338
8	Radioactive chemicals	444
9	Safety with chemical engineering operations	463

Volume II
10	Hazardous chemical processes	491
11	Safety in laboratories and pilot plants	533
12	Safety in chemical-process plant design and operation	653
13	Safety in marketing and transportation of chemicals	769
14	Waste disposal	821
15	Control of hazards from large-scale installations	864
16	Legislative controls	915
17	Management of safety and loss prevention	966
18	Chemical safety information	1031
	Index	1071

Preface

This text is intended to assist in the safe use of chemicals whether in the laboratory or factory, in offices or workshops, in schools and colleges, on the land, or elsewhere. While it should be of value to anyone associated with the chemical and process industries, the breadth of coverage is aimed in particular at the technical user of chemicals who is not involved in their manufacture and therefore not necessarily aware of their properties and safe handling procedures. Thus the audience includes plant operators, technicians and their line management plus occupational medical officers and their nursing staff, safety officers, chemists, hygienists, chemical engineers, safety representatives and staff with responsibility for health and safety in laboratories and pilot plant in industry and in academic institutions. It should also be of assistance to students of occupational safety and health in the broadest sense. Although the reader is introduced to the hazards and control of radioactive chemicals, the book does not cover the special risks associated with the nuclear industry.

While some reference is necessarily made to UK legislation and codes of practice, the aim throughout is to identify hazards commonly arising in chemicals' storage, transport, handling and disposal; and to summarise measures which result in safe plant, a safe working environment and a safe system of work. It represents best current practice at the time of compilation in the mid 1980s. Thus, although there will inevitably have been some recent changes in legislation, and even advances in technology, the general principles remain unchanged and are not restricted in anyway to the UK.

Brief cameo case histories of incidents are used to exemplify some problems which can arise with chemicals, engineering and system-of-work failures, etc. Presented in a typeface different to text typeface they are included to help relate theory and principles to practice and to be useful, together with other case histories given in the sources referred to in Chapter 18, in the preparation of training materials.

Where possible the examples chosen for these case histories, and to explain fundamental principles, are restricted to a relatively small selection of common, simple chemicals and processes. These are supplemented by additional examples in numerous tables to increase the reader's awareness of the scope. Even so, while the book contains a significant amount of information in tabular form this is intended to be illustrative rather than encyclopaedic. Thus the absence of data in a table, e.g. on a specific chemical, should not be interpreted necessarily as having significance and the reader must consult the literature (such as the sources listed in Ch. 18) for information relating to specific enquiries.

The main thrust is to describe techniques for the recognition of chemical hazards and to quantify and control the risks so that chemicals can be handled safely in practice. The hazards, preventive measures, legislation, and sources of information discussed are based upon current knowledge. So far as the general fire and explosion characteristics of common chemicals are concerned, except where very large vapour clouds are possible, this is reasonably well established. With regard to the toxicity of chemicals there tends to be more uncertainty, particularly when considering chronic effects or the possibility of disease following long latency periods, for example with chemical carcinogens. Thus occupational exposure limits/threshold limit values, from which 'permissible' levels of workplace exposure are derived, are constantly under review. It is a basic tenet of chemicals handling, as discussed in detail in the appropriate chapters, that occupational exposures and releases to the environment generally should in all cases be as low as is reasonably practicable.

Acknowledgements

Books, published papers and other sources of information which have been drawn upon have generally been referenced in the text. However, we also acknowledge those sources, e.g. of case histories, which have – perhaps understandably in a book of this size and scope, but quite unintentionally – escaped due reference.

We thank numerous colleagues within Unilever and Aston University for helpful discussions during the preparation stages. In particular, recognition is due to those who took time to read and offer constructive comments on different parts of the text, namely Messrs N. Burak, G. R. Cooper, G. R. Cliffe, S. Evans, A. Jarman, J. G. Lyons, and Drs D. M. Roberts, M. J. How and T. C. Pestell (all of Unilever) and R. C. Keen of Bristol Polytechnic. Gratitude is also expressed to the Management of Unilever Research Laboratory for generously making available typing and administrative facilities.

Finally we express special thanks to our respective families without whose encouragement this book could not have been written.

P. A. C.
C. J. M.

CHAPTER 1

Introduction

The hazards in modern industry are mechanical, electrical, energy radiation (ionising and non-ionising), biological, ergonomical and physical. In addition, in numerous operations and manufacturing processes, there are hazards arising from the use of 'chemicals'. Thus in the first column of Table 1.1 the most common causes of lost-time accidents in industry generally are listed in descending order.[1]

Table 1.1 Common causes of industrial accidents

Accident cause	Total reported accidents for factory premise (%)[1]	Chemical laboratory/pilot plant accidents (requiring first-aid treatment) (%)[2]
Handling	28.7	27
Others	24.6	—
Machinery	15.6	1
Slipping and falling	15.7	17
Striking against	8.4	14
Struck by	6.0	3
Chemicals	1.0*	38†

* Fires or explosions of combustible material, poisonings, gassings, or explosion of molten metal.
† All chemical accidents including acid splashes, gassings, etc.

However, what constitutes a 'chemical' accident may be ill-defined and furthermore, as discussed in Chapter 5, prolonged or intermittent exposure to certain chemicals can result in an occupational disease, e.g. dermatitis, a pneumoconiosis or systemic poisoning.

The hazards of chemicals stem from their inherent flammable, explosive, toxic, corrosive, radioactive, carcinogenic, or chemical reactive properties, or a combination thereof.

> A fire started in a warehouse in which were stored large quantities of combustibles such as charcoal, wickerware, wood and garden furniture

together with 10,700 tonnes of bagged ammonium nitrate, compound fertilisers and potassium nitrate. The warehouse was destroyed and corrugated iron and roofing material were projected into neighbouring premises by the explosions which occurred. Dense clouds of smoke containing toxic fumes drifted over an area of 1 km^2 which led to the temporary evacuation of more than 750 of the surrounding population downwind.[3] This incident illustrates the risks associated with a combination of hazards such as fire, explosion, exposure to toxic chemicals and the dangers arising when incompatible materials are mixed.

Safety, therefore, must be a primary consideration whenever chemicals are manufactured, stored, transported, used or disposed of. Otherwise the producer, transporter, consumer, third parties such as contractors, intermediate handlers (e.g. dockers, ship's personnel) or innocent bystanders may be put at risk. The main thrust of this book is to describe how chemicals can be used in industry with minimal risk of accident or disease by a consideration of hazard recognition, risk evaluation and appropriate control measures.

The second column in Table 1.1 lists causes of accidents requiring first-aid attention for one large, typical, research laboratory. Figures are based on data collected over a 10-year period.[2] This demonstrates the importance of attention to chemical handling in avoiding accidents in a chemical/chemical engineering pilot-plant/laboratory complex involving small-scale experimentation and production – as distinct from large-scale manufacturing processes – and hence increased opportunity for exposure to chemicals. Because of the particular hazards associated with laboratory work, and since laboratories of varying size and sophistication are so commonplace and represent a microcosm of industry in general, the handling of chemicals at the bench-scale is discussed fully in Chapter 11. However, 'chemicals' are widely used in agriculture; in offices, laboratories, shops and engineering workshops; in the home; in manufacturing and process industries and in medicine. They are also generated either inevitably, or accidentally, in processes such as welding, effluent treatment, metal treatment, combustion and chemical cleaning. In all these situations the ability to identify any potential exposures to toxic chemicals or potentially hazardous situations, e.g. due to a possible fire or explosion, or pressure release, or arising from the use of 'scientific' apparatus, e.g. lasers, is a prerequisite for control. Such control is based, wherever possible, on quantification of the exposure/risk.

Chapter 3 outlines basic physico-chemical principles to assist in predicting whether hazardous conditions are likely to develop in a particular situation. Since the monitoring of atmospheric concentrations of toxic or flammable substances is essential in quantifying and control of risk from these materials a complete chapter is devoted to this.

Reduction of risk is not limited to occupational safety and health considerations but extends into the realms of consumer safety, incorporating such aspects as packaging, provision of information, testing and quality control. This subject is discussed in Chapter 13. It also encompasses loss prevention, i.e. the protection of a company's or an individual's assets (e.g. plant, building, materials) from damage or deterioration. This relates to the minimisation and control of fire and explosion, toxic release, materials spillage or escape, etc. Considerable losses are incurred each year due to fire. For example, in the UK alone, in 1983 losses in the chemical and allied industries (including textiles and petrochemicals) amounted to around £30 million[4] and the majority are avoidable on any assessment by the use of reasonably practicable precautions. The hazards of flammable chemicals and precautions for their control are covered in Chapter 4 and particularly hazardous processes are considered in Chapter 10.

Furthermore, the manner in which a chemical is used, and the control measures inside the workplace, should not give rise to a hazard to the outside environment. This necessitates consideration of (Ch. 14):
1. Effluent discharges.
2. Methods of solid and liquid waste collection, treatment and disposal (including transportion).
3. Emissions to atmosphere, from the processes and from exhaust ventilation and gas treatment plant.

It is also necessary to allow for the consequences of 'inadvertent' discharges, for example from pressure-relief systems, blow-down systems and flare stacks. Inadvertent contamination with chemicals by transfer, for example on footwear or overalls or by the wheels of vehicles, is a further consideration.

Very considerable benefits have accrued from chemicals, e.g. in drugs, fertilisers, foods, pesticides, plastics, etc., but, due partly to the complexity and rapid growth rate of the chemical industry, increasing concern has been expressed in recent years about a variety of possible health and safety problems. Thus, in the past ten years 4 million new chemicals have been prepared and identified (though more than 75% are cited but once in the literature). In 1950 the total world production of synthetic organic chemicals was 7 million t per annum which rose to 63 million t by 1970. By extrapolation of this growth rate it has been estimated that the total annual world production will be 250 million t by 1987.[5] With annual sales of around £17,000 m, the UK chemical industry is fifth largest in the world (excluding the Eastern Bloc).[6] It has a very good safety record overall, for example the FAFR (fatal accident frequency rate) of 4 is as shown in Table 1.2, comparable with that of UK industry as a whole.[7,8]

However, the search for improved safety and health standards in the manufacture and use of chemicals follows in the wake of increasing knowledge. For example Table 1.3 summarises estimated rates of diseases attributed to different types of chemical or physical exposure.[8]

Table 1.2 Fatal accident frequency rates (FAFR)* for various occupations and activities[7,8]

Occupation/activity	FAFR
British industry (i.e. all premises covered by the Factories Act)	4
Clothing and footwear	0.15
Vehicles	1.3
Timber, furniture, etc.	3
Metal manufacture, shipbuilding	8
Agriculture	10
Coal mining	40
Railway shunters	45
Construction erectors	67
Staying at home (men 16–65)	1
Travelling by train	5
Travelling by car	57

* FAFR $\equiv$ number of fatalities in 10^8 man-hours $\equiv$ number of deaths from industrial injury in a group of 1,000 workers during their working lives.

Table 1.3 Estimated rates of fatality (or incidence) of disease attributed to types of chemical or physical exposure[8]

Occupation	Cause of fatality	FAFR
Shoe industry (press and finishing rooms)	Nasal cancer	6.5
Printing trade workers	Cancer of the lung and bronchus	10
Workers with cutting oils		
Birmingham	Cancer of the scrotum	3
Arve District (France)	Cancer of the scrotum	20
Wood machinists	Nasal cancer	35
Uranium mining	Cancer of the lung	70
Coal carbonisers	Bronchitis and cancer of the bronchus	140
Viscose spinners (ages 45–64)	Coronary heart disease (excess)	150
Asbestos workers		
Males, smokers	Cancer of the lung	115
Females, smokers	Cancer of the lung	205
Rubber mill workers	Cancer of the bladder	325
Mustard-gas manufacturing (Japan 1929–45)	Cancer of the bronchus	520
Cadmium workers	Cancer of the prostate (incidence values)	700
Amosite asbestos factory	Asbestosis	205
	Cancer of the lung and pluera	460
Nickel workers	Cancer of the nasal sinus	330
(employed before 1925)	Cancer of the lung	775
β-Naphthylamine manufacturing	Cancer of the bladder	1,200

Improvements in working practices and conditions during the last decade are not, however, reflected in these data.

Processes and products involving vinyl chloride, isocyanates and potentially carcinogenic hair dyes are examples of where the risks have recently been questioned. The hazards associated with asbestos and aromatic amines have become more widely acknowledged. The adverse effects of chemicals on health are described in Chapters 5 and 8. General strategies for handling toxic substances are given in Chapter 6 with a brief introduction in Chapter 8 to the special precautions for using radioactive substances.

In the UK this upsurge in interest in the control of safety with chemicals covered the period in which the Flixborough disaster occurred and in which the Roben's report and The Health and Safety at Work, etc., Act 1974 were published. Sometimes concern regarding the risk is confounded and confused by emotional and political aspects and it is important that these factors be identified. Risk is associated with all activities in life, including work, and nowadays the concept of an occupational health strategy is based on levels of risk which are *acceptable*. However, debate surrounds these 'levels' of acceptability and the mechanisms adopted for arriving at such values.

In a limited number of cases individual 'chemical' plants, or storages of, for example, liquefied petroleum gas, are so large as to represent a credible – albeit demonstrably small – risk to other factories, houses, and in a few situations people, in the immediate vicinity. These special risks are considered in Chapter 15.

Nowadays, a more scientific and sophisticated approach must be adopted for assessing danger than in the past. In assessing the level of risk two factors must be considered:
1. The probability of an accident, or over-exposure to a chemical in the form of solid, liquid or vapour occurring.
2. The likely consequences of such a mishap.

Clearly, therefore, before considering control measures it is necessary to recognise the potential hazard and, if possible, to quantify it. Furthermore with chemicals-related accidents, as with other types of accident, experience suggests that a relationship will exist between the frequency of incidents causing minor losses and the losses or fatalities.[9] This is exemplified in Fig. 1.1. The corollary is that measures taken during design or operation to minimise the number of small incidents/exposures or 'near misses' will also tend to reduce the frequency of major incidents. Safety with chemical engineering operations is discussed in Chapter 9 while the design and operation of chemical process plants is covered in Chapter 12.

In developed countries management, and indeed each individual, will have common law and statutory duties to comply with as regards safety

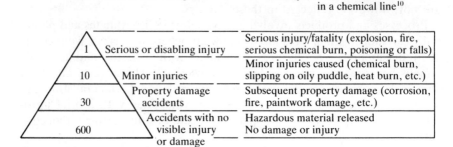

Fig. 1.1 Relationship between non-injury/non-damage incidents and more serious accidents

with chemicals. The common law duties of an employer, which in the UK are also statutory requirements under The Health and Safety at Work Act 1974, are to provide a safe place of work, information, training, proper plant and appliances and a safe system of work.

While a brief review of UK and US legislation is given in Chapter 16 and occasional reference is made to it elsewhere, the general approach in this text is to explain the main hazards associated with 'chemicals', interpreted in a broad sense, and either to summarise best current practice to minimise these hazards (so far as is practicable) and/or to refer to where this is described in detail. The recommendations are hence universal in application, although of course in specific cases it is always necessary to check for any relevant legislation.

Since chemicals are not used in isolation other hazards coexist (e.g. due to electricity, machines, material movement) and these are dealt with briefly in Chapter 2.

Clearly co-operation between all interested parties is the cornerstone of a successful safety programme. However, as described in Chapter 17, ultimate responsibility for minimising risk and preventing loss rests with management. This necessitates the setting up of safe systems of work and the training of employees with regard to both the operation of such systems and the consequences of non-compliance.

Because of the breadth of safety and health technology, the coverage of certain areas in the text has been restricted to basic information. Indeed a substantial volume of information in some areas – such as radioactive hazards, specific hazardous chemical reactions, loss prevention measures related to specific industries, the detailed analysis of potential bulk chemical hazards, etc. – had to be omitted.

Moreover, except in Chapter 14 on waste disposal and in Chapter 13, with regard to chemicals testing, discussion of the impact of chemicals upon the environment – from either normal occasional or prolonged use, or misuse, of toxic or non-biodegradable chemicals – has been restricted.

Therefore, in addition to full referencing throughout the text, a final chapter deals with sources of chemical safety information.

References

(*All places of publication are London unless otherwise stated*)
1. King, R. and Magid, J., *Industrial Hazard and Safety Handbook*. Newnes-Butterworth 1979 (Appendix C).
2. Private Communication.
3. Health and Safety Executive, *Fire at Cory's Warehouse, Toller Road, Ipswich, 14 October, 1982*. HSE report, HMSO Mar 1985.
4. Perry, B., *Fire Prevention*, 1985 (177), 12.
5. Langley, E., *Chem. Ind.*, 1978 (14), 504.
6. Chemicals Industries Association.
7. Kletz, T. A., *Hazard Analysis – A Quantitative Approach to Safety, Instn Chem. Engrs Symp. Series*, No. 34, 1971.
8. Kletz T. A., 'The application of hazard analysis to risks to the public at large', in Koctsier (ed.) *World Congress of Chemical Engineering on Chemical Engineering in a Changing World*. Elsevier, Amsterdam, 1976.
9. Heinrich, H. W., *Industrial Accident Prevention*. 4th ed. McGraw-Hill, New York, 1959.
10. Ling, K. C., 'Loss control – a practical application of the Health and Safety at Work etc., Act 1974', in *Process Industry Hazards, Accidental Release Assessment Containment and Control. Instn Chem. Engrs Symp. Series*, No. 47, 1976.

CHAPTER 2

General occupational safety

Common causes of industrial accidents and diseases include those associated with:
- Fire
- Chemicals
- Physical dangers associated with slips and falls, material handling, and the use of machines
- Sources of electricity
- Exposure to non-ionising radiation such as heat, light and noise, and to ionising radiation.

Fire, chemical, and ionising radiation hazards and their control measures are dealt with elsewhere in the book. This chapter concentrates on other sources of occupational accidents. A detailed treatise is inappropriate but reference is made to more general texts on occupational health and safety,[1,2] and to other works cited at the end of the chapter.

Fall hazards

Falls account for a significant proportion of all industrial accidents as illustrated by Table 2.1 which quotes accident statistics for UK factories in a single year. They arise from people either falling, or being struck by falling objects (see Fig. 2.1): a further breakdown of the figures to illustrate how persons fell is given by Table 2.2.

Table 2.1 Falls of persons and total accidents in UK factories

	Fatal	Total
Total accidents	290	219,001
Falls of persons	49	33,631
Struck by falling objects	27	13

General occupational safety 9

Fig. 2.1 Complete collapse of scaffold during the extension of a chemistry research laboratory. The cause was removal of reveal bars which tied the scaffold structure through the windows. These ties were taken out to enable plasterers to 'finish' plaster work around window frames

Table 2.2 Analysis of the ways in which persons fell in factories

Type fall	Fatal	Total
On or from stairs	3	4,511
On or from ladders	6	3,309
From one level to another	37	8,136
Falls on the same level	3	17,675
Total falls of persons	49	33,631

Falls from same height

The main reasons for persons falling on the same level are slipping and tripping hazards. Prevention should concentrate on removal of the danger; when this is not immediately feasible it should be suitably fenced off and, where relevant, well lit. Reliance on warning signs, lights, etc., should be a last resort. Regular formal inspection of work areas is essential to ensure hazards are continuously identified and rectified; all employees should be encouraged to report situations which may result in persons falling.

Struck by falling objects

Injuries from being struck by falling objects may prove fatal. Use of personal protection such as hard hats, safety shoes and other protective clothing reduces injuries. Additional precautions involve instituting measures to discourage any of the workforce from walking under loads suspended from cranes, etc., from throwing objects from aloft, or from placing objects high up near edges, and encouraging them to stack loads securely, etc.

Falls from heights

Falls to lower levels frequently involve falls through floor openings and into pits, trenches, vessels, and falls from ladders and other elevated workplaces.

The latter is aggravated by poor means of access, unfinished work areas, fragile surfaces (e.g. felt, asbestos, glass) or surfaces that have deteriorated due to age or weather, sloping surfaces such as roofs, and by adverse climatic conditions during the work.

Statutory controls normally apply to roof work and working at heights and relevant guidance is provided, e.g. in the UK as in references 4–11.

Advice for safe working at heights is given in Table 2.3.

Table 2.3 Guidelines for working at heights

General
- Safe means of access (see below).
- Work at heights must only be undertaken by persons who have the knowledge, experience, training, and resources necessary for the work to be completed in safety.
- The person responsible for arranging the work must consider the hazards involved in the job and how they can be eliminated.
- The person responsible for the planning of the job must consider the necessary equipment, the experience of the workmen, access required, protection of staff, the instruction and supervision required together with any relevant documentation, e.g. permits to work.
- For work of short duration it will still be necessary to identify hazards and arrange for appropriate safeguards to be implemented.
- All equipment brought on site must be in good condition, be properly maintained and be used correctly.

Table 2.3 (continued)

Safe access to place of work

Laboratories, rooms, stores, libraries
- Step ladders, access platforms 'kick-stools' and step stools should be available for use by persons requiring access to high shelves, equipment, etc. Improvised access, e.g. chairs or bench stools should be avoided.

Ladders
- All ladders must be in good condition and be suitable for the work to be done and for the access required. Ladders should always be inspected by the user before use. (Defective ladders should be reported and not used.) The base of the ladder should be placed 1 m from the wall for every 4 m in height. They must extend at least 1.1 m above the uppermost landing platform. Ladders should be securely fastened near to the top and close to the base before work commences. Extension ladders should not be over-extended – an overlap of at least one-third is required.
Ladders used on felt roofs should not bear directly on to the roof.

Scaffolding
- Scaffolds must satisfy all legal requirements and all components of the scaffold must be examined by a Competent Person before it is erected. There should also be an inspection of the complete scaffold by a Competent Person before use. Scaffolding should thereafter be inspected weekly and after any modification or after severe weather conditions. Figure 2.1 shows the consequences of incompetent modification to falsework.
The scaffold platform for roof/access should be positioned below the fascia board in order to provide a reasonable step-up distance.[6]

Mobile self-erect scaffolds
- Self-erect scaffolds should only be erected by persons who have been given adequate training.
- Outriggers or means to prevent the scaffold from overturning must be used when the height of the working platform exceeds three times the narrowest width of the scaffold.

All wheels of a mobile scaffold must be securely locked wherever persons or materials are on any of the working platforms.

Hand rails or means to prevent a person from falling must be fitted to all mobile/self-erect scaffolds.

Roof work
- No contractor or employee should generally be allowed to go on the roof of any of the company's buildings without permission (e.g. from the Engineer, Safety Officer, etc.).
- Any roof that is to be used as a means of access or a place of work must be examined to identify any areas covered with fragile materials. Prominent warning notices must be displayed at any fragile roof covering.
- All debris, rubbish and rubble must be removed from the roof in an approved manner and not thrown down to ground level.
- Hot work, e.g. asphalting, etc., must not be commenced before prior discussion with the Engineer and the necessary hot-work permits are issued (see Ch. 17).

Work on flat roofs and roofs up to 10° pitch
- Guard rails and toe-boards must be provided at the unprotected edge of a flat roof or at an opening in the roof where a person could fall. Methods of guarding edges are shown in reference 7.
- If an opening is covered instead of being fenced it should be of such construction as to prevent persons or materials falling through. The covering should be securely fixed and the area clearly marked.

Table 2.3 (continued)

Work on sloping roofs over 10° pitch
- Under no circumstances should any person walk on a sloping roof except on a roof ladder or when wearing a safety harness which is properly anchored.
- Proper roof ladders or crawling boards are required for work on all sloping roofs. Details of roof ladders can be found in reference 7.
- On all sloping roofs at *least two* roof ladders or crawling boards are required. Where materials are likely to be stored on a roof while awaiting installation then a third roofing ladder is required for the materials.
- The anchorage should whenever possible bear on the opposite slope by means of a special ridge iron or similar device and where necessary be secured by a rope as an added precaution. Eaves and gutters should never be used to support roof ladders.

Work on or near fragile roofs
- Prior to work commencing all roofs must be examined by the Engineer to identify any areas covered with fragile material.
- Specific warnings and instructions must be given to the men engaged on the work and adequate supervision applied whilst the work is in progress.

Lifting appliances
- Suitable guard-rails and toe boards must be provided at the edge of the roof or platform where a lifting appliance is being used. Guard-rails and toe-boards should, if possible, be kept in position when raising or lowering materials. If they have to be removed then persons near the edge should wear a safety belt or a safety harness.[7]

Storage of materials on the roof
- The storage of materials on the roof should be limited to the amount that the structure will carry safely, or to the day's requirement.

Safety nets
- Where safety nets are used they should be installed as close to the working level as possible.[9]

Safety harnesses
- A suitable anchorage point must be provided where safety harnesses are to be used.
- The anchorage point must be capable of sustaining the anticipated shock load. Excessive shock loads on the anchorage should be avoided and for this reason the free fall distance should not be more than 2 m where a safety harness is worn. Details of the design and construction of harnesses are given in BS 1397[10] and anchorage points in BS 5845.[11]

Window cleaning
- Additional precautions are given in reference 8.

Work in loft spaces
- Prior to commencing work people working in a room underneath who may be affected by the activities should be notified and their agreement obtained for the work to proceed.
- Wherever possible work should be restricted to be within reach of permanent walkways.
- Light-weight portable access platforms should be used where a permanent access platform is not provided.
- Warning notices should be displayed at access points to areas with fragile ceiling materials.
- All access equipment must be adequately maintained and used correctly.
- Smoking should be prohibited.

'Material handling' hazards

'Goods handling' represents the largest single cause of occupational injury, accounting for about a quarter of all reported industrial accidents. Unless materials are handled with skill and due care, problems can be encountered both when materials are man-handled or moved with the aid of mechanical devices.

Man-handling

Many materials (e.g. sacks, kegs, drums, boxes, items of plant) are man-handled on a regular basis often with little thought of the dangers. Hazards arise from the weight, size, shape and nature of the load. Common accidents are contact with chemicals (e.g. from spillage of vessel contents, contamination of container), cuts on sharp edges, abrasions by rough surfaces, crushing injuries, hernia and back injury.

> Back pain is one of the fastest-growing and most expensive occupational injury problems. In the UK alone costs of £300 million are incurred annually through medical care, sickness benefit and lost production. On any one day 94,500 employees are absent from work because of back pain, amounting to 30 million working days lost each year.[12]

Man-handling accidents are largely avoidable by attention to simple rules. In general the necessity for manual handling of goods should be eliminated by design of the plant or the work layout, and by the use of mechanical lifting aids such as fork-lift trucks, hoists, etc., wherever possible.

When loads must be moved by hand the weight to be lifted should be limited. Several guides exist on maximum weights to be lifted, but no universally accepted limits have been set, mainly because of the complications of coping with wide human variations and adaptability. Thus the Chartered Society of Physiotherapists recommends a maximum weight to be lifted of 120 lb (54 kg). The ILO recommends between 80 and 110 lb (36–50 kg) as a maximum which may vary according to the physique and age of the operator. Others have suggested care is needed when lifting a weight more than half one's own weight. Some experts suggest the limits listed below, while the TUC proposes stricter limits. In the UK comprehensive regulations have been drafted for comment but have yet to be ratified.[13]

Age	Limit – lb (kg)	
	Men	Women
16–18	44 (20)	26 (11.8)
20–35	55 (25)	33 (15)
> 50	35 (15.9)	22 (10)

The use of proper personal protection also reduces the opportunity for injury. Thus when lifting heavy weights safety footwear should be worn and protective gloves, for example of chrome leather, are advisable when handling equipment with sharp or jagged edges. When moving containers of chemicals appropriate eye protection and protective clothing should be worn (Ch. 6).

While limitation of the load is important in preventing accidents, most back strains result from poor lifting techniques. Thus operators should receive both on- and off-the-job instruction in human kinetics, the elements of which are given in Table 2.4. Excellent training films are available[14] and the use of video recordings of employees at work (e.g.

Table 2.4 Simple rules when lifting and moving objects

- Examine the load to be lifted for hazards and study how best to tackle the problem.
- Assess the load and if it is likely to be too heavy or awkward for comfort obtain assistance from a colleague, preferably of similar physique.
- Wear appropriate personal protection.
- Bend the knees and hip joint then grip the object with the palms of the hand, not finger tips.
- Keep the back straight to prevent uneven stress on the discs and back muscles.
- Position feet about 30 cm apart, one slightly in front of the other to provide balance and a strong lifting base.
- Tuck the chin in to protect the top of the spine and help straighten the back.
- Tuck arms well into the body.
- Breath in.
- With a strong thrust forward and upwards off the back foot, lift the load by straightening the legs and keeping the spine straight – jerking should be avoided since it may cause severe strain to arms, back and shoulders.
- Before moving with the load ensure it does not obstruct the view.
- Carrying a load under one arm and supported by the hip can cause local strain; such a load may be better carried on one shoulder. To change direction the whole body should turn not just the weight.
- Do not change the grip while carrying; it is better to rest the load on a ledge, etc., in order to change the grip.
- Standing holding a heavy weight should be avoided whenever possible – a bench or platform or other support of suitable height will reduce lowering and lifting movements and avoid strain during waiting periods.
- Use the body weight not 'brute force' when moving heavy or cumbersome objects about their centre of gravity. Upending and overturning large filled drums requires use of mechanical lifting aids or two persons. To roll a barrel push with the hands not the feet.
- Sacks should be lifted by grasping opposite corners and supporting the load on the shoulder before walking.
- When lowering the object bend the knees and keep the back straight. Place one corner or edge down first to avoid trapping fingers.
- Loads should not be pushed up on to stacks above chest height. A sturdy platform should be provided for this purpose.

in stores, engineering workshops, packing lines) is a useful aid in demonstrating both poor and correct lifting techniques.

Mechanical handling

Common mechanical aids for lifting and moving material include jacks, handtrucks and barrows, conveyors, powered fork-lift trucks, lifting tackle such as chain and rope slings and hoists, yokes, derricks, cranes, etc. A selection is briefly discussed below. Clearly a more detailed appreciation of the safety requirements and legal obligations associated with these aids will require reference to the literature – see references 1 to 3, 15, 16.

Conveyors

The three common types of conveyor are belt, roller or screw conveyors. The main hazard is trapping of fingers, limbs or clothing in in-running nips. Belt conveyors should have guards at the head and tail of the belt to a distance of at least 42 in (107 cm) from trapping points. For roller conveyors attention to in-running nips is essential. The injury potential associated with screw conveyors necessitates guards to prevent access at all times. Maintenance should only be permitted with the drive motor locked off and consideration should be given to the need for a permit to work.

Barrows and handtrucks

Two-wheeled trucks and barrows should be equipped with knuckle guards to protect hands. To ensure the truck and not the operator carries the load, heavy objects should be placed to keep the centre of gravity low and near the wheels. Loads must be stacked in a stable manner and they should not obstruct the operator's view. Operators should keep clear of the wheels and should use leg muscles to raise the truck, keeping the back straight. Overhanging edges or sacks should also be kept clear of the wheels. Handtrucks should be pushed rather than pulled except when going up an incline; the operator should never walk backwards. The trucks should be used at safe speeds for avoiding collision with pedestrians, obstructions such as doors and corner walls. Care is needed when running wheels off bridge plates, planks, etc. Trucks should be parked in allocated areas and not left in passage ways.

Lifting tackle

Lifting gear such as block and tackle, jacks, etc., should only be used by trained operators. Other persons in the vicinity should be protected from falling or moving loads. Portable equipment should be stored, and issued only to authorised personnel. All lifting tackle must be inspected and tested regularly and records kept of the date and results of such examinations. Equipment with faults such as stretched links, frayed ropes and undue wear must never be used and should be repaired promptly or destroyed. The safe working load must be prominently displayed on all lifting equipment and this value must never be exceeded.

Chain hoists are preferred to rope tackle on account of their greater strength and durability. Screw-geared and differential hoists are self-locking and will hold a load in position automatically, whereas spur-geared versions are free-running and require a crane-like brake. Chain hoists should be of high quality welded steel with a safety factor of at least 5 and should be chosen to have a capacity in excess of the normal working load. These should preferably be hooked permanently on to a monorail trolley itself designed for at least the maximum load of the hoist. The load chain should be free of twists and never reeved around the load. Lifting must always be performed vertically, or as near as vertical as possible. A chain block should never be thrown or dropped from a height or used for purposes for which it was not designed.

Jacks should be in good condition and adequate to withstand the load to be lifted. The jack should be correctly centred and its head should make full-face contact with the load and precautions taken to prevent it slipping. As the load is raised, blocks should be placed underneath on either side of the jack. Operators must wear appropriate safety footwear. Jacks require regular maintenance and inspection and records must be kept.

Lift trucks

One of the most common means of lifting and moving loads in industry is the lift truck, which can be driver-operated or pedestrian-controlled. Powered versions are electric, or fuelled by diesel or LPG. However, their use requires skill to avoid accidents. (In the UK there are about 20 deaths and 5,000 injuries annually from accidents involving lift trucks, almost half of which are the result of operator error.) Sources of danger include overturns, loss of loads, moving parts, batteries, normal driving hazards, etc.

Trucks are built to withstand heavy loads and are counter-balanced. Overloading will cut down the normal drive capacity and increase the chance of it overturning. The risk of loads falling from the truck can be minimised by proper stacking on pallets, use of appropriate attachments (fork extensions, drum clamps, etc.). Overhead guards are needed to protect the driver.

As with any machine, dangers can be created by moving parts. Hazards with battery-operated trucks include splashing with acid during 'topping up' and the build-up of hydrogen during charging. Lift trucks also represent sources of ignition and even non-motorised versions can generate static via wheels.

> A company recovered solvent from contaminated highly flammable liquids. Failure of the cooling-water supply to the condenser of the pot still, in which 6,000 litres of contaminated hexane were being distilled, resulted in escape of vapours into the flameproof room. A violent explosion occurred killing one man, injuring another, demolishing a two-storey building and causing damage to nearby residential property. A possible source of ignition was the fork-lift truck found in the doorway to the flameproof room and which was not explosion-protected.[17]

As an example of the detailed rules which may be necessary in order to minimise material-handling hazards – however simple in concept – safety precautions with lift trucks are summarised in Table 2.5. To illustrate the importance of inspection, Fig. 2.2 depicts some significant fork-lift truck defects found by engineers.

Table 2.5 Rules for use of lift trucks

General
- Separate routes, designated crossing places and suitable barriers at danger spots should be provided to keep pedestrians away.
  Notices should be posted to warn pedestrians that they are entering a lift-truck area.
- Roads, gangways and aisles should have sufficient width and overhead clearance for the largest lift trucks using them to do so safely whether loaded or unloaded and, if necessary, to allow vehicles and loads to pass each other in safety.
  If ramps are used in factory roads to reduce the speed of other traffic, a bypass should be provided for use by lift trucks.
- Flexible doors of transparent, or translucent, material may reduce risks where vehicles have to pass through (but if it is appreciably darker on one side of the door than on the other, the material will instead reflect like a mirror).
- The surfaces used by lift trucks should be as level and firm as is reasonably practicable, preferably surfaced with concrete or other suitable material.
- Signs showing directions should be clear and distinctive.
- Gradients should be kept as low as possible.
- Lighting should be adequate.
- All accidents must be reported, however minor, whether they involve injury or not. A check is necessary that the truck is not damaged and still functions correctly.
- Chemicals, oils or flammable materials must not be handled in open-topped containers.
- Home made attachments must not be used.
- Lift trucks should not be used as a means of access without the proper attachment.

Table 2.5 (continued)

Driver selection and training
Persons are normally permitted to drive lift trucks only if they:
  (i) are 18 years old, or over;
  (ii) are reasonably intelligent;
  (iii) are medically fit;
  (iv) have good eyesight (with glasses if necessary);
  (v) have good colour vision;
  (vi) have good hearing and reflexes;
  (vii) hold a current Driving Licence (if the fork lift is to be driven on the public highway);
  (viii) are trained in the operation of the type of fork lift they will be expected to drive (see later);
  (ix) are of a physical stature to enable them to operate the fork lift safely and efficiently;
  (x) have sufficient knowldge of the fork lift to see that it is in full working order before use;
  (xi) are familiar with relevant communication systems or signals.

Banksmen (i.e. overlooker)
Persons who are going to act as Banksmen should be:
  (i) aged 18 years or over;
  (ii) medically fit, with particular regard to good eyesight, hearing and reflexes;
  (iii) familiar with relevant communication systems, signals, etc.
  (iv) sufficiently trained in the workings of the lift truck to enable them to direct the driver as necessary.

If the operator is ill, or cannot for any reason operate his truck, the supervisor should be told. The operator should not allow others to operate his truck unless they are authorised drivers instructed to do so by management.

In addition to pre-employment medicals, all operators should be screened for fitness at five-yearly intervals in middle age, and also after sickness or an accident.

Training
The special training requirement of lift truck drivers are exemplified in Table 17.24.

Inspection and maintenance
- Detailed checks should be made of the lift truck by the operator with the assistance of appropriate tradesmen, e.g. tyres, controls, horn, reversing lights, hydraulics, attachments, battery.
- Weekly checks of electrics should be made by a qualified electrician.
- Mechanical checks should be made regularly by a qualified fitter, e.g. forks, hydraulics, carriage and load guard, brakes, hoist chain, mast and carriage tracks, steering, tyres, etc.
- Trucks, and all safety guards and devices, should be maintained by qualified and authorised personnel according to the supplier's instructions.
- Cleaning procedures should follow supplier's instructions (e.g. avoiding the use of steam lines, air lines or water hoses for cleaning electric trucks).
- As with all lifting equipment, there should be an annual inspection for insurance purposes. Recommended adjustments and repairs should be undertaken by a third party lifting equipment contractor.

Table 2.5 (continued)

Operational

Loads in general
(a) A plate should be on the vehicle stating the maximum load capacity
(b) Do not attempt to lift loads that exceed the truck's rated capacity as shown on the data plate.
(c) Never add counterweight to a truck in order to increase its load carrying capacity.
(d) Do not pick up a load if someone is standing close to it. Stop people from walking underneath the load. If the load appears to be unsuitable or the pallet unsound, it should be left alone, and its condition reported to the supervisor.
(e) Do not attempt to handle loads using two trucks at the same time unless you are under the direct supervision of an expert responsible for handling operations.
(f) Never lift with one fork only.
(g) Never operate with an insecure or unsafe load or use defective or damaged stillages or pallets.
(h) Keep the load right back against the fork carrier.
(i) If you are carrying divisible loads, make sure that they are not higher than the fork carriage or backrest extension. The backrest extension or the overhead guard should never be removed.
(j) Always cross stack your load.

Loading sequence
(a) Make sure bundles and parcels are correctly stacked and tied and are therefore secure before transporting or hoisting on to a pallet. The load should be balanced.
(b) Always open the forks to ensure an even spread of the load on the forks. Know the weight of articles to be lifted so that the loads are within the rating of the truck.
(c) Adjust the forks to suit the load being lifted.
(d) Position forks level and fully inserted under load – *not* tilted forward
(e) Take weight.
(f) Tilt mast back, and position forks about 15 cm (6 inches) from the floor.
(g) Move truck to position in front of stack where lift is to be made. Apply parking brake.
(h) With mast still tilted back, raise the load to the top of the stack.
(i) Tilt the load forward to level.
(j) Move the truck forward so that the load is positioned accurately.
(k) Apply parking brake, and lower to stack.
(l) Lower forks to clear pallet.
(m) Look to make sure that nobody is behind.
(n) Move back to clear the forks 1.2 to 1.8 m (4 to 6 ft) from the stack.
(o) Apply brake. Tilt mast back and lower the forks to travel position.

Non-routine loads (i.e. loads which the operator would handle infrequently)
(a) Before lifting non-routine loads, ensure a Banksman is present.
(b) Handling awkward loads, e.g. roof trusses and timber or scaffold tubes of irregular length needs special care. A banksman should always work in association with the driver in difficult loading situations.
(c) When about to carry very wide loads, meticulous care must be taken to ensure lateral integrity is maintained by critical approval and maintenance of machine approach runs. Before travelling with a wide load, it must be ascertained that both sides of the access way are free from obstruction. Wide loads should be packed up close to the centre of gravity, with items strapped tightly together.

Table 2.5 (continued)

Stack/destacking in general
(a) Stacking and destacking should not be attempted on inclines since the lift truck could overturn as the load is lifted, and the stack itself may be unstable. With goods on pallets, make sure you know what weight each pallet will carry, so that the one at the bottom does not collapse when others are stacked on top of it.
(b) When stacking materials either palletised or loose, ensure that the floor is of sound construction and strong enough to carry the whole stack.
(c) Do not stack directly against walls, in doorways, or too close to emergency exits, fire points or first aid posts, trench walls, excavations, etc.
(d) Always use pallets or dunnage strips, bond the layers and bind the tiers to each other, rejecting damaged pallets or stillages.
(e) Post or box pallets containing 1000 lb (454 kg) each should only be stacked two high; 500 lb (227 kg) each, four high and so on.
(f) When stacking sacked goods, ensure that the ears of the sacks face inwards. Stack each layer squarely on top of the preceding layer and bind the top stack.
(g) Stack the pallet loads as closely together as possible allowing room for the forks to protrude beyond the back of the pallet.
(h) Beware of overhead pipes and never stack near to steam or high temperature water pipes.
(i) Beware of electric cables, lamps, roof trusses and do not stack within three feet of sprinkler heads, if they are of the type that spray out sidewards.
(j) Lower loads as soon as they are clear of the stack.

Stacking sequence
(a) Always approach the stack with the load low and tilted backwards.
(b) At the face of the stack, slow down and stop.
(c) Apply brakes and reduce backward tilt to an amount just enough to stabilise the load.
(d) Raise the load to the desired stacking height and when it is clear of the top of the stack, move forward slowly, being careful not to dislodge the loads in adjacent stacks.
(e) When the load is over the stack, stop, apply brakes and bring the mast to the vertical position and lower the load on to the stack.
(f) When the load is securely stacked, lower the forks until free of the pallet or dunnage strips and after ensuring that the way is clear, withdraw by reversing the truck.
(g) At this position, a slight forward tilt may be helpful.
(h) It should seldom be necessary to use forward tilt otherwise. When clear of the stacks, lower the forks to not more than 6 inches above the ground before moving on.

Destacking sequence
(a) Stop at the face of the stack and apply the brakes.
(b) Bring the mast to the vertical position.
(c) If it is necessary, adjust the fork spread to suit the width of the load and make sure that the weight of the load is within the truck's capacity.
(d) Keeping the mast vertical, raise the forks to a position allowing clear entry into pallet or dunnage strips.
(e) Fully insert forks by moving forward very slowly and apply the brakes.
(f) Lift the load until it is clear of the stack ensuring that the load is stable.
(g) When it is clear of the top of the stack, make sure that the way is clear and reverse slowly to clear the face of the stack. (Be careful not to dislodge the loads in the adjacent stacks.)
(h) Apply the brakes, then carefully lower the load to the correct travelling position and apply backward tilt before moving off.

Table 2.5 (continued)

Stacking with side-loading lift trucks
(a) Come towards the stack with the load firmly positioned on the deck of the truck and, if fitted, use the backward tilt.
(b) Stop when the load is in line with and the truck parallel to the stack and apply the parking brake and engage neutral. If the truck has stabilising jacks and they are needed (i.e. if the load is greater than the unjacked capacity), lower them and make sure that they are hard down on firm ground.
(c) To bring the load level, neutralise the tilt.
(d) Lift the load to the desired height and traverse it out until it is over the stacking position.
(e) Be careful not to dislodge loads in any adjacent stacks.
(f) Lower the load onto the stack adjusting the tilt slightly, if necessary.
(g) Lower the forks until they are free from pallet or dunnage strips and traverse mast fully in when the load is securely stacked. Lower or raise the forks to just below deck level, raise the stabilising jacks if they are used.

Destacking with side-loading lift trucks
(a) Stop in line with, and parallel to, the face of the stack, apply the parking brake and neutralise the engine.
(b) If they are fitted and required, lower the stabilising jacks.
(c) Raise or lower forks to a position which allows clear entry into pallet or between dunnage strips using tilt, if necessary.
(d) Traverse mast out until forks are fully inserted and lift load clear and apply backward tilt, if provided.
(e) Traverse load fully in.
(f) Be careful not to dislodge loads in adjacent stacks.
(g) Neutralise the tilt to bring the load parallel with deck and lower the load onto the deck.
(h) Raise the stabilising jacks if they have been used.

Manoeuvring general
(a) Do not operate your truck with greasy hands or shoes
(b) Never engage reverse when a vehicle is moving forward.
(c) Always look in the direction of travel, and travel at a speed consistent with conditions; reduce speed if the surfaces are wet, slippery or loose. Drive slowly when approaching pedestrians.
(d) Always carry your loads as near as it is practicable to the ground.
(e) If your load obscures your forward vision, drive in reverse.
(f) Always cross railway lines slowly and only at authorised points, and, if possible, diagonally.
(g) Always stop before doorways – sound your horn and go through slowly.
(h) Never cross:
   — soft ground;
   — cables or flexible pipes, etc., that are on the floor unless they are suitably protected;
   — objects such as dunnage;
   — bridges and ramps until you are sure they are safe, which means you must have a good idea of the weight of your truck, loaded or unloaded.
(i) Use the controls smoothly and avoid making fast starts, jerky stops and quick turns, especially when materials are stacked high.
(j) Be careful when braking, as braking violently when loaded may cause the load to fall off or the truck to tip.
(k) When driving loaded uphill, your load must lead, and when driving downhill, you must drive in reverse with the load and forks trailing. (This stops load from slipping on forks.)

Table 2.5 (continued)

(l) When travelling unloaded uphill, drive in reverse with the forks trailing behind. When travelling unloaded downhill, drive forwards with the forks leading.
(m) Always drive at a safe braking distance from any vehicle in front (i.e. at least three vehicle lengths behind), and never overtake another vehicle travelling in the same direction.
(n) Keep to the left when passing oncoming vehicles; drive with the forks about 15 cm (6 inches) off the ground, and always travel with the mast vertical or tilted back.
(o) If the fork truck is used outside the factory on the highway, it must be fitted with registration plates and lights, and driver must have a driving licence.
(p) Sound horn at corners, cross-roads, and if vision is obstructed.
(q) Be conscious of overhead obstructions and width limitations.
(r) The arms or hands should be outside the truck body or cab *only* when signalling.
(s) Never allow anyone to stand or walk under the elevated forks. Passengers should not be carried.
(t) Before entering lifts, road vehicles or other wagons, ensure that they will support the weight of the truck and load, and that the truck's mast will enter.
(u) When approaching a lift, slow down; then with the load first and using a bridge plate, if necessary, enter the lift squarely. When inside apply the brakes and fully lower the forks, shut off the power and neutralise the controls.
(v) When using a bridge plate, make sure that it will support the weight of the truck and load, and never drive over it unless it is securely fixed.
(w) Always keep away from anyone working at a bench, near a wall or other fixed objects – there may be no way of escape for them.
(x) When working with trailers or lorries – make sure the handbrake is on, the engine is stopped, and that the lorry driver knows what you are doing.
(y) On elevated work such as maintenance, the truck should be equipped with an approved working platform. If you are going to tow, drag or push other vehicles or loads, make sure that proper attachments are fitted to the trucks and use them.
(z) Never remove the backrest extension or overhead guards.

Steering
(a) Fork-lift trucks steer on the rear wheels. At corners, move the steering when the front wheels come level with the beginning of the corner. The front of the truck will pivot round the nearside front wheel, and the back wheels will swing out.
(b) When going backwards, the rear of the truck should pass the corner and the front wheels be level with the turn before moving the steering.
(c) Take care when manoeuvring near pits, trenches, excavations and scaffolding.
(d) Use roadway wherever possible and avoid steep, slippery or loose gradients.

Pedestrian trucks
(a) Operators should always face the direction of travel and never walk backwards.
(b) Never walk directly in front of the control handle.
(c) Always keep feet clear of the truck's chassis when reversing.
(d) When reversing from a stack make sure that there is sufficient standing space between the control handle and the stack, rack or wall behind.
(e) In an emergency completely release the control handle, thus putting the 'dead man's brake' in action or otherwise use the stop button 'cut-outs'.

Parking
(a) Do not park across fire-fighting equipment, lifts or stairways.
(b) Before truck is left, even for a short period of time, ensure that the forks are fully lowered, power is shut off, controls are in neutral, brakes are applied and the starter key is removed.
(c) If forced to park on a gradient, apply the brake and check the wheels.
(d) Never park closer than 2.5 m (8 ft) to railway lines.

General occupational safety 23

(a)

Fig. 2.2 Some significant fork-lift truck defects found by engineer surveyors
(a) The counterweight (arrowed A) on this truck was loose and its detachment would have resulted in loss of stability of the truck with the possibility of serious damage or injury. Both rear supports (arrowed B) for the canopy are fractured, seriously affecting the strength of the canopy, which is provided to protect the driver should the load be displaced

24   The Safe Handling of Chemicals in Industry

(b)

(c)

Fig. 2.2 (continued)
(b) The fracture bracket (arrowed) attached the steering base to this fork-lift truck and affected the ability of the driver to steer the truck effectively.
(c) If the wear at the heels of these forks is allowed to progress, failure of the forks due to overstress is inevitable (Courtesy National Vulcan Engineering Insurance Group Ltd)

Machinery hazards

As indicated in Chapter 1, a significant number of occupational accidents involve machines, despite legislative controls, e.g. in the UK ss. 12–17 of the Factories Act 1961. Probably three-quarters of all machinery accidents could have been prevented by reasonably practicable means, the responsibility for which rests with the employer and employee. The subject is considered in detail in reference 2.

Mechanical hazards with machines can be classified according to the form of motion (e.g. rotary, reciprocating), type of machine (abrasive wheels, circular saw), component (e.g. mixer arms, pulleys), location and severity of injury (e.g. amputation of toe), or the way by which harm arises (e.g. entanglement). The last classification is most constructive and encompasses:

- *Traps* – components whose movements could lead to trapping points where the limbs are either drawn into in-running nips or simply trapped by a closing or passing movement.
- *Impact* – parts which due to their speed of movement could cause injury if a person gets in the way.
- *Contact* – parts which because they are sharp, abrasive, hot, cold, or electrically live, could cause injury as a result of touching them (contact injuries may involve both moving and stationary machinery).
- *Entanglement* – moving components of a machine which can cause hair, rings, gloves, clothing, etc., to become entangled.

A lady in a chemical manufacturing company suffered a leg injury when her skirt became caught in gearwheels on a packing machine. She was dragged into the wheels but a workmate stopped the machine before more serious injuries could occur. There was a guard for the machine but a fitter who had been working much earlier in the day had failed to replace it on the machine.

- *Ejection* – hazards from components or material being worked upon being thrown out of the machine.

Good supervision, co-operation, training and constant attention by the operator have an important role in preventing machinery accidents but they are no substitute for engineering safeguards. Thus machines should ideally be of an inherently safe design. Where this is not achieved they should be equipped with proper guards to eliminate, or reduce, the hazard before access to the danger zone can be achieved, unless the machine is considered safe by virtue of its position. Even here, however, safety cannot be guaranteed and where the risk of access and possible injury is reasonably foreseeable safeguarding should be provided. The assessment of risk must take cognisance of the agent of danger, of the need and ease of access and of the possibility of human error. Frequently guards are defeated by operators, sometimes with apparent acceptance by management, e.g. to improve productivity. Design considerations should ensure the need to override the safety feature is removed and the effort required far outways the perceived benefits.

A youth crushed his hand and lost a finger in a power press. The machine was fitted with a gate-guard which was supposed to be in position before the foot pedal could be operated. However, the safety device could be easily bypassed; indeed several employees defeated the guard because it was easier and faster on production.[18]

The most effective means of preventing machinery accidents is a combination of high-quality guards (with a regular inspection and maintenance programme), administrative controls (e.g. permits to work) and training.

Types of guard

Within the UK advice on machine guarding is available as guidance notes (refs 19–28) and safety booklets (refs 29–34), and from the British Standards Institution (refs 35–41).

Several types of guard are available as summarised in Table 2.7: choice will depend upon the level of risk and the nature of the machine and its mode of operation. Selection of materials for constructing guards should consider durability, strength, stiffness, possible adverse effects on machine reliability (e.g. overheating) and visibility. Attention to the design of the guard and its operating mechanisms is crucial in affording the adequate degree of protection. Even the mode of fixing the guard can influence its effectiveness, reliability and ease of defeat. Methods of fixing guards are given in Table 2.6.[2]

Table 2.6 Hierarchy of means of fixing machine guards

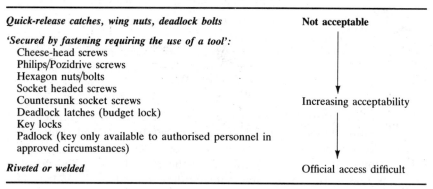

Quick-release catches, wing nuts, deadlock bolts	**Not acceptable**
'*Secured by fastening requiring the use of a tool*': Cheese-head screws Philips/Pozidrive screws Hexagon nuts/bolts Socket headed screws Countersunk socket screws Deadlock latches (budget lock) Key locks Padlock (key only available to authorised personnel in approved circumstances)	Increasing acceptability
Riveted or welded	Official access difficult

A trimming machine was fitted with a guard designed to protect the operator from 36-cm long knives by switching off the machine automatically as the guard was raised. However, screws were missing from the cut-off switch which prevented this from happening. As a result a dangerous machine was insecurely fenced.[42]

Table 2.7 Common types of machine guard

Guard type	Description
Fixed guard	Physically prevents all access to dangerous components. Design and installation should not introduce additional hazards by creating traps between the guard and moving (but otherwise safe) machine components. Effectiveness is mainly determined by the size of any openings and their distance from the hazard, coupled with their ease of removal (see Table 2.6).
	A worker sustained a serious arm injury on a cable handling machine. The front of the machine was protected by a wire mesh with a hole in it big enough for the cable to go through. However, the hole was too big because it had previously been used for larger cables. The guard should have been changed depending upon the size of the cable. There was a switch to stop the machine but because of the nature of the accident the man had not been able to reach it.[43]
Interlock guard (provided when the guard needs to be moved)	The machine is rendered incapable of starting until the guard is closed and brings the machine to rest when the guard is opened. The interlock should be connected to the start controls to prevent uncovenanted motion.
	On the first day of use a power press made an uncovenanted stroke which over-rode the interlocking machanism of the guard and caused serious injuries to both hands of the operator. It was discovered that a vital lubrication line was blocked with paint and also part of the machine which should have been supplied with oil through this line had been installed with inadequate clearance.[44]
	They should be of fail-to-sate design and difficult to over-ride. Types of interlocks used include direct manual switch or valve, mechanical, cam-operated limit switches, trapped and captive key interlocks, magnetic switches and time-delay devices. Attention to design details and frequent inspections are essential.
	A youth was injured while operating a horizontal paper baling machine. His left arm was amputated at the elbow as it became trapped between the ram and a machine frame. Investigation revealed that the postively-operated interlock switch on the top hinged cover of the baler was out of position; the baler ram was also defective.[45]
Automatic guard	The guard operates automatically as the machine starts and it prevents the machine from moving until the guard is in a safe position.
Distance guard	A simple barrier erected around a danger zone. Ergonomic requirements will be dictated by level of risk, etc.
Trip guard	Trip bars and wires, photoelectric devices, pressure sensitive mats, etc., arranged so as to render the machine inoperative as the operator approaches beyond a safe distance. They require careful adjustment and attention to the machine's brake system. They should not be easy to over-ride.

Table 2.7 (continued)

Guard type	Description
	A machine operator lost parts of two fingers when they were trapped in a hydraulic press. Earlier that day a magic eye safety system, designed to cut out the press if a hand passed through the machine, had been switched off to overcome a problem. Since trouble was encountered on the automatic setting and the fault could not be easily rectified the operator was allowed to use the machine with the guard switched off.[43]
Adjustable guard	A fixed guard with an adjustable component. Maintenance, operator training and supervision are essential.
	A worker needed 16 stitches in his hand after being cut by an unguarded power saw. The firm subsequently admitted failing to keep an adjustable guard around machinery and failing to keep machinery securely fenced.[46]
Self-adjusting guard	Prevents access to danger zone until the guard is forced open by the work piece.
Two-handed controls (for machines with cyclic operations where the work is introduced and the machine activated)	The control circuit is arranged so that two controls must be activated simultaneously to start the cycle, and release of either button stops or reverses the machine. Generally of value if only one operator is involved and if they are difficult to defeat.

Electrical hazards

Accidents arise from misuse of purpose-generated electricity and from that produced as an unwanted side effect (e.g. static). These dangers can lead to human injury directly (e.g. electric shock) or indirectly. Examples of the latter include ignition of flammable material, production of toxic thermal degradation products, failure of critical apparatus to operate when required, or unintentional start-up or shut-down of equipment.[43-46]

> A foreman and a director were changing the casting moulds on a machine. Although the electricity supply should have been cut off this was not done. The foreman slipped and grabbed onto the machine's handle; this activated the machine and his hand became caught. Fortunately the director acted swiftly and switched off the power in time to save the hand from being completely crushed.[47]

Problems associated with failure of critical equipment are self-evident. Prevention relies on regular inspection, testing of alarms, detectors, etc., provision of back-up facilities, locking off, etc. The hazards of electrical ignition (including static) are discussed in Chapter 4. The present section concentrates briefly on the hazards and precautions relating to electric shock.

Electric shock

In 1869 a carpenter in Lyon was electrocuted by an a.c. current of 250 volts from a dynamo. This was the first reported death from electric shock in industry.

Many fatalities are caused by occupational exposures to electricity but the rise in industrial accidents has not paralleled the rise in consumption of electricity. For example, in the UK in 1974 there were 804 reported electrical accidents of which 25 were fatal (compared with a total of 290 fatal industrial accidents from all causes that year).[48] Currently there are about 50 industrial fatalities and about 1,000 injuries annually resulting from electrical accidents.[2]

The body is a good conductor of electricity. Thus, if it becomes part of an electrical circuit (e.g. as a result of contact with poorly insulated equipment) current will flow through the body to earth. This can result in injury by chest muscle contraction, temporary paralysis of the nerve centre (both of which lead to interference with breathing), ventricular fibrillation (causing failure of blood circulation), suspension of heart action, haemorrhage and, because of the current's heating effects, tissue damage and burns. Also, flash/arc burns caused by, e.g. opening switches, or removing fuses from energised circuits, etc., can be deep and slow to heal or even fatal.

The severity of injury caused by current flowing through the body is influenced by a variety of factors such as amperage, duration of contact, nature of the current, and current pathway through the body. The effects of current at various magnitudes are summarised in Table 2.8. This demonstrates that mains frequency a.c. is more dangerous than either d.c. or high frequency a.c.

Generally a longer contact time will produce more damage than a short one. A particular hazard with shock from a.c. is tetanic contraction of voluntary muscles such as in the hand or forearm (most pronounced in

Table 2.8 Effects of electric current on a man's body

Current in milliamps			Effect
a.c. 50 Hz	d.c.	a.c. 10,000 Hz	
0–1	0–5	0–9	No sensation
1–8	6–55	10–55	Mild shock
9–15	60–80	60–80	Painful shock
16–20	80–100	80–100	Some loss of muscular control
20–45	100–350		Severe shock and loss of muscular control
50–100	400–800		Possible heart failure (ventricular fibrillation)
Over 100	Over 800		Usually fatal

the 15–150 Hz frequency range) which 'freezes' the victim in contact with the electrified object thereby increasing the contact time. The maximum safe 'let-go' current is about 6–9 mA. However, even 'safe' currents in the range 2–10 mA produce an unpleasant shock sensation and can cause the victim to fall if working at heights.

In most industrial accidents involving electricity the current flows from hands to feet near to the heart. Internal resistance of the body is poor though, if dry, the skin possesses good electrical resistance. As an insulator, however, the skin is poor because of its ability to heat up easily and it undergoes thermal degradation at relatively low voltages.

Precautions

In the UK, in addition to the duties under the Health and Safety at Work etc. Act (HASAWA) current specific statutory requirements are as in

Table 2.9 Summary of Electricity (Factories Act) Special Regulations*

Regulation	Topic
1	Apparatus and conductors to be of sufficient size and power for the work and to be constructed, installed, protected, worked and maintained as to prevent danger so far as is reasonably practicable.
2,6	Insulation and protection of conductors, points and connections.
2–5	Construction, protection, etc., of switches, circuit breakers and fuses.
7,11 and 12	Provision of means of cutting-off electricity supplies.
8	Protection from excess current.
9,10	Special precaution for dealing with a conductor of a system which is earthed or bare.
13	Particular requirements for portable equipment.
14–18	Special requirements for switchboards.
19	Protection of parts of motors, etc.
20	Protection when voltages are transformed.
21	General earthing requirements.
22–24	Personal protection.
25	Adequate working area and access.
26	Adequate lighting.
27	Precautions for electrical apparatus in hostile (including flammable) environments.
28	Requirements as to accompanied work.
29	First aid for victims of electric shock.
30	Substations.

* The Electricity (Factories Act) Special Regulations 1908 and 1944.

Table 2.9. Three consultative documents containing new proposed regulations and attendant codes of practice on the safe use of electricity at work have been published by the Health and Safety Commission.[49] These seek to remedy shortcomings of existing legislation and will apply to all works and work activities.

Equipment liable to become charged with electricity should be adequately earthed to allow sufficient current to flow from the fault so as to blow the fuse or operate the circuit breaker. The earth current should be at least three times the rated current in order to ensure the fuse blows instantly.

Advice on earthing is provided by the British Institution,[50] and the Institute of Electrical Engineers.[51] A blown fuse indicates a fault which requires immediate attention. Fuses should be properly rated and constructed, guarded, and placed so as to protect against overheating or flying hot metal when they blow.

Circuit breakers can be voltage operated, or installed to electromagnetically detect excess current and cut off the electricity supply automatically. They are available to detect overcurrent and earth-leakage current. Use of sufficiently sensitive earth-leakage circuit breakers can prevent electric shocks, i.e. they tend to protect people whereas fuses are designed to protect equipment.

Electrical conductors must be adequately insulated with material of high resistance. Insulations should be inspected periodically for faults and only competent persons should work with live circuits.

> When an assistant, operating a machine for wrapping garments in a dry-cleaning establishment, adjusted an inefficiently insulated thermostat she touched the 240 volt live conductors. The current went across her chest and she suffered finger burns and became unable to release herself. But for the prompt action of a colleague, coupled with the fact that she was a fairly fit person, the accident could easily have proved fatal.[52]

Before work commences on equipment the electricity supply should be turned off and locked off (or fuses removed) to isolate it. A voltmeter or test lamp (which has recently been shown to work on a live circuit) should be used to confirm the circuit is dead. Electrical accidents occur often when work is undertaken on 'live plant' when it could equally have been performed on 'dead plant'.

For work on electric motors which drive pumps, fans and mixers, the coupling should first be broken or other means used to ensure that the motor cannot be inadvertently driven mechanically. If necessary the circuit should be identified and separated from adjacent live circuits by barriers or insulating sheets. Danger notices should be attached to adjacent live equipment plus cautionary notices on equipment controlling the supply to the circuit.

> A fitter was fatally electrocuted as he was putting a new drive belt on a bakery's automatic wrapping machine. The machine should have been isolated and a warning 'DANGER' board attached before the work was started. However, the isolation switch was still 'ON' when the man was found dead under the machine, and the warning sign was still in his locker.[53]

In high-risk situations (e.g. involving high voltages) additional precautions will be required and a permit to work issued to confirm that relevant apparatus is dead, properly isolated, discharged of electricity and earthed. The permit should be issued by an authorised person who should arrange for any testing necessary and who ensures the recipient is conversant with the work and the precautions.

> Working 'under pressure' to repair a ship's alternator an electrical engineer touched a test probe to a terminal and was knocked flat in a ball of flame. The accident resulted because the engineer was given insufficient time to take out the high-voltage coils and test them with low voltage on a workbench. (After recovery, the man suffered from depression and was unable to work again for more than 2 years.)[54]

Substations in the UK are covered by the general requirements under the HASAWA together with the legal obligations to control access under the Electricity Regulations. They should be of substantial construction and arranged so as to prevent unauthorised entry. They should be under the control of an authorised person and access should be restricted to the authorised individual, or a person under his immediate supervision. Access doors should be kept locked and substations should not be used for other purposes (e.g. as a store).

Portable electrical equipment especially electrical tools, handlamps, instruments, etc., often pose a hazard as a result of abuse, misuse, neglect and incorrect wiring. Any person using such equipment should wear rubber gloves or stand on rubber mats, or the instrument should be well insulated externally. Alternatively, the appliance can be double insulated. The cable must be connected to the supply using an approved plug to which it is properly wired and the sheath firmly clamped, and the equipment must be efficiently earthed. Electrical leads should be wires of standard colours, i.e. live = brown, neutral = blue, earth = green and yellow.

> A shop fitter died as a result of electrocution because an electric till had been wired up incorrectly: the positive wire had been connected with the earth terminal, and the earth wire was connected to the live terminal. When a chargehand touched him he too received an electric shock and ordered that power supply be switched off.[55]

Hand-held portable tools may suffer an undetected break in the earth connection and any fault in the wiring can then put the metal casing at

240 V. Defects should be reported immediately and not 'patched up' as a temporary measure.

> A roofing worker was electrocuted while using a portable electric saw connected to a 240 V supply. Several weeks prior to the accident he had accidentally cut through the supply cable which he repaired by twisting the conductors and taping the joint. When he was working from a ladder the joint pulled apart and the live wire touched his body electrocuting him. The deceased should have returned the faulty saw to his employer for proper repair. Equally the employer should have provided a low-voltage system. Use of a double-wound isolated transformer, with secondary voltage of not more than 110 V, with the centre tap earthed is advised.[56]

Only standard equipment should be used, with its use restricted to that for which it was designed[57]: only low-voltage all-insulated handlamps should be used for inspection purposes. The equipment should be kept dry and the tools themselves should be prevented from coming into contact with electric wiring which it could damage. Electric cables must be protected against mechanical damage, water, corrosive chemicals and solvents.

> A worker was electrocuted while cleaning external paintwork on a wall using an electrically-driven water spray machine. The accident arose from the combination of water and electricity. While the plugs on the machine and extension lead were earthed the socket outlet to which the extension was connected was not earthed so that any leakage currents could not be discharged to earth. This was confirmed by an earth loop impedance test which indicated infinity. The fatality would probably have been prevented by the provision of a sensitive, current-operated, earth leakage circuit breaker.[58]

Wherever possible cables should be slung or fixed at least 7 ft (2.1 m) above the floor, platform, or working level to prevent them from being either run over or creating a tripping hazard.

> A man was blown into the air after he tripped over a mains cable while erecting scaffolding. As he grabbed the metal bar to stop himself falling 6 m to the ground he again fell on to the wire and could not release his grip: the electric shock locked the muscles in his arm. The man was rescued by colleagues who beat the chest because his breathing had apparently stopped. Fortunately, he survived but suffered burns to his arms, waist and legs.[59]

Regular inspection of fuse boards, portable electric hand-tools and lamps (plus their cables and plugs), and rubber gloves and mats is essential as is the frequent testing of the current-carrying capacity of the earth conductor between metal clad tools and their plug, and circuit breakers. Some electrical defects found by engineers are shown in Plate 1 (page 80–81).

Noise and vibration hazards

Effects of exposure to noise[60-73]

Noise can be considered as unwanted sound. Sound is generated when fluctuations in air pressure are radiated away from a source of vibration. Besides being a nuisance, noise can reduce working efficiency, particularly where tasks demand high degrees of concentration. It can also contribute to accidents if communication is hampered, or if warning signals and alarms cannot be heard. Furthermore, exposure to excessive levels of noise can impair hearing, e.g. occupational deafness is a prescribed disease in the UK. In order to understand how this irreversible damage arises an awareness of the hearing mechanism is helpful by reference to Fig. 2.3.

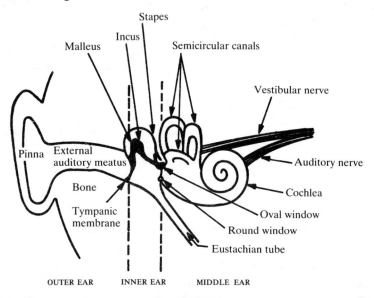

Fig. 2.3 Diagrammatic representation of the ear

Fluctuations in air pressure cause the tympanic membrane (eardrum) to vibrate. This motion is amplified via ossicles (hammer, anvil and stirrup bones) in the middle ear and transmitted to the oval window. From here the air-pressure waves are transformed into hydraulic waves in the cochlea of the inner ear. This snail-shell-like structure is filled with fluid and contains a membrane covered with tiny hair cells. Movement of these hair cells creates electrochemical activity which transmits signals to the brain for interpretation. Irreversible fatigue of these cells by age (termed presbyacusis) or by exposure to excessive noise levels, leads to a loss in hearing. Occupational deafness can develop unnoticed over a period of years. Initially, excessive noise exposure causes a temporary threshold shift, a difference in the hearing sensitivity measured before and after exposure to sound. After a period of hours away from intense sound the individual's pre-exposure hearing level returns. However, repeated insults of excessive noise eventually transforms this threshold shift permanently. Noise-induced deafness does not amount to total silence but affects the frequency range of 500–4,000 Hz which covers the speech range and the consonants in particular. The consequences are that the sufferer has difficulty in interpreting conversation which appears as a garbled string of vowels.

In addition to hearing loss, noise can cause tinnitus. This is a noise heard in the absence of any real sound, often described as 'ringing in the ears'. The condition may be permanent in a proportion of those suffering occupational hearing loss but may also occur in deafness due to old age and various ear disorders.

Effects of vibration

Adverse health effects can result from exposure to vibrations at frequencies below the audible range. Pronounced mechanical physiological and psychological effects may result. Whole body vibration can be experienced, for example, in a tractor cab while segmental vibration is associated with various tasks, for example Raynaud's syndrome (white finger) in hands used to guide vibrating tools (see Table 16.7).

Units of noise measurements

The ear is sensitive to a wide range of sound-wave frequencies (the pitch or sound quality) and amplitudes (which influence the degree of loudness or sound quantity). Frequency is measured in cycles per second or hertz (Hz) and amplitudes in newtons per square metre (N/m^2). For the average healthy ear the audible frequency range is 20 Hz to 20,000 Hz and the audible amplitudes span 2×10^{-5} N/m^2 to 20 N/m^2. With such a range of amplitude a more convenient logarithmic decibel (dB) scale is used to quantify sound pressure levels (SPL):

$$SPL = 20 \log_{10} P_1 P_0.$$

where: P_1 = pressure amplitude of the sound; P_0 = the reference pressure, 2×10^{-5} N/m^2.

However, the ear is not equally sensitive to all frequencies; it is most sensitive to the 2 kHz–5 kHz range and least sensitive at extremely high and low frequencies. Noise measurements are therefore usually taken with an instrument equipped to mimic the response of the ear and levels are recorded as dB(A) units. Table 2.10 lists some familiar surrounding and typical noise levels. Because of the logarithmic nature of the scale every 3 dB(A) increase represents a doubling of the sound intensity. Thus, two machines each capable of producing 90 dB(A) in isolation would, if switched on together, produce noise levels of 93 dB(A) not 180 dB(A).

Table 2.10 Examples of noise levels of different environments

Sound pressure level (dB(A))	Sound pressure (N/m^2)	Typical environment	Average subjective description
140	200	Jet aircraft	Threshold of pain
130	63		Intolerable
120	20	Pneumatic chipper	
110	6.3	Loud car horn (dist 1 m)	
100	2	Newspaper press	
90	0.63	Heavy vehicles at 6 m/inside bus	Very noisy
80	0.2	Busy road	
70	0.063	Busy office/vacuum cleaner	Noisy
60	0.02	Conversational speech	
50	0.0063	Quiet office	
40	0.002	Living room	
30	0.00063	Library/country lane	Quiet
20	0.0002	Quiet chuch/bedroom at night	
10	0.00006	Broadcasting studio	Very quiet
0	0.00002		Threshold of hearing

Assessing noise risk

Risk assessment of noise exposure requires a knowledge of the levels and frequencies of sound to which operatives are exposed together with the duration of their exposure. A wide selection of instruments of various degrees of sophistication is available to measure sound levels.

The basic sound-level meter comprises a microphone amplifier–attenuator circuit and an indicating meter. Most are equipped with fast and slow meter response facilities for measuring sustained noise. The fast response is used to measure, with reasonable accuracy, noise levels that do not change significantly in periods of less than 0.2 seconds. The slow setting gives a more sluggish response which is useful to average-out meter fluctuations which would otherwise be impossible to read; however, this setting would prove inaccurate if sound levels change substantially in less than 0.5 seconds. Thus the only noises which cannot be measured directly with ordinary sound-level meters are impact noises where there are significant changes in less than 0.2 seconds, such as that associated with drop forging or 'explosive' noises, like cartridge hammers.

Impulse noises are generally measured either directly from a calibrated oscilloscope, or more conveniently with a peak-reading impact noise analyser attached to the output of the sound-level meter.

General purpose sound-level meters are often fitted with three filters or frequency weighting networks termed A, B or C, to reflect the approximate response characteristics of the ear at various sound levels. Of these options, the A-weighting is now used almost exclusively. A more detailed assessment of the frequency characteristics of the sound pressure can be obtained by feeding the output of the sound-level meter to a suitable analyser such as an octave band analyser.

In reality, noise exposures tend to fluctuate with time and space. This variation in exposure can be accounted for using noise-average meters or personal dosimeters which are worn by the operator and which produce a single figure known as the 'Leq'. This is equivalent to that continuous noise level delivering the same amount of sound energy as the fluctuating level.

Whether the sound-level meter is of an 'industrial' standard, or is a precision grade instrument, calibration and proper use are essential for obtaining reliable data. More detailed advice on noise monitoring and the interpretation of data is given in references[63,65,70,71,73]. The following guidance summarises the main rules when taking measurements:
- Study the manufacturer's literature.
- Choose equipment appropriate for the measurements to be taken with consideration of grade of meter and frequency content of the noise.
- Calibrate the equipment according to the supplier's instructions.
- Ensure batteries are in good order.
- Set speed and weighting network.
- To assess operator exposure place microphone of sound-level meter (and noise-average meter) close to the operator's ear position at the normal work station.
- To assess noise levels in general point the microphone at arm's length away from the body towards the source of sound. When sound comes from a variety of directions pan the area to obtain average sound levels.
- Keep away from reflecting surfaces.
- Check background noise levels.
- Use wind shields where wind or draft can blow across the microphone.

- Report noise levels/exposure levels; and where relevant operator's name, machine reference number, location of study, date, time, serial number of meter, settings and method of calibration. Noise topographs (or noise contour maps), can be useful for indicating sound distribution patterns and, in particular, zones of noise danger.

To monitor the overall effectiveness of a hearing conservation programme, the hearing acuity of exposed personnel may be monitored by measurement of pure tone air-conduction hearing thresholds. This requires accurately calibrated audiometers, special quiet test rooms, and staff well trained in audiometry.

Occupational noise limits

Exposure to noise should be reduced to as low as is reasonably practicable, and in any event continuous 8 hr exposure to noise should not exceed 90 dB(A) per day. To ensure that this noise dose is not exceeded when noise levels are above 90 dB(A) the duration of exposure must be adjusted accordingly as in Table 2.11. In the UK the concept of equal

Table 2.11 Maximum permissible duration of exposure at different noise levels

Noise level (dB(A))	Maximum exposure time (hr)
90	8
93	4
96	2
99	1

energy or equivalent continuous sound level normalised to 8 hr is the basis of the Department of Employment Code of Practice[65] later adopted in the Woodworking Machine Regulations (1974). While no general legislation currently exists in the UK relating to occupational hearing protection, draft regulations propose the use of hearing protection for employees exposed to noise levels in excess of 90 dB(A) (Leq) and audiometry for those exposed to more than 105 dB(A) (Leq). Draft EEC proposals centre around an acceptable workplace noise level of 85 dB(A) but this value is not internationally accepted. (Notwithstanding the absence of specific legislation on the subject, prosecutions under the HASAWA can be brought against employers who subject employees to unacceptable levels of noise.)

Noise control

Noise control strategies include engineering control, personal protection and administrative measures. Where it is reasonably practicable, noise should be reduced by engineering procedures. Where this is not the case a combination of approaches is usually required.

Engineering control

The best means of preventing noise is control at source by design. Frequently economically attractive and worthwhile reductions in noise levels can be accomplished by small changes to the design of a machine or careful choice of materials of construction. Control of noise from existing plant is normally achieved by palliative measures such as soundproof enclosures.

Noise is initially generated either as mechanical vibrations or as air turbulence; it can penetrate through machines and structures as well as through air and is capable of being transferred between the different pathways. Choice of appropriate measures for controlling noise, therefore, depend upon the sound sources and the transmission paths. Ideally, to reduce noise levels, consideration should first be given to replacement of a noisy machine by a quieter version or adoption of quieter processes (e.g. welding rather than riveting, use of electric trucks in place of trucks with internal combustion engines). Where this is not practicable the main engineering tactics available include vibration isolation, damping, silencers, noise insulation and noise absorption.

Vibration isolation

Mechanical vibrations pass through rigid structures. Suitable resilient mounts, such as springs or rubber cushions, are used to isolate vibrating machines (or their vibrating components) from radiating surfaces. Mounts should be selected so that the natural frequency of the system is low relative to the frequencies to be isolated.

Damping

Damping of noise-radiating surfaces involves the conversion of vibrational energy into heat energy which reduces the amplitude of vibration and hence the intensity of radiated sound. The technique is useful for reducing impact noise and can be introduced into structures by change in material of construction (e.g. cast iron has more damping properties than steel), by stick-on or spray-on palliative damping substances, or by use of layered materials such as cork sandwiched between steel.

Silencers

Mufflers reduce the transmission of sound while allowing free flow of gas and are used to reduce air turbulence created by intakes or exhausts of engines; compressors, air lines, and from ventilation systems. The accoustical characteristics are achieved by the shape of the silencer or by use of sound absorbent materials.

Noise insulation

Noise insulation involves use of a barrier to reduce airborne noise transmitted from one point in space to another. Good insulation is achieved by airtight heavy, limp, impervious, well-damped materials with no air spaces. Transmission loss at different frequencies depends for a given barrier material on its stiffness, mass and resonance frequency.

Noise absorption

Absorption is achieved by the dissipation of airbone noise energy into heat and damping. Absorbent materials mounted on walls or suspended from ceilings reduce reflected sound and minimise the build-up of reverberant sound.

Materials which are suitable for thermal insulation tend to be poor accoustic insulators but efficient sound absorbents. Lists of sound absorption coefficients of materials and sound transmission loss of materials and structures are to be found in references included at the end of the chapter, (e.g. ref. 63).

Soundproof enclosures

The design of these enclosures is based on many of the techniques described above. The insulating properties of each part of the enclosure should be balanced accoustically, the internal surfaces should be absorptive, and machines inside the enclosure should be mounted on vibration isolators.

Summary

In summary, excluding control at source, the main engineering measures for control of noise from different transmission pathways are as follows:

Source/pathway:	Remedy:
Structure-borne noise	Isolate vibrations from radiating surfaces
Radiating surfaces	Prevent resonance by damping
Airborne noise	Use a barrier to insulate noise from operator, or use a porous blanket to adsorb noise

Table 2.12 lists typical sound-level reductions achievable by various applications or combinations of control measures (after ref. 67). Table 2.13 lists some suitable methods for controlling noise from specific plant items or processes.

Table 2.12 (after ref. 67) Reduction in machine noise levels with various control applications

Application \ Typical sound level reduction (dB(A))	Datum	2–10	0–10	2–10	0–10	10–20	10–30	15–40	30–60
Original machine	✓								
Machine modification		✓							
Partial cover			✓						
Partial screen				✓					
Vibration isolation					✓		✓	✓	✓
Sealed enclosure						✓	✓	✓	✓
Absorption (within enclosure)								✓	✓
Double walls									✓

Table 2.13 Sources of noise and its control

Plant	Noise source	Control measures
Electric motors	(1) Windage from air turbulence in the cooling fan	– Redesign fan – Fit ducts containing sound absorbent material
	(2) Mechanical noise caused by bearings and poor balance	– Maintenance – Incorporate vibration isolators in motor and drive shafts
	(3) Chatter of brushes on the commutator segments	– Stiffen brush holder – Improve commutator finish
	(4) Vibration arising from fluctuating magnetic forces between rotor and stator	– Avoid sharp changes in flux density by chamfering rotor claws and skewing armature slots
Hydraulic systems	Pump whine and valve noise caused by turbulence from sharp changes in flowrate creating pressure fluctuations in fluid which is transmitted to structure and pipework	– Attention to internal shape and size of components – Mount pump on resilient supports – Incorporate isolating vibration breakers in pipes – Use two small pumps in place of one large pump where feasible – Enclose existing systems

Table 2.13 (continued)

Plant	Noise source	Control measures
Air compressors	(1) Mechanical components of both drive motor and compressor	– Use close shield enclosures with attention to isolation and sealing
	(2) Air intake and pressure output	– Fit silencers
Pneumatic tools, etc	(1) Turbulence from high velocity exhausts	– Silencers (except where blockages could arise, e.g. debris in air line, ice)
	(2) Radiating impact noise from drill bit	– incorporate damped metal inserts into bit
Fettling, chipping, riveting and grinding	(1) Noise from electrical and pneumatic tools	– As above
	(2) Resonant noise excited and radiated by the work	– Difficult but examples include: dampening work piece with magnetic damping sheets, or clamp in felt or rubber. Minimise numbers at risk by arrangement of work layout, use of partial enclosures and clad walls and roofs with absorbent materials
Gears	Tooth impact	– Improve precision of manufacture
		– Routine maintenance
		– Consider helical rather than spur gears or gears made of well-damped materials (e.g. nylon) rather than steel or rubber belts
Circular saws and band saws	Vibration of saw blade caused by impact of blade and work piece	– Stiffen or dampen blade
		– Attention to angle of the cutter blade
Lathes	(1) Mechanical noise (gears, etc)	– See above
	(2) Impact noise in indexing mechanism	– Incorporate rubber buffers in mechanism
	(3) Impact between stock and stock tube	– Use double-skin shock tubes incorporating resilient sandwich
	(4) Chatter between workpiece and tool	– Stiffen tool post, and stiffen work piece with rubber lined clamp
	(5) Pneumatic exhaust noise	– Adapt splash guard to form acoustic close shield
		– Remove exhaust from operator
Presses	Clutches, gears, pneumatic exhausts, and impact noise	– As above

Table 2.13 (continued)

Plant	Noise source	Control measures
Material handling	Impact between product and chutes, hoppers, conveyors, pallets, bins, etc.	– Minimise drop heights – Absorb impacts with tough rubber linings (protected by sheet steel if necessary) – Apply damping treatments to noise radiating surfaces

Personal protection

Ear defenders are no substitute for effective noise control and should only be regarded as an interim measure while more suitable means of reducing noise exposures are under consideration.

Forms of hearing protection available include muffs and both re-usable and disposable plugs. Selection must ensure adequate protection is afforded to the wearer by study of the supplier's data on sound attenuation achievable at the relevant frequencies prevailing in the area under consideration. Users must be instructed in the need for personal protection and in its proper use, with some personal choice from a selection of appropriate equipment. Ear defenders should be protected from contamination during use and storage and be carefully inspected regularly.

Administrative control

Administrative procedures are an essential component of a hearing conservation programme. They include:

- Regular inspection and maintenance of all noise control measures including arranging noise monitoring where necessary.
- Training and supervision of staff with regard to use of procedures provided.
- Warning signs prominently displayed in hazardous areas particularly where personal protection is required.
- Reducing duration of exposure to noise by job rotation, rearrangement of work schedules, etc.
- Provision of noise refuges.
- Maintaining records of noise levels around the site and individual records of all users of personal protection.

Electromagnetic radiation

The electromagnetic spectrum is shown in Fig. 2.4. Implications of exposures to ionising radiations such as gamma- and X-rays are discussed

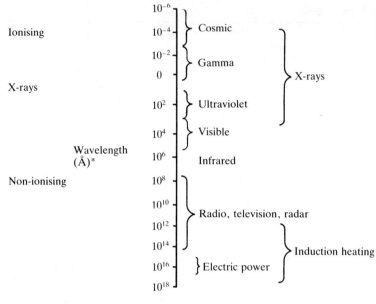

*Å = Angstrom where 1 Å = 10^{-8} cm

Fig. 2.4 The electromagnetic spectrum

in Chapter 8. The present discussion centres around health and safety aspects associated with non-ionising sources such as visible, ultraviolet (UV) and infrared (IR) radiation.

Visible radiation[74-78]

Adequate artificial lighting is essential for all environments in which natural light is absent, e.g. work at night; underground; in blast-proof, windowless control rooms of chemical plants, etc. Even where natural light exists it usually requires supplementing with artificial illumination. For example the illuminance indoors from natural light at a desk next to a window may only be 20% of the value outside and working further away from the window reduces values further to only 1–10% of outdoor levels.

Visible light is one component of the electromagnetic spectrum bounded on one side by UV radiation and on the other by IR. The implications of insufficient illumination are obvious. Generally hazards from exposure to visible light are considered relatively unimportant though ergonomic considerations are crucial.

The quantity of light emitted by a source is termed the luminous flux and the light density is the luminous flux per unit area of surface, whose units are lux (illuminence per square metre of surface), or footcandles 1 lux = 10.7 footcandles.

The aim should be to provide sufficient light to illuminate the area or task without introducing glare, which results when some part of the visual field is extensively bright in relation to the general level of brightness. The degree of glare depends on the surface to be viewed, illumination levels, position of light sources within the visual field, and the general level of background lighting. Broad guidance is:
- Keep light sources outside the field of vision.
- Use matt finishes for worktops, etc.
- Arrange contrast ratio of 10 : 3 : 1 for task : immediate surrounds: general background level, respectively.

In general, discharge and fluorescent lamps are more efficient than incandescent bulbs.

Recommended illumination levels for a variety of different tasks, areas and situations are published by the Illuminating Engineering Society,[75] as exemplified by the selection given in Table 2.14. Local legislation should also be consulted as illustrated in the UK by the Factories Act

Table 2.14 Levels of interior illumination currently recommended by the IES

Area	Footcandles on Tasks*	Dekalux† on Tasks*
Chemical works		
Hand furnaces, boiling tanks, stationary driers, stationary and gravity crystallisers	30	32
Mechanical furnaces, generators and stills, mechanical driers, evaporators, filtration, mechanical crystallisers, bleaching	30	32
Tanks for cooking, extractors, percolators, nitrators, electrolytic cells	30	32
Chemical laboratory	50	54
Materials handling		
Wrapping, packing, labelling	50	54
Picking stock, classifying	30	32
Loading, trucking	20	22
Inside truck bodies and freight cars	10	11
Soap manufacturing		
Kettle houses, cutting, soap chip and powder	30	32
Stamping, wrapping and packing, filling and packing soap powder	50	54
Woodworking		
Rough sawing and bench work	30	32
Sizing, planing, rough sanding, medium quality machine and bench work, gluing, veneering, cooperage	50	54
Fine bench and machine work, fine sanding and finishing	100	110

* Minimum on the task at any time for young adults with normal vision and better than 20/30 corrected vision
† The dekalux is an SI unit equal to 0.929 footcandles; 1 dekalux = 10 lux.

1961, the Chemical Works Regulations 1922, the Electricity Regulations 1908, the Horizontal Milling Machines Regulations 1928, etc.

Ultraviolet light[79-82]

Natural and man-made sources of UV radiation are employed for a variety of purposes including sterilisation, promoting the production of vitamin D, increasing the blood supply to the tissues (assisting healing), creating fluorescence for chemical analysis, UV spectroscopy, free radical initiation in chemical processes, photocopying machines, etc. Often it is also the by-product of processes (e.g. welding). Workers at potential risk from UV light include dentists, food and drink irradiators, hairdressers, laboratory workers, lighting technicians, lithographic and printing workers, nurses, paint and resin curers, physiotherapists, photographic workers, plasma-torch operators, spectroscopists and welders.

Ultraviolet radiation is radiation with wavelengths between the blue region of the visible spectrum and the X-ray region from 10 to 400 nanometres (nm) and is non-ionising.

The sun is a source of electromagnetic radiation which radiates rays over a wide spectrum from the harmful cosmic, gamma- and X-rays, to the less harmful UV rays, visible light and infrared rays.

Ultraviolet radiation is generally produced either by the heating of a material, e.g. carbon, to an incandescent temperature, or by the excitation of gas or vapour at discharge, e.g. mercury vapour lamps. The UV spectrum is classified into regions on the basis of wavelength and the hazard potential is a function of the wavelength of the source plus its intensity. UV radiation is governed by the inverse square law, i.e. the intensity of the rays varies inversely as the square of the distance from the source. The shorter wavelength UV radiations are severely attenuated by all common substances, including air, and hazards to health are derived in general from wavelengths longer than 200 nm. Though vacuum radiations which occur at 100–200 nm are absorbed by air, they can cause chemical reactions whereby harmful gases, such as ozone and oxides of nitrogen, are formed. At wavelengths 150–200 nm phosgene can also be formed when chlorinated hydrocarbons are irradiated (e.g. carbon tetrachloride) which makes adequate ventilation important in confined spaces.

Many sources of UV radiation also emit copious quantities of visible light, but some do not. For example, germicidal lamps which generate large amounts of UV radiation give off only a faint visible glow. Therefore, it is unsafe to judge hazard potential solely by brightness. Scattered light, particularly from black-light sources, can present an additional hazard. For example, reflections of light of wavelength 254 nm (invisible) may not be noticed from a low-power mercury source – especially if it is fitted with a Corning glass filter which cuts out the visible light.

Hazards

The main effects arising from exposure to UV radiation are summarised by Table 2.15 and discussed below. In addition, there may be physical or electrical hazards associated with some equipment containing UV sources.

Table 2.15 Physical properties of ultraviolet radiation

Effect	Wavelength (nm) range	(max. effect)
Production of NO_x; HCl; $COCl_2$	130–190	
Ozone production	< 250	(185)
Keratitis	250–310	(288)
Erythema (i)	240–275	
(ii)	290–310	(greater effect)
Carcinogenesis	290–310	
Skin pigmentation	270–365	(340)

Effects on the skin

Acute effects on the skin are familiar as sunburn (the redness produced is termed 'erythema') which is caused mainly by wavelengths shorter than 315 nm. The severity of the effect depends on the duration and intensity of exposure. Chronic effects of repeated exposures include premature skin ageing and cancer; these effects may not manifest themselves for years or even decades. Again, as with erythema, the most important wavelengths appear to be those below 315 nm. The areas of skin usually at risk are the backs of the hands, the forearms and the face and neck. Hands can be protected by gloves. The arms should be covered by long sleeves of material of low transmission in the range 280–315 nm. In general, materials which are visibly opaque are suitable. The face can be protected by a face-shield which can also give protection for the eyes.

Photosensitisation

Sensitisation of the skin to UV radiation can occur. There are many possible causes and these fall into two broad categories: chemical and genetic.
1. *Chemical photosensitisation* occurs only in a small proportion of those exposed but unfortunately there are no easy means of identifying those, who will react in this way. The speed and severity of the reaction largely depends on how the chemical reaches the cells of the skin. Medicaments, cosmetics, and industrial chemicals, such as tar compounds, organic solvents, anthracene and synthetic dyes, which

are in direct contact with the skin, may produce a burning sensation and erythema within an hour or two in the presence of UV radiation. However, therapeutic agents introduced orally or by injection will not produce symptoms until effective concentrations of the chemical have built up within the skin. This may take several days, or even weeks. A large number of drugs are known to be capable of producing temporary photosensitivity, but the reaction seems most common with certain antibiotics and tranquilisers.
2. *Genetic causes* of photosensitisation are illustrated by a rare group of metabolic diseases known as the porphyrias. People who suffer from these diseases produce excessive quantities of a sensitising chemical which may result in skin lesions and aversion to the light.

Effects on the eyes

The principal effect on the eyes of excessive exposure to UV light is kerato-conjunctivitis. The symptoms are pain, discomfort similar to that resulting from grit in the eyes, and an aversion to bright light; the cornea and conjunctiva show inflammatory changes. As with erythema of the skin, the severity of kerato-conjunctivitis depends on the duration, intensity and wavelength of the UV radiation exposure. The length of time between exposure and the occurrence of symptoms for both erythema and kerato-conjunctivitis is similar. The symptoms tend to disappear after about 36 hours and permanent corneal damage is rare. The maximum effect on the eyes for a given exposure is produced at a wavelength of 270 nm. The elimination of harmful UV radiation must be given attention because it can produce injury to the eyes *without warning*. It is only after the damage has been done (some 4 to 6 hours later) that their effects begin to appear (cf. 'welders' flash'). Although a small amount of UV may not produce permanent injury to the eyes, the only safe procedure is to completely exclude all harmful UV rays by containment, e.g. placing protective cabinets around sources of emission or, where necessary, to provide filters and wear approved spectacles and protective clothing. (Table 2.16 summarises the grade of glass for welders' face-shields as recommended in BS 679.)[82]

Table 2.16 Grades of glass for welders' face-shields[82]

Arc welding currents (A)	Grades of glass	Shade no.
Up to 100	EW	8 or 9
100–300	EW	10 or 11
300–500	EW	12, 13, or 14
Over 500	EW	15 or 16

17 out of 50 employees complained of sore, irritable eyes – though there had been no change in production methods or materials, and no local epidemic in the general population. More personnel eventually became affected and some of the treated employees suffered relapses. Eventually one of the affected persons was diagnosed at an eye hospital as suffering from radiation exposure, probably from an ultraviolet source.

Workplace investigations subsequently revealed that a proprietary fly-insect killing device installed in the messroom, and commonly found in canteens and food production areas, had been fitted with incorrect fluorescent tubes. Instead of emitting 'black light' (UV-A), at 357 nm (which is supposed to attract insects) the equipment had been fitted with UV tubes intended for sunbed use, and persons in the messroom were exposed to UV-B without protection. Subsequent eye irritation – similar to welders' flash (arc-eye) – was easily prevented by replacing the incorrect tubes with ones designed for use with this equipment; there being no further complaints.

Indirect effects/toxic gases

An important indirect effect of UV light is to initiate free radical reactions, sometimes with formation of toxic products. For example, wavelengths below 250 nm can dissociate oxygen molecules to produce ozone and those between 190 and 130 nm break N—O and N—N bonds respectively, producing oxides of nitrogen. Homolytic fission of Cl—Cl and C—C bonds occurs below 487 and 313 nm respectively. Irradiation of chlorinated hydrocarbons such as carbon tetrachloride and trichlorethylene with light in the far UV range will decompose them with the formation of hydrogen chloride, phosgene or other acid chlorides. (Hence the need to avoid arc welding near chlorinated hydrocarbon degreasing vats.) The hazards from ozone can be avoided by ensuring adequate ventilation in the area around the source. Very intense short-wavelength sources may require an extraction system to remove ozone.

Implosion/explosion

Ultraviolet lamps often operate at pressures below or above atmospheric pressure. In some lamps, the pressure can be as high as 110 atmospheres. Care must be taken to ensure that persons using such lamps or engaged in their maintenance, are appropriately protected in the event of breakage of the tube.

Hygiene standards

In assessing the hazards of exposure to UV radiation, both the wavelength and the intensity must be considered. Compliance with the standard which is based on the biological effects of ultraviolet radiation, including most known data on the production of erythema and kerato-conjunctivitis will protect most individuals from these acute effects. The maximum permissible exposures (MPE) are expressed in radiant exposure units of joules per square metre ($J\ m^{-2}$), the total UV energy falling on 1 m² of surface, or in irradiance units of watts per square metre ($W\ m^{-2}$), the average radiant exposure per second over the exposure period.

Wavelength range 400–315 nm
1. The total irradiance on unprotected eyes and skin for periods of greater than 1,000 seconds should not exceed $10\ W\ m^{-2}$.
2. Total radiant exposure on unprotected eyes and skin for periods of less than 1,000 seconds should not exceed $10^4\ J\ m^{-2}$.

Wavelength range 315–200 nm The radiant exposure on the unprotected eyes and skin should not exceed, within any 8-hour period, the values given in Table 2.17. These apply only to sources emitting essentially monochromatic UV radiation, and a calculation must be made to assess the effective irradiance of a broad-band source. This is carried out as follows. The MPE for a broad-band source is calculated by summing the relative contributions from all its spectral components, each contribution being weighted by the relative spectral effectiveness, S_λ. Values of S_λ for the range 200–315 are also given in Table 2.18.

$$E_{\text{eff}} = \Sigma E_\lambda S_\lambda \Delta_\lambda$$

where E_{eff} = effective irradiance relative to monochromatic wavelength 270 nm ($W\ m^{-2}$)
E_λ = spectral irradiance at wavelength ($W\ m^{-2}\ nm^{-1}$)
S_λ = relative spectral effectiveness
Δ_λ = band width employed in the measurement of calculation E_λ (nm)

Table 2.17 Maximum permissible exposures in an 8-hour period

E_{eff} ($W\ m^{-2}$)	Maximum permissible exposure
10^{-3}	8 hours
8×10^{-3}	1 hour
5×10^{-2}	10 minutes
5×10^{-1}	1 minute
3	10 seconds
30	1 second
3×10^2	0.1 second

Table 2.18 Relative spectral effectiveness (S_λ)

Wavelength (nm)	MPE (J m^{-2})	S_λ
200	1,000	0.03
210	400	0.075
220	250	0.12
230	160	0.19
240	100	0.30
250	70	0.43
*254	60	0.5
260	46	0.65
270	30	1.00
280	34	0.88
290	47	0.64
300	100	0.3
305	500	0.06
310	2,000	0.015
315	10,000	0.003

* Mercury lamp – resonance emission line

The maximum permissible exposure, expressed in seconds may be calculated by dividing the MPE for 270 nm radiation (30 J m^{-2}) by E_{eff} (W m^{-2}).

Precautions

Where sources of UV light are powerful enough to be a hazard, protection against over-exposure may be achieved by a combination of measures such as personal protection (previously mentioned), administrative and engineering control measures.

Personal protection
- Ideally, hazards should be contained (e.g. by engineering control measures). Where this is inappropriate, personal protection should be worn to protect the eyes and skin against UV radiations, explosion and implosion, etc. Table 2.19 indicates the various types of available eye protection for absorbing UV light.

Administrative control measures
- Access to an area where equipment emits UV radiation should be limited to those persons directly concerned with its use.
- All persons concerned with the use of equipment which emits UV radiation should be made aware of hazards involved.

Table 2.19 Types of eye protection for absorbing UV light

Description	% transmission at wavelength	Comments
Black-Ray UVC 303	$<10^{-6}$ at 254 nm and 365 nm Cuts off at 400 nm	Fit comfortably over spectacles. Uncomfortable for extended viewing.
Photochromatic safety spectacles	10% at 380 nm 30% at 392 nm (light) 21% at 392 nm (dark)	Recommended for regular exposure to low amounts of UV. Can be fitted with prescription lenses.
Clip-on lenses: (a) CO/400/F – clear (b) CO/400/F – green	10% at 395 nm >1% at 392 nm 6% at 390 nm 17% at 350 nm	Clip on to spectacles.
NS 623J green spectacles	Zero transmission.	

- Hazard warning signs should be used to indicate the presence of a potential UV radiation hazard (e.g. on lamp housings). Warning lights may be used to show that the equipment is energised.
- The user should keep as far away from the source of UV radiation as is practicable. (The radiation falls off as the square of the distance from the source.)
- The exposure time should be kept to the minimum, and the maximum exposure limits should not be exceeded: where necessary the levels of UV radiation should be monitored.
- Maintenance work must be undertaken by suitably qualified engineers.

Engineering control measures
- Indiscriminate emission of UV radiation must not be allowed. This can be prevented either by carrying out the process within a sealed housing or providing a screened area.
- Wherever possible, the radiation should be contained within a sealed housing. If observation ports are required they should be made of suitably absorbent materials such as certain grades of acrylics, PVC and window glass.
- Where the exposure process takes place external to the source housing, a screened area should be provided where it may be carried out.
- Where the frequency and intensity of stray UV radiation is hazardous, interlocks should be fitted to the source-housing to prevent excessive and unnecessary exposure. All interlocks should be fitted in an appropriate manner, i.e. to BS 5304.[36]
- To reduce the intensity of reflected radiation, surfaces should be painted in a dark, matt colour.

- Where toxic, gaseous by-products are involved, the equipment should be used under ventilated conditions.
- As with any electrical equipment, cables should be in good condition and arranged in a safe and tidy manner (see page 33).

Lasers[83–91]

Lasers are optical devices which emit intense beams of monochromatic, highly collimated coherent light. A diagrammatic representation is given in Fig. 2.5. (The word LASER is an acronym for 'light amplification by the stimulated emission of radiation'. This name describes fairly well the general mode of action of a laser.) A He/Ne laser is a typical example. Basically it consists of a glass tube containing helium (acting as the catalyst) and neon (which does the work) at a pressure of a few millimetres Hg.

Helium atoms are raised to an excited state in the tube by an electric discharge. The atoms subsequently transfer this energy to the neon through hard-sphere collisions. This collision-energy transfer produces a population of excited atoms which is greater than that in the ground state – the so-called 'population inversion'. The consequence of this is the phenomenon of negative absorption and the emission of radiation. The tube is maintained between two mirrors which may be curved or flat depending on the laser design. The emitted radiation travels along the tube via the Brewster windows and is reflected back along its own path if incident on the mirrors at 90°, otherwise it is reflected out of the cavity. In this way a standing wave is built up by continuous reflections between both mirrors with the transmission of ~1% of the power through the output coupler, as a laser beam.

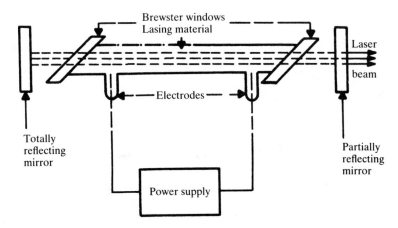

Fig. 2.5 Diagrammatic representation of a laser

Depending on the material undergoing excitation in the laser, the radiation emitted may be of almost any wavelength in the UV, visible or IR region of the electromagnetic radiation spectrum. (NB The damage potential can vary with the wavelength.) Some lasers emit their radiation in shorter pulses and others like a He/Ne produce continuous beams or waves.

Hazards

It is the coherence of laser emission that distinguishes laser radiation from most other types of light and this permits a unique degree of collimation. The potential hazard depends on power output and on wavelength. The high energy content means that the amount of laser light arriving at a target may damage it and the beam collimation ensures that the intensity of the beam is maintained over large distances. In addition there are also hazards from fire formation of toxic gases and also the electrical components of associated equipment.

Electrical hazards In addition to the normal hazards associated with electrical apparatus (see page 28) the high tension associated with lasers needs special consideration because all laser systems have potential for causing serious injury. Often, laboratory models of lasers have open electrical circuits which are hazardous. Direct current models carry sufficient power at the cathode to cause injury and care should be taken when protective covers are removed from such equipment. The latter should only be undertaken by authorised personnel. All electrical circuits should be housed in positively grounded cabinets to limit the possibility of shock from a floating ground potential.

Proper maintenance of cables, connectors, cabinets and switches is essential. Where practicable an interlock system should be provided to ensure that all connections are made with power supplies disconnected. Components not carrying current including instrumentation, chassis, etc., should be grounded.

Toxic gas hazards Ozone may be produced by ionisation of the air surrounding high voltage electrical discharges as could occur with laser systems. Where lasers are used in confined areas, ozone levels should be monitored when the laser is first installed and after any significant electrical alterations or servicing. Consideration should be given to suitable ventilation. Laser radiation of some materials, e.g., chlorocarbons, can result in the formation of toxic by-products (particularly with pulse lasers).

Fire hazards Visible lasers are not potential sources of ignition except when focused on to absorbing flammable material. With some instruments the metal outer casing can become hot and attain temperatures which may be sufficient to ignite certain organic solvents, etc.

Biological hazards The two parts of the body at greatest risk to laser damage are the skin and eyes. In general, the hazards associated with lasers are determined by the power/wavelength of the light and range from the obvious dangers associated with high power lasers which can burn holes through armour plate in seconds, to the lower power version, which cannot cause skin burns and for which only entry of the beam into the eyes need be prevented. Because of its high focusing power ($\sim \times 10^6$) the eye is susceptible to laser radiation and different parts of the eye may be damaged depending on the wavelengths of the radiation. Of course, light from any source is concentrated very considerably at the retina, but the image produced by a laser beam is about 200 times smaller and thus more concentrated than that of the sun.

> Direct sunlight of about 0.025 W/cm^2 at the cornea rapidly damages the retina. Thus, dividing by 200 and then applying a safety factor of 100, demonstrates that laser light does not become completely safe for continuous viewing until its power density is reduced to about 10^{-6} W/cm^2.

The eye does not focus UV and most IR radiations so that when emitted by lasers they are no more hazardous than radiation of these wavelengths from non-laser sources. However, the IR radiation lying just beyond the red end of the visible spectrum is focused and possesses the added danger of being invisible and therefore without prior warning.

Figure 2.6 shows a diagrammatic horizontal section of a left eye. The eye may be separated into two parts, the front (anterior) chamber being bounded by the cornea, iris and lens; and the back (posterior) part, which is bounded by the retina, containing the vitreous humour. Since the eye, or any other part of the body, can only be damaged when radiation is absorbed, it follows that for lasers, the areas of the eye which absorb visible radiation are of prime concern, e.g. the retina.

The retina (Fig. 2.6) is a complex layer structure which forms a continuous lining around the posterior chamber of the eye. Two important areas of this structure are the fovea, which is responsible for high quality vision and colour awareness, and the optic disc which is the end of the optic nerve, connecting the eye to the brain. The remainder of the retina has little optical resolving power and this coupled with the relative size of this area makes damage to it less important.

The optic disc is of such a small size that the chances of it being hit by a laser beam are extremely low, but if a direct hit with sufficient energy were to occur, then the whole eye could be made completely inoperative. The fovea (Fig. 2.6) is somewhat different to the rest of the retina in several respects; the most important of these being the presence of macular pigment. This pigment is highly absorbent in the green/blue

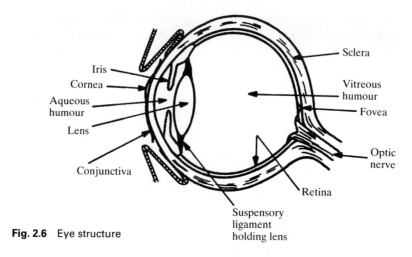

Fig. 2.6 Eye structure

region of the spectrum. The effect of wavelength on eye damage has been established by animal experiments showing that He/Cd lasers (wavelengths c. 442 nm) are potentially more hazardous than He/Ne lasers (wavelengths c. 633 nm) of the same power rating. Damage to the fovea, if it does occur, may result in 'blind spots' which if repeated could cause the eye to be virtually inoperative.

Laser damage to the eye may be divided into two types, thermal and explosive, both producing retinal lesions. Thermal damage is caused by absorption of radiation in the pigment epithelium, thereby generating heat which is conducted away with destruction of nearby cells. Explosive effects are usually caused by a short pulse of high energy radiation which, although again absorbed in the pigment epithelium, produces very rapid heating which cannot be conducted away quickly enough. The lesions produced by these processes involve complete destruction of the membrane structure in the receptor cell layer of the retina, which may not only damage the immediate irradiated area, but also may cut the optic nerves from surrounding parts.

Laser radiation to the eye is not believed to be cumulative, but any damage caused is almost certainly irreversible. Coloured races are more at risk than Caucasians because of the selective adsorption of laser energy by pigmented ocular structures.

Apart from those factors previously considered, there are several which might have been expected to control the radiation exposure to the eye. Of these, the coherence of the radiation and defocusing of the light beam are apparently of little value for safety purposes. The eye itself is of greater significance in that it has two main protective devices: the lid reflex, taking ~ 0.02 seconds and pupil constriction which usually may take 0.1 to 0.5 seconds.

When using optical instruments it is imperative to ascertain the relative importance of all the apertures which limit the spectral power density on the retina. Factors which affect the power density include image size, beam convergence (or divergence) and the use of polarising optics. Each appliance will involve different combinations of these parameters so that each situation must be considered on its merits, taking into account all of the above factors.

Safety standards

Recent publications considerably simplify safety guidance by classification of lasers according to their maximum damage potential, which depends on their maximum accessibility power levels and their wavelengths – accessible emission limits (AEL) being defined for each class – Table 2.20.

In the UK laser safety goggles or shields are required under the Protection of Eyes Regulations (1974) 'where there is a reasonable risk of injury to the eye'. The wearing of goggles is recommended when non-enclosed lasers of power >500 mW (class 4) are being used. At lower

Table 2.20 Classification of lasers

Class	Reason for classification	AEL example (visible continuous working lasers)
1. Exempt	Output is so low that they are intrinsically safe under all conditions.	< 1 μW
2. Low power	Eye protection is normally afforded by aversion responses including the blink reflex and with not staring into the beam.	< 1 mW
3A. Medium power	Protection for the unaided eye should be afforded by aversion responses but looking directly into the laser, along the beam, must be avoided.	< 5 mW
3B. Medium power	Hazards are associated with direct viewing and also from specular reflections. Under certain conditions they may safely be viewed via diffuse reflector.	< 500 mW
4. High power	Hazardous diffuse reflection with sufficient outputs to cause flesh burns. Total enclosure with indirect viewing (e.g. closed-circuit TV) should be considered.	> 500 mW

powers goggles should be worn when class 3B lasers are being set up/manipulated. Nevertheless, the wearing of goggles can never be a substitute for beam enclosure and good safety practice. Great care must also be taken to ensure that appropriate goggles for one laser are not inadvertently used for another of different wavelength or power density.

In addition to this classification scheme, there is a complex system of maximum permissible exposure levels (MPEs or MPELs). However, these are difficult to apply and the use of the classification scheme (which basically takes these values into account) renders such a strategy superfluous.

Safety methods adopted in laser laboratories in the UK have demonstrably led to a satisfactory degree of containment and ophthalmological control for laser workers. Acceptance of the more stringent codes now under consideration is likely to make this environmental control even more effective.

Precautions

In addition to the guidance given above the following precautions are worthy of emphasis.

- **Never look into a laser, along the beam, either directly or by reflection.**
- Prior to working with lasers, all operators should have visual acuity checks by opticians. Retinal photography is of dubious value. Authorisation to work with lasers will depend also on intelligence, knowledge, experience, etc. Persons with monocular vision should not be permitted to work with lasers. A register should be kept of laser operators and any laser incidents.
- Medical examination and retinal photography is important following accidental exposure to laser beams or on development of vision defects.
- Safety and health aspects of using lasers should be considered at the design stage of any new project involving these devices.
- Prominent warning notices should be displayed (e.g. on the equipment or as illuminated signs outside the room as appropriate) and entry prohibited unless authorised by the person in charge.
- Optical components should be rigidly fixed to the bench to prevent accidental deflection of the beam. For the higher power devices, the laser should be isolated in special areas/rooms.
- In normal usage, the light path of the laser should be totally enclosed wherever practical and terminated with a heat-resistant non-reflecting beam stop (black card is unsuitable for all but the lowest power lasers). For higher risk lasers (including He-Cd and Ar-ion lasers), where practicable, the enclosure should be interlocked in a fail-safe manner

with the sources so that it cannot be energised until the enclosure is in position.
- A high level of general illumination is advisable in the laser room to constrict the pupil and thereby limit the energy which may enter the eye.
- Looking into the primary beam or at specular reflections of the beam must be avoided particularly if the power energy density is above the maximum/permitted level. Eye protectors cannot be assumed to give complete protection. **It is prudent to avoid this practice altogether.**
- When not in use, the mains supply to the laser should be switched off.
- Eye protection filters are no substitute for good practice and beam enclosure.
- The ozone concentration in the working environment of laser rooms should be assessed on commissioning of new equipment and when significant electrical alterations or servicing are undertaken to ensure operator exposure is well within the Hygiene Standard. (The results will determine the frequency of monitoring and whether modification, such as increased ventilation, is required.)
- An interlock system should be used to ensure all circuit alterations are made with power supplies disconnected.

Visual display units

VDUs and microfiche readers are a feature of most modern workplaces. VDUs are cathode-ray tubes resembling television sets and are an integral part of computers, word-processor units, etc. Their widespread use has led to concern about possible health hazards such as radiation exposure, epileptogenic properties, cause of eye-strain and other alleged adverse health effects. In the UK a selection of important sources of information is given in references 92–97.

Radiation levels from VDUs have been shown to be far lower than permissible levels for continuous exposure. While it is acknowledged that VDUs possess epileptogenic potential the likelihood of operators experiencing such effects as a result of working with VDUs is extremely remote. In the main, problems with VDUs stem from inadequate attention to ergonomical and psychological considerations. Thus workplace layout and environmental conditions such as levels of background lighting and noise are important. Provision of detachable keyboards, well-designed seats, and reader-holders help prevent problems arising from poor posture. Operators should be presented with stable, clear images and consideration should be given to the repetitive nature of the task together with the duration that operators are expected to work at VDUs without a break.

Thermal hazards[98-103]

Heat

To function normally the body must maintain a core temperature of 36–38 °C. It is therefore equipped with an efficient thermoregulatory system, to make adjustments to fluctuations in heat generated by metabolism (determined by work load) and exposure to differing ambient thermal environments.

The fundamental thermodynamic process involved in heat exchange between the body and its milieu may be described by

$$M \pm R \pm C - E = \Delta H$$

where M = heat gain from metabolism; R = heat exchange by radiation; C = heat exchange by convection; E = heat loss by evaporation and ΔH = change in body heat content. (Heat transfer by conduction is negligible and therefore ignored.) Under conditions of thermal equilibrium, heat generated within the body by metabolism is completely dissipated to the environment and $\Delta H = 0$.

When heat loss fails to keep pace with heat gain the body temperature begins to rise; this triggers certain physiological mechanisms into action. Thus, blood vessels of the skin and subcutaneous tissues dilate with diversion of cardiac output to those superficial regions. Concomitantly with this the volume of circulating blood is increased by contraction of the spleen and by dilution of the circulating blood with fluids drawn from other tissues. Cardiac output is increased. These adjustments result in enhanced heat transport from the body core to the surface; the sweat glands are also activated.

Heat transfer by convection relies on a current of moving air. Its effectiveness in cooling depends on wind speed and the temperature differential between the air and skin. Clearly clothing will reduce the air movement next to the skin.

Sweating is an important heat dissipation process involving evaporation. Normally an individual at rest will sweat at a rate of 1 litre per day. This can rise to 1 litre per hour under heavy exertion and/or high temperatures, and as much as 10 g of the body's salt will be lost with the water. If the sweat can be evaporated as rapidly as it is produced, and if the heat stress is not such as to exceed the maximum sweating capability of the body, the body maintains the necessary constant temperature. However, the rate of evaporation will depend upon the moisture content, temperature and movement of ambient air.

Heat sources such as furnaces, fires, etc., radiate electromagnetic energy in the IR region of the spectrum. Air movement provides no relief and protection must involve shielding of the source from the operator with material highly reflective in the IR (e.g. aluminium foil or sheet).

When the body's ability to shed sufficient of the heat load causes the thermoregulatory defence mechanism to become overburdened, heat stress results. Tolerance to heat is influenced by the size/weight of individuals, age and degenerative disease, physical fitness, water and salt balance, alcoholic intake and degree of acclimatisation. After about 10 days of relatively short daily exposures to heat a worker becomes acclimatised which enables him to perform the job with less physical stress than a person unaccustomed to the hot conditions. Such beneficial effects are retained over a weekend but lost if there is no exposure to heat for a period of about 2 weeks.

Physical effects

Physical disabilities caused by excessive heat exposure together with predisposing factors, treatment and prevention are summarised in Table 2.21.[98] The most important examples are discussed below.

- *Heat rash* (prickly heat) – results in hot–humid environments because of sweat gland dysfunction. The orifices of sweat ducts become blocked by a swollen moist keratin layer of skin which leads to inflammation of the glands. In addition to discomfort, heat rash greatly diminishes the workers' capacity to tolerate heat.
- *Heat cramps* – occur after prolonged exposure to heat with profuse perspiration and loss of salt. Signs and symptoms include pain in muscles of abdomen and extremities.
- *Heat exhaustion* – may result from physical exertion in a hot environment when vasomotor control and cardiac output are inadequate to meet the demand put on them. Signs and symptoms include pallor, lassitude, dizziness, syncope, profuse sweating and cool, moist skin.
- *Heat stroke* – a serious condition resulting from excessive physical exertion in extreme thermal environments. Signs may include dizziness, nausea, severe headache, hot dry skin, cessation of sweating, high core temperature, collapse, delirium and coma.
- *Heat cataract* – many years of exposure to IR radiation may result in the classic eye lesion of posterior cataract. Because of its incidence this heat cataract, which is a Prescribed Disease in the UK, is often referred to as glassblowers' or chain-makers' cataract. Furnace-men and smelters, etc., may also be exposed to a significant degree of IR radiation. The provision and use of appropriate eye protection is an important precaution.

Measurement of the thermal environment

Detailed discussion of this topic is given in references 70 and 99. Depending upon the nature of the environment, factors which need to be considered include wind speed, clothing index, mean radiant temperature, humidity, work rate and air temperature. Monitoring hardware will

Table 2.21 Heat illness – classification, medical features and treatments, and preventive measures (after ref. 92)

Category	Clinical features	Predisposing factors	Prevention
Temperature regulation			
Heat stroke Heat hyperpyrexia	Heat stroke: (1) *Hot dry skin*: red, mottled or cyanotic. (2) *High and rising* T_e, 40.5 °C and over. (3) *Brain disorders*: mental confusion, loss of consciousness, convulsions, coma as T_r continues to rise. Fatal if treatment delayed. Heat hyperpyrexia: milder form. T_e lower; less severe brain disorders, some sweating.	• Sustained exertion in heat by unacclimatised workers. • Lack of physical fitness and obesity. • Recent alcohol intake. • Dehydration. • Individual susceptibility. • Chronic cardiovascular disease in the elderly.	Medical screening of workers. Select based on health and physical fitness. Acclimatise for 8–14 days by graded work and heat exposure. Monitor workers during sustained work in severe heat.
Circulatory hypostasis			
Heat syncope	Fainting while standing erect and immobile in heat.	Lack of acclimatisation.	Acclimatise workers. Intermittent activity to assist venous return to heart.
Salt and/or water depletion			
Heat exhaustion	(1) Fatigue, nausea, headache, giddiness. (2) Skin clammy and moist. Complexion pale, muddy or hectic flush. (3) May faint on standing with rapid thready pulse and low blood pressure. (4) Oral temperature normal or low; rectal temperature usually elevated (37.5–38.5 °C). *Water restriction type*: Urine volume small, highly concentrated. *Salt restriction type*: Urine less concentrated, chlorides less than 3 g/l.	• Sustained exertion in heat. • Lack of acclimatisation. • Failure to replace water and/or salt lost in sweat.	Acclimatise workers using a breaking-in schedule for 1 or 2 weeks. Supplement dietary salt only during acclimatisation. Ample drinking water to be available at all times; to be taken frequently during work day.
Heat cramps	Painful spasms of muscles used during work (arms, legs, or abdominal). Onset during or after work hours.	• Heavy sweating during hot work. • Drinking large volumes of water without replacing salt loss.	Adequate salt intake with meals. In unacclimatised men, provide salted drinking water.

Table 2.21 (continued)

Category	Clinical features	Predisposing factors	Prevention
Skin eruptions			
Heat rash (miliaria rubra; 'prickly heat')	Profuse tiny raised red vesicles (blister-like) on affected areas. Pricking sensations during heat exposure.	Unrelieved exposure to humid heat with skin continuously wet with unevaporated sweat.	Cooled sleeping quarters to allow skin to dry between heat exposures.
Anhidrotic heat exhaustion (miliaria profunda)	Extensive areas of skin which do not sweat on heat exposure, but present goose flesh appearance, which subsides with cool environments. Associated with incapacitation in heat.	Weeks or months of constant exposure to climatic heat with previous history of extensive heat rash and sunburn.	Treat heat rash and avoid further skin trauma by sunburn. Periodic relief from sustained heat.
Behavioural disorders			
Heat fatigue – Transient	Impaired performance of skilled sensorimotor, mental, or vigilance tasks, in heat.	Performance decrement greater in unacclimatised, and unskilled men.	Acclimatise and train for work in the heat.
Heat fatigue – Chronic	Reduced performance capacity. Lowering of self-imposed standards of social behaviour (e.g. alcoholic overindulgence). Inability to concentrate, etc.	Workers at risk come from homes in temperate climates, for long residence in tropical latitudes.	Orientation on life abroad (customs, climate, living conditions, etc.)

also depend upon the hazard and a range of instruments is available. For example for air temperature measurements mercury in glass, alcohol in glass, thermocouple and thermistor thermometers are available; for air velocity measurements swinging vane, rotating vane, hot wire and heated thermocouple versions of anemometers exist along with Kata thermometers; humidity measurements are made with wet bulb thermometers or sling hygrometers. Each of the foregoing has particular strengths and weaknesses: all are of value for various situations.

To determine the risk associated with exposure to a thermal environment monitoring data are interpreted with the aid of nomograms, charts, etc., for comparison against appropriate indices such as Wet-Bulb Globe Temperature Index, Heat Stress Index of Belding and Hatch, Effective Temperature and Corrected Effective Temperature, Predicted Four-Hour Sweat Rate, etc.[99]

Precautions

Exposure to heat in industry often exceeds that experienced in even the most hostile of natural climates. 'Hot' industrial jobs frequently also demand heavy work. This combination of exposure to thermal environments with production of high levels of metabolic heat may exceed the worker's physiological capacity to regulate body temperature. In such situations control measures need to be introduced to prevent heat strain. The two conditions under which heat becomes troublesome are warm–moist and hot–dry. For the former, heat is mainly convected and work becomes difficult because of the humidity which hampers evaporative cooling. Amelioration is primarily ventilation and air conditioning. In hot–dry environments ventilation is inappropriate and amelioration relies on reducing heat falling on the individual.

Where practicable these hazards should be designed out at the planning stage or controlled by engineering methods. Thus to remove radiant heat the strategy is to reduce the temperature or thermal emissivity of the IR source where possible and apply conductive heat insulation to sources whose temperatures cannot be lowered. This reduces surface temperature and hence diminishes radiant heat emitted. Where heat is an inherent part of the process and temperatures cannot be lowered (e.g. molten glass) then shields should be installed to provide IR shadows for operators. These should be chosen to function either:

1. By reflecting IR (e.g. aluminium foil or sheet). These must not absorb radiant heat and become hot.
2. Exchanging heat, e.g. heat absorbed by screens of iron or steel is removed by water-cooling. Ventilated reflective booths/control rooms also provide a useful haven.

Personal protection made of IR reflecting material is available for operators in extremely hot environments. Such protection is usually non-porous and the body can experience difficulty 'breathing'. Full-quilted suits faced with aluminised cloth and fitted with a helmet with IR reflective glass view plate and self-contained supply of compressed air has been used in extreme situations. Refrigerated suits are also available.

Other methods for controlling heat exposure, depending on the nature of the hazard, include increased air movement, evaporative cooling, supply of fresh outdoor air, air-conditioned work areas, air-conditioned rest areas, proper acclimatization to heat, intake of salt, environmental monitoring, minimisation of exposure period, medical screening and other administrative controls.

Table 2.22 lists some occupations with potential for heat exposure. A selection of chemical processes which may pose a potential hazard from heat stress, with mention of the appropriate control measures adopted, is given below (after ref. 103).

Table 2.22 List of occupations with potential heat exposures

Animal-rendering workers	Glass manufacturing workers
Bakers	Kiln workers
Boiler men	Laundry workers
Cannery workers	Miners in deep mines
Chemical plant operators and maintenance workers working near hot containers and furnaces	Outdoor workers during hot weather
	Paper makers
	Sailors passing hot climatic zones
Cleaners, e.g. boilers	Shipyard workers when cleaning cargo holds
Coke oven operator	
Cooks	Smelter workers
Foundry workers, e.g. casting and 'knocking-out' operations	Steel and metal forgers
	Textile manufacturers, e.g. in de-sizing
	Rubber workers, e.g. on calenders

Bottle glass production Heat stress can pose a problem in production areas to a degree dependent upon the size of the operation, the crowding of equipment, etc. Heat stress is controlled by isolation of employees, design of work practices, administrative and engineering controls.

Cellulose acetate manufacture During the washing of the product to free it of acetic acid large amounts of water are involved and the relative humidity in the room reaches 100%. On hot days this can be particularly problematical since there is little opportunity for workers to lose heat by sweat evaporation. Spot cooling or air-conditioned remote control rooms aid comfort and efficiency.

Cement manufacture Operators can be exposed to stressful combinations of radiant and convection heat from the kiln (for clinker formation) and the clinker quenching operation. Control of heat exposure in the kiln area of the plant involves isolation of employees in control room and/or cooling-off areas, reducing duration of exposure, education (including recognising symptoms of heat stress), and use of personal protection.

Chlorine production Heat stress may be encountered in the mercury electrolytic cell process. Stripping of sodium–mercury amalgam in the decomposer is an exothermic reaction liberating considerable amounts of heat.

Chromic acid production Ore is roasted to oxidise the chromium from valency 3 to 6, and to initiate the reaction with soda ash. Heat stress can be a problem around kilns and driers.

Ethyl alcohol from hydration of ethylene Heat in and around the boilers (used to raise the temperature of ethylene feedstream) combined with strenuous work or warm climatic conditions may result in the working environment becoming thermally stressful. Maintenance operations in the boiler region may also expose personnel to excessive heat. Chilled air, shielding from radiant heat, protective clothing (including protective suits fitted with Vortex tubes), the use of fans to move air, administrative measures, provision of cooling-off areas and regular monitoring may be required to control heat exposures.

Glass fibre production During the fibre-drawing stage of production, which is a labour intensive operation, the warm, humid atmosphere on the forming platform may result in discomfort rather than heat stress. The main control measure is air movement or air-conditioning in the forming areas. An added complication stems from potential noise problems, since the high humidity and warm environments makes it difficult to enforce the wearing of ear protection.

Nickel refining Matte containing nickel and impurities is handled in solution through to separation and packaging. This ensures dust problems are not encountered. During the leaching of impurities, heat stress can be a consideration in this process particularly in hot summer months. Roof fans ventilate the building to relieve the heat.

Nylon-6 fibre manufacture Heat can be a difficulty around the chip melting and fibre extrusion facilities. Exposure to heat in chip melting tends to be limited to service and maintenance activities. However, the spinnerettes demand more continuous attention and require a proper programme of heat control.

Petroleum refining Heat sources at crude fractionation units are in and around furnaces, heat exchangers, hot feed pumps and associated equipment. With the exception of on-stream maintenance, exposures of operators to heat are usually of short duration except in hot climates, e.g. the Middle East, where appropriate work rotation is necessary.

Significant heat stress sources such as furnaces are also encountered during reforming and cracking.

Polyester resin production Heat is used in the polymer production process to accommodate reaction requirements and temperatures in adjacent work areas may be high. The situation is exacerbated during hot summer conditions accompanied by high humidity. Control is achieved by a programme of acclimatisation, provision of fluid replacements, cooling-off areas, plus engineering and administrative measures.

Steel production Heat exposures vary considerably throughout the process. The main problem is exposure to intense heat for short intervals of time. Opportunities exist to retreat to cooler, often air-conditioned areas. Aluminised clothing may be required to reduce the high radiant heat load, e.g. at the teeming platform, charging floor, during removal of slag, etc. Training and thermal monitoring programmes should be implemented.

Textile industry Heat stress situations in this industry differ from those upon which most criteria are based. In finishing operations, there can be very high humidities coupled with very low air movement. Radiant heat loads from tenter frames complicate the situation.

Cold

The body possesses physiological mechanisms to ensure thermal homoeostasis is maintained. These tend first to limit heat loss and then to increase heat production. Vasoconstriction of skin vessels, reduced blood flow, and hunching (to reduce surface area) decrease heat lost by the body. As vasoconstriction becomes inadequate to maintain the body temperature muscular hypotonus, and shivering become important to raise heat production. Inactivity can strain the mechanism of heat conservation by vasoconstriction and the extremeties such as fingers and toes may 'freeze'. Even before this, hands and fingers become numb as temperatures fall below 15 °C, resulting in increased probabilities of malfunction and accidents. In general, cooling stress is proportional to the thermal gradient between the skin and the environment. The combined effect of wind and temperature produce 'windchill' as shown by Fig. 2.7.[100] Loss of heat by evaporation of sweat is insignificant below 15–20 °C.

Harmful effects of exposure to cold conditions include frost-bite, trench-foot, hypothermia and vascular abormalities.

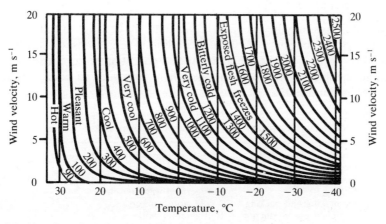

Fig. 2.7 Heat loss from body in kcal hr^{-1} m^2 for various temperatures and wind velocities

Physical effects

Frostbite This occurs when freezing takes place in tissue with mechanical disruption of cell structures. It occurs at −1 °C and is facilitated by forced convective cooling by wind. After onset it progresses rapidly. Frost-bite can also result at warm temperatures if the skin comes in direct contact with objects of sub-zero temperatures. Usually warning is given by a prickling sensation but numbness may permit freezing to develop with the sufferer being unaware. Injury ranges from superficial redness of the skin to deep tissue freezing and thrombosis, cyanosis and gangrene.

Trench-foot This may result from long continuous exposure to cold conditions, combined with persistent dampness. This condition arises from local tissue anoxia and severe cold causing damage to capillary walls. Oedema, tingling, itching or even severe pain may develop followed in some cases by blistering and ulceration.

General hypothermia This extreme acute problem results from prolonged cold exposure and heat loss. Individuals fatigued during physical activity are most prone to heat loss. As exhaustion is reached the vasoconstriction mechanism is overburdened and vasodilation occurs with rapid loss of heat.

Vascular abnormalities Conditions such as chilblains, Raynaud's disease (see page 35) acrocyanosis and thromboangiitis obliterans can be precipitated or aggravated by cold exposure. Workers suffering these ailments therefore need to adopt special precautions to avoid chilling.

Precautions

Cold conditions may be encountered in industry because of failure of the heating system, because of climatic conditions (e.g. transport workers in winter) or because of the inherent process conditions (e.g. refrigeration).

Table 2.23 lists some occupations where exposure to cold conditions may be encountered.

Table 2.23 Occupations with potential for cold exposures

Diving	Outside working during cold weather
Dry ice handling	Offshore oil-rig workers
Fire fighting	Packing house working
Fishing	Refrigerated warehousing/
Ice making and packing	cold store workers
Liquefied gas production	Refrigeration workers

General occupational safety 69

In general, however, exposures to low temperatures tend to be less of a problem in industry than do high temperatures, mainly because extra insulation in the way of clothing can be readily provided. Even so, the face will be exposed and workers may need to limit their exposures for relief. Also because the chilling effect of wind is often overcome to a certain extent by use of impermeable clothing, sweat evaporated from the body will condense and clothing will require to be dried out periodically.

Precautions in a cold working environment include physical fitness, use of several layers of clothing (with some venting of perspiration) with personal protection where necessary. Metal tools should be used with care since metal conducts heat rapidly from the body. In more extreme conditions (at sub-zero temperatures) workers should be cleared by the medical department, wear appropriate protective clothing (e.g. parkas, gloves, boots, half-face masks, etc.), avoid perspiring before entering the cold room, breath through the nose rather than mouth, and take care not to slip on frosty floors or stairs. Administrative controls should be introduced to ensure visual or telephone contact is maintained and duration of exposures is minimised. Table 2.24 lists time limits that have been recommended for work in a variety of thermal conditions.[102]

Table 2.24 Low-temperature time limits

Temperature range (°F)	Maximum daily exposure
30 to 0	No exposure time limit, if the person is properly clothed.
0 to −30	Total cold-room work time: 4 hours. Alternative 1 hour in and 1 hour out of the chamber.
−30 to −70	Two periods of 30 minutes each, at least 4 hours apart. Total cold-room work time allowed: 1 hour. (*Note*: Some difference exists among individuals: one report recommends 15-minute periods – not over 4 periods per work shift; another limits periods to 1 hour out of every 4, with a low chill factor, i.e. no wind; a third says that continuous operation for 3 hours at −65 °F has been experienced without ill effect.)
−70 to −100	Maximum permissible cold-room work time: 5 minutes over an 8-hour working day. For these extreme temperatures, the wearing of a completely enclosed headgear, equipped with the breathing tube running under the clothing and down the leg to preheat the air, is recommended.

References

(*All places of publication are London unless otherwise stated*)
1. King, R. & Magid, J., *Industrial Hazard and Safety Handbook*. Newnes-Butterworths 1979.
2. Ridley, J., *Safety at Work*. Butterworths 1983.

3. Anderson, P. W. P. (ed.), *Safety Manual for Mechanical Plant Construction*. Klewer Publishing 1984.
4. *The Construction (Lifting Operations) Regulations*. HMSO 1961
5. *The Construction (General Provisions) Regulations*. HMSO 1961.
6. Health and Safety Executive, *Guidance Note GS 15: General Access Scaffolds*. HMSO 1982.
7. Health and Safety Executive, *Guidance Note GS 10: Roof Work: Prevention of Falls*. HMSO 1979.
8. Health and Safety Executive, *Guidance Note GS 25: Prevention of Falls to Window Cleaners*. HMSO 1983.
9. British Standards Institution, *BS C993: The Use of Safety Nets on Constructional Work*. HMSO 1972.
10. British Standards Institution, *BS 1397: Specification for Industrial Safety Belts, Harnesses and Safety Lanyards*. HMSO 1979.
11. British Standards Institution, *BS 5845: Specification for Permanent Anchors for Industrial Safety Belts and Harnesses*. HMSO 1980.
12. Watkin, B., *Health and Safety at Work*, 1983, **5**(10), 29.
13. Health and Safety Commission, *Consultative Document: Health and Safety (Manual Handling of Loads) Regulations and Guidance 1983*. HMSO.
14. Royal Society for the Prevention of Accidents, '*Moving On*' (a film in 3 parts). RoSPA film library.
15. Dickie, D. E. & Short, E., *Lifting Tackle Manual*. Butterworths 1981.
16. Dickie, D. E., *Crane Handbook*. Butterworths 1981.
17. Health and Safety Executive, '*The Explosion and Fire at Chemstar Ltd, 6th September 1981*'. HMSO 1982.
18. Royal Society for the Prevention of Accidents, *RoSPA Bull*. 1983 (July), 2.
19. Health and Safety Executive, *Guidance Note PM 1: Guarding of Portable Pipe Threading Machines*. HMSO 1984.
20. Health and Safety Executive, *Guidance Note PM 2: Guards for Planing Machines*. HMSO 1976.
21. Health and Safety Executive, *Guidance Note PM 6: Dough Dividers*. HMSO 1979.
22. Health and Safety Executive, *Guidance Note PM 10: Tripping Devices for Radial and Heavy Vertical Drilling Machines*. HMSO 1977.
23. Health and Safety Executive, *Guidance Note PM 23: Safety in the use of Woodworking Machines* HMSO 1981.
24. Health and Safety Executive, *Guidance Note PM 23: Photoelectric Safety Systems*. HMSO 1981.
25. Health and Safety Executive, *Guidance Note PM 35: Safety of Bandsaws in the Food Industry*. HMSO 1981.
26. Health and Safety Executive, *Guidance Note PM 35: Safety in the Use of Reversing Dough Brakes*. HMSO 1983.
27. Health and Safety Executive: *Guidance Note PM 40: Protection of Workers at Welded Steel Tube Mills*. HMSO 1984.
28. Health and Safety Executive, *Guidance Note PM 41: The Application of Photo-Electric Safety Systems to Machinery*. HMSO 1984.
29. Health and Safety Executive, *Guarding of Hand-Fed Platen Machines*. Health and Safety at Work Booklet 11: HMSO.

30. Health and Safety Executive, *Safety in the Use of Mechanical Power Presses*. Health and Safety at Work Booklet 14: HMSO.
31. Health and Safety Executive, *Drilling Machines – Guarding of Spindles and Attachments*. Health and Safety at Work Booklet 20 (2nd edn). HMSO 1974.
32. Health and Safety Executive, *Safety in the Use of Guillotines and Shears*. Health and Safety at Work Booklet 33. HMSO 1974.
33. Health and Safety Executive, *Safety in the Use of Woodworking Machines*. Health and Safety at Work Booklet 41. HMSO 1970.
34. Health and Safety Executive, *Guarding of Cutters of Horizontal Milling Machines*. Health and Safety at Work Booklet 42. HMSO.
35. British Standards Institution, *BS 4999: Specification for General Requirements for Rotating Electrical Machines. Part 20: Classification of Types of Enclosure*, 1972.
36. British Standards Institution, *BS 5304: Code for Safeguarding of Machinery*, 1975.
37. British Standards Institution, *BS 5498: Specification for the Safety of Hand-Operated Paper Cutting Machines*, 1977.
38. British Standards Institution, *BS 5669: Specification for Continuous Mechanical Handling Equipment – Safety Requirements. Part 18: Conveyors and Elevators with Chain Elements – Examples for guarding of Trip Points*, 1979.
39. British Standards Institution, *BS 5933: Specification for Overhead Guards for High Lift Rider Trucks*, 1980.
40. British Standards Institution, *BS 5945: Specification for Guards and Shields for Earth-Moving Machinery*, 1980.
41. British Standards Institution, *BS 6491: Electro-Sensitive Safety Systems for Industrial Machines. Part 1: Specification for General Requirements*, 1984.
42. Royal Society for the Prevention of Accidents, *RoSPA Bull.*, 1980 (Sept.), 2.
43. Royal Society for the Prevention of Accidents, *RoSPA Bull.*, 1983 (June), 2.
44. Royal Society for the Prevention of Accidents, *RoSPA Bull.*, 1984 (Mar.), 3.
45. Royal Society for the Prevention of Accidents, *RoSPA Bull.*, 1984 (Nov.), 2.
46. Royal Society for the Prevention of Accidents, *RoSPA Bull.*, 1983 (Aug.), 2.
47. Royal Society for the Prevention of Accidents, *RoSPA Bull.*, 1981 (Apr.), 2.
48. *H. M. Chief Inspector of Factories Annual Report 1974*. HMSO.
49. Health and Safety Commission:
    (a) *Electricity at Work Other Than in Mines and Quarries*. Draft Regulations, Approved Code of Practice and Guidance. HMSO 1984.
    (b) *The Use of Electricity at Mines*. Draft Regulations and Approved Code of Practice. HMSO 1984.
    (c) *The Use of Electricity at Quarries*. Draft Regulations and Approved Code of Practice. HMSO 1984.
50. British Standards Institution, *BS CP 1013: 1965 'Earthing'* 1965.

51. Institution of Electrical Engineers, *Regulations for Electrical Installations* (15th edn). IEE 1981.
52. Royal Society for the Prevention of Accidents, *RoSPA Bull.*, 1983 (Aug.), 3.
53. Royal Society for the Prevention of Accidents, *RoSPA Bull.*, 1984 (Jan.), 3.
54. Royal Society for the Prevention of Accidents, *RoSPA Bull.*, 1984 (June), 2.
55. Royal Society for the Prevention of Accidents, *RoSPA Bull.*, 1980 (Dec.), 3.
56. Health and Safety Commission, *Newsletter No. 39*. HMSO 1984 (Dec.), 6.
57. British Standards Institution, *BS 2769 Portable Electrical Motor-operated Tools*, 1964.
58. Health and Safety Commission, *Newsletter, No. 39*. HMSO 1984 (Dec.), 4.
59. Royal Society for the Prevention of Accidents, *RoSPA Bull.*, 1984 (Oct.), 4.
60. Health and Safety Commission, *Consultative Document on Protection of Hearing at Work*. HMSO 1981.
61. Petrusewicz S. A. & Longmore, D. K. (eds), *Noise and Vibration Control for Industries*. Elek Science 1974.
62. Beranek L. L. (ed.), *Noise and Vibration Control*. McGraw Hill, New York 1971.
63. Michael, P. L., 'Industrial noise and conservation of hearing', in *Pattys Industrial Hygiene and Toxicology*, Vol. 1 (3rd edn). Wiley, New York 1978.
64. Atherley, G. R. C. & Booth, R. T., *Labour Research*, 1974 (July)
65. Department of Employment: *Code of Practice for Reducing the Exposure of Employed Persons to Noise*. HMSO 1972.
66. International Labor Office, *Protection of Workers Against Noise and Vibration in the Working Environment Code of Practice*.
67. Health and Safety Commission, *Some Aspects of Noise and Hearing Loss: Notes on the Problem of Noise at Work and Report of the Health and Safety Executive Working Group on Machinery Noise*. HMSO 1981.
68. Health and Safety Executive, *100 Practical Application of Noise Reduction Methods*. HMSO 1983.
69. Jensen, P. & Vokel, C., *Industrial Noise Control Manuals*. National Institute for Occupational Safety and Health 1979.
70. Gill, P. S. & Ashton, I., *Monitoring for Health Hazards at Work*. Grant McIntyre Medical and Scientific 1982.
71. Botsford, J. H., 'Noise Measurement and Acceptability Criteria', in *The Industrial Environment – its evaluation and control*. National Institute for Occupational Safety and Health, Washington 1973.
72. Hill, V. H., 'Control of noise experience', in *The Industrial Environment – its evaluation and control*. National Institute for Occupational Safety and Health, Washington 1973.
73. Health and Safety Executive, *Guidance Note EH 14: Level of Training for Technicians Making Noise Surveys*. HMSO 1977.
74. Brief, R. S., *Basic Industrial Hygiene: a training manual: Section 15 Visible*

Light. American Industrial Hygiene Association.
75. Illuminating Engineering Society, *IES Lighting Handbook* (5th edn). New York 1972.
76. Creuch, C. L. 'Lighting for seeing', in *Patty's Industrial Hygiene and Toxicology* (3rd edn), Vol. I. Wiley, New York 1978.
77. Illuminating Engineering Society, *Office Lighting Studies*. New York 1977.
78. Kaufman, J. E., 'Illumination', in *The Industrial Environment – its evaluation and control*. National Institute for Occupational Safety and Health, Washington 1973.
79. Hughes, D., *Hazards of Occupational Exposure to Ultraviolet Radiation*. Occupation Hygiene Monograph No 1 (2nd edn). Science Reviews, Leeds 1982.
80. National Radiological Protection Board, *Protection Against Ultraviolet Radiation in the Workplace*. HMSO 1978.
81. Dutt, G. C., *Effects of Ultraviolet Radiation on Man*. National Research Council of Canada 1978.
82. British Standards Institution, *BS 679: Specification for Filters for Use During Welding and Similar Industrial Operations*, 1959 (Amd 1977).
83. British Standards Institution, *BS 4803: Radiation Safety of Laser Products and Systems*, 1983.
84. Cox, E. A., *Safety in the Use of Lasers on Site*. Chartered Institute of Building, Ascot 1983.
85. Laney, J. C., 'Lasers in construction', in *Site Safety*. Construction Press 1982, Ch. 20.
86. Marshall, J., *Annals Occup. Hyg.*, 1978, **21**, 69.
87. Harlen, F., *Annals Occup. Hyg.*, 1978, **21**, 199.
88. Friedmann, *Annals Occup. Hyg.*, 1978, **21**, 277.
89. American Conference of Governmental Industrial Hygienists, *A Guide for Control of Laser Hazards* (3rd edn). ACGIH, Cincinnati 1981.
90. Health and Safety Executive, *Guidance Note PM 19: Use of Lasers for Display Purposes*. HMSO 1980.
91. D. Sliney and Wolbarsht, *Safety with Lasers and Other Optical Sources*. Plenium Press, New York 1980.
92. Association of Scientific, Technical and Managerial Staffs, *Guide to Health Hazards of VDUs – An ASTMS Policy Document*. ASTMS 1979.
93. Association of Scientific, Technical and Managerial Staffs, *Hazards of Microfiche Viewers – Prevention of Occupational Eyestrain*. ASTMS 1980 (revised 1981).
94. 'Prevention of occupational eyestrain – hazards of VDUs, health and safety information', *ASTMS Monitor*, 1981 (Feb.).
95. Chartered Institution of Building Services, *Lighting for a VDU*, Technical Memoranda No. 6. London 1981.
96. Health and Safety Executive, *Human Factors Aspects of VDU Operation*, Research Paper No. 10, HSE 1980.
97. Health and Safety Executive, *Visual Display Units*, HSE 1983.
98. Minard, D., 'Physiology of Heat Stress', in *The Industrial Environment – its evaluation and control*. National Institute for Safety and Health, Washington, 1973.

99. Hertig, B. A., 'Thermal Standards and Measurement Techniques', in *The Industrial Environment – its evaluation and control*. National Institute for Occupational Safety and Health, Washington 1973.
100. Gill, F. S., 'Heat', in H. A. Waldron, & J. M. Harrington, (eds) *Occupational Hygiene*. Blackwell Scientific Publications 1980.
101. The British Standards Institution, *BS 5429: Code for Safe Operation of Small-scale Storage Facilities for Cryogenic Liquids* 1976.
102. National Safety Council, *Cold Room Testing of Gasoline and Diesel Engines*. Data Sheet 465, 1958.
103. Cralley, L. V. & Cralley, L. J., *Industrial Hygiene Aspects of Plant Operations*: Vol. 2, *Unit Operations and Product Fabrication*. Macmillan, New York 1984.

CHAPTER 3

Physico-chemical principles and safety

The extent of the risks associated with chemicals depends upon the properties of the materials stored and processed, the inventory, and the processes to which they are subjected, e.g. high pressures, oxidation, hydrogenation, etc. These are discussed in Chapters 4, 5, 8 and 10.

However, the chemical engineering operations, material transport and reaction processes involved all depend upon elementary physico-chemical principles. These principles enable the manner in which materials will behave (e.g. flow, change phase, react or decompose, exert pressures, release or absorb heat, mix or stratify, and expand or contract) to be predicted with sufficient accuracy for equipment to be designed and operated. Furthermore, the practice of chemical engineering uses as a basis certain fundamental concepts regarding heat transfer (whether by conduction, convection, radiation or some combination of these mechanisms) and mass transfer. The latter is concerned with how, and at what rates, material is transferred across a phase boundary, e.g. in absorption or distillation. Operating variables, such as agitation, or turbulent flow, or increased interfacial area, significantly affect these phenomena.

Application of the same principles enables hazards due to fire, various types of 'explosions', release of toxic or corrosive chemicals, or exposure to atmospheres containing toxic contaminants to be identified. Some of these principles are reviewed in this chapter (after ref. 1). Reference is made to incidents which have arisen time and time again, either because the principles were overlooked, or a safe design was not introduced, or a safe operating procedure was circumvented. Examples are drawn from large plant situations to small laboratory, or domestic, incidents; the purpose is to demonstrate that scale is not the determining factor – it is the physical chemistry and chemical engineering principles which determine the behaviour of the chemicals involved.

The chemical engineering and management required to avoid the types of incident exemplified are covered in detail in Chapters 6, 9, 10, 12, 13 and 17.

Vapour pressure relationships

The vapour pressure of a substance is a measure of its volatility at any specific temperature. Except at very high pressures, it is largely independent of pressure. Thus a reasonable approximation of the vapour pressure p^1 of a pure substance is given by the equation,

$$\ln p^1 = \frac{A}{T} + B \qquad [3.1]$$

when: A and B are empirically determined constants; T is the absolute temperature.

It is clear from this that the vapour pressure of a substance increases markedly with temperature; the relationship for a variety of compounds is demonstrated by Fig. 3.1.[2]

When a component 1 is present in a mixture of vapours its partial pressure p_1 is the pressure that would be exerted by component 1 at the same temperature if present in the same volumetric concentration.

The total pressure is equal to the sum of the partial pressures,

$$P = \Sigma p \qquad [3.2]$$

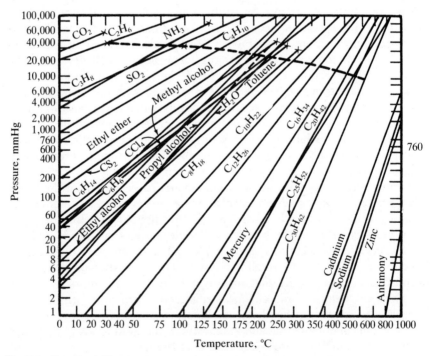

Fig. 3.1 Cox chart. Vapour pressure–temperature plot
Key: X = critical state

Thus with a mixture of two components,

$$P = p_1 + p_2$$

or if an inert gas is also present,

$$P = p_1 + p_2 + p_{inert}$$

In an ideal mixture the partial pressure p is proportional to the mol fraction y_a (the volume fraction) of the component in the gas phase

$$p_a = y_a \cdot P \qquad [3.3]$$

For an ideal mixture the partial pressure of component 'a' is related to the concentration in the liquid phase, expressed as a mol fraction x_a, by

$$p_a = p_a^1 x_a \qquad [3.4]$$

where p_a^1 is the vapour pressure of a component 'a' at the prevailing temperature.

It follows from eqns [3.2] and [3.4] that the total pressure P of a mixture is then

$$P = p_a^1 x_a + p_b^1 x_b + p_c^1 x_c \qquad [3.5]$$

Clearly the concentration of a component in air is directly proportional to its vapour pressure and increases with temperature. Put another way, exposure of any flammable or toxic material to atmosphere will, at equilibrium, result in concentrations in the atmosphere, which are directly proportional to its vapour pressure.

One important effect of the situation expressed by eqn [3.5] is the lowering of the flashpoint of liquids by the presence of traces of volatile contaminants, e.g. the flashpoint of a gas-oil, normally 60 °C, can be lowered to below ambient temperature by the presence of 3% of petrol.[3] This is an important consideration when dealing with mixed solvents, e.g. in waste disposal. The main area of concern, however, is probably that of keeping fuels free from contamination, or the avoidance of confusion of the tank contents, during tanker filling and discharge.

A fire occurred in a 1,000 m³ storage tank partly filled with a hydrocarbon normally having a flashpoint of 35 °C. Air was present because the tank was vented to atmosphere. Contamination of the hydrocarbon with acetone was found to have lowered the flashpoint to below ambient temperature.[4] The source of ignition was not firmly identified.

The roof of a gas-oil tank blew off while it was being filled with gas-oil from a ship. Two men who were on the roof were killed and (due to fire-fighting problems and lack of adequate water supplies) the fire escalated to involve other tanks and burned for 38 hours.[5] The gas-oil was contaminated with petrol which leaked through a partition between tanks on the ship. It was known that the oil was contaminated but the hazard was not appreciated.[6] The ignition source was provided by a static discharge. (The tank was being splash filled and dipped using a steel dip tape.)

In general the concentration of any component in the vapour phase will, in the presence of free liquid, be significantly affected by temperature. The consequences of increasing temperature are described below.

(a) Creation of a flammable vapour in air concentration

This may arise due to a significant increase in ambient temperature, e.g. by transportation and use of a flammable solvent in hotter parts of the world where the flashpoint (refer to Ch. 4) may be exceeded. Alternatively, the application of open flames to diesel fuel tanks to thaw them out in winter, or to grease or bitumen barrels to lower the viscosity of the contents for emptying, may generate a flammable vapour–air mixture and also provide an ignition source. The latter also occurs if hot-cutting, welding or soldering operations are performed without adequate precautions. (These operations and appropriate safeguards are discussed later in Ch. 17.)

> A man was engaged upon cutting open a drum that had contained a waxy solid, flammable ester. The heat from the torch caused residual material sticking to the inside of the drum to vaporise while the steel bands around it were being cut. An explosive concentration had formed inside the drum by the time he had cut off the top; this was ignited by the flame of the torch and the man was killed.[7]

The prediction of such events may be complicated by more volatile components being produced as a result of 'cracking', i.e. thermal breakdown, of the residues.[8]

In general the problem associated with operations which result in relatively involatile materials being raised in temperature until their flashpoint in air is exceeded is not widely appreciated.

> A 20-litre oil can, said to have contained lubricating oil, was drained by being suspended upside down for some time. A handle was to be soldered to it. Shortly after an oxyacetylene burner was applied to the surface, the can exploded.[9]

Moreover the use of 'empty' drums which have contained flammable materials as a temporary support for piping or steelwork subjected to flame-cutting or welding can pose a serious hazard. Used drums cannot be treated as empty unless they have been cleaned and/or washed free of all residue and purged free of all vapour.

(b) Creation of excessive concentrations of toxic vapours in air

> A workman was engaged for 5 hours, without protection, on the recovery of

mercury which had spilled from a mercury vapour boiler. The liquid mercury was still warm but not too hot to handle. He subsequently developed acute mercury pneumonitis and died 3 days later.[10]

Absorption of mercury by inhalation is the main potential risk from work with this chemical (although it may also enter the body by ingestion or by skin absorption). The vapour pressure between 0 °C and 150 °C may be calculated from eqn 3.6.[11]

$$\log p^1 = -3212\, T + 8 \qquad [3.6]$$

where: p^1 is the pressure in mmHg; T is the absolute temperature.

Typical values demonstrate the significant effect of temperature on vapour pressure, e.g.

Temp.(°C)	Vapour pressure (mmHg)
15	0.0007
25	0.0016
125	1.0

Thus around ambient temperature a 10 °C temperature rise results in a doubling of the vapour pressure. Furthermore, if air at 20 °C is passed over a 10 cm^2 surface of mercury at 1 litre/minute the outlet concentration of mercury in air is approximately 3 mg/m^3; experience has shown that at normal room temperatures the concentration of mercury vapour in the air of the workroom may reach 1 mg/m^3 if the ventilation is inadequate.[11] The current Time-Weighted Average Occupational Exposure Limit and Threshold Limit Value (see Chs 5 and 7) for mercury is by comparison 0.05 mg/m^3.

> Laboratory workers have commonly been liable to chronic mercury poisoning due to the inhalation of low concentrations of mercury vapour over long periods, generally as a result of 'bad housekeeping' with liberal sprinklings of mercury droplets in uncleaned corners.[12]

(c) Increased total pressure in closed equipment, resulting in failure due to over-pressure

Because of this phenomenon, under no circumstances should gas cylinders or liquefied gas cylinders be warmed to aid discharge (except with the specific approval of the suppliers).

> Direct heat in any form must not be applied to liquid chlorine containers to assist liberation of gas.[13]

Over-pressurisation may be serious and result in rupture of the equipment, unless pressure relief, e.g. a properly-sized, rupture-disc or relief-valve, is provided. Gas cylinders exposed to fire, or aerosol cans heated excessively are typical examples. Typically the latter result in boiling liquid expanding vapour explosions (BLEVEs).

> A portable bitumen boiler was in use on a flat roof for the application of roofing felt. The heat to the boiler was provided by a propane burner connected to a cylinder standing alongside. Through overheating the bitumen boiled out of the pot, enveloped the cylinder and was ignited by flames from the burner. Shortly afterwards the overheated cylinder exploded and three men were badly burned.[14]

> In November 1978, 37 cars of a 92-car frieght train were derailed outside Canyon, Texas. Ten of these cars were tankers, five of which contained LPG, one fuel oil, one petroleum naphtha and one petroleum distillate. Firefighters found one LPG tanker and the naphtha tanker burning when they arrived. Because of the isolated location and lack of available water the decision was made not to fight the fires and the area around the site was evacuated. Less than 20 minutes after the fire team's arrival one of the LPG tanks ruptured; it did not blow apart but released its contents through a two-feet square hole in the top, the flames extending 131 metres into the air. A second LPG tank ruptured in a similar manner one hour and five minutes after derailment and, after 3 hours, a 149.8 metres^3 LPG tank car BLEVEd with a portion of the tank being projected over 450 metres and passing through a house.[15]

The phenomenon of BLEVEs is discussed further in Chapter 4 and examples given in Chapter 15.

Over-pressure may also be caused by ingress of more volatile contaminants into vessels or equipment.

> Pentane was stored in a large refrigerated tank. A leak in the low molecular weight refrigerant line allowed refrigerant into the pentane. The sum of the partial pressures (eqn [3.5]) unduly stressed the tank – apparently because a relief valve failed to operate. The top was blown off the tank and after rising to 45 to 60 metres the vapour cloud 'slumped' over the surrounding area where after about 30 seconds it was ignited.[16]

(d) Changes in pressure in sealed containers, resulting in partial ejection of the contents on opening

Changes in pressure may occur, for example, in drums or Winchester bottles, due to the difference between filling temperature and ambient temperature. If the ambient temperature is highest, this may result in ejection or splashing of the liquid contents or vapour release on opening.

> An operator needed a quantity of concentrated nitric acid for cleaning purposes and therefore placed a drum on its side on a pallet. Assisted by another man he began to open the side bung of the drum when internal pressure forced the acid out, spraying both men.[17]

(a)

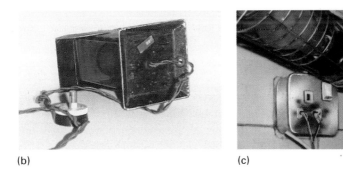

(b)　　　　　　　　　　　　　　(c)

Plate 1 Some electrical defects found by engineer surveyors
(a) The earth continuity to this motor depends on the steel wire armouring of the cable, which has been terminated too short and become disconnected from the gland. After detecting failure of earth continuity, the surveyor pulled back the plastic shrouding to locate the defect
(b) This metal clad illuminating lamp should have been wired with three-core flex so that the metalwork could be earthed, but a two-core flexible cable had been used which is already showing signs of wear
(c) This was found during an inspection of a garage:

1. Mains voltage hand-lamps should never be used and hand-lamps should always be of reduced voltage and separated from the mains by a transformer, preferably with an earthed centre tapping on the secondary winding.
2. Any electrical appliance with metal parts, unless it is specifically designed for double insulation, must have an earth connection.
3. Proper plugs should always be used. The use of wires directly into a square pin 13 amp plug is particularly dangerous as the fuse in the plug is necessary to give overload protection on the portable apparatus. (Courtesy National Vulcan Engineering Insurance Group Ltd)

Sometimes a similar effect is produced by deliberate heating of a sealed container.

> A glass flask of 250 ml capacity and containing 100 ml of 98% formic acid was fitted with a glass stopper held in position by a twisted copper wire. It was placed in an oven at 105 °C. After removal from the oven the stopper was taken off; the sudden reduction in pressure resulted in rapid evaporation of the liquid and the research student concerned was splashed on the neck and face.[18]

Despite being repeated elsewhere in the text, it is worth while noting here the precautions to be taken in such work, viz.:
- Heated flasks containing volatile corrosive liquids should be allowed to cool before opening.
- Work should be performed inside a fume cupboard, when practicable.
- Suitable protective clothing and face protection should be worn.
- The neck of the flask should be directed away from the laboratory worker during opening.

Furthermore in this particular incident it was fortuitous that a second person was present in the laboratory to render first aid since the student was temporarily blinded by the vapour.[18]

With immiscible liquids the total pressure is the sum of the individual vapour pressures.

$$P_{total} = p_1' + p_2' \ldots \quad [3.7]$$

The consequences of this are considered separately later.

Gas–liquid solubility relationships

For dilute solutions the partial pressure exerted by a liquid solute 1 in a liquid solvent is given by,

$$P_1 = Hx_1$$

where: H is Henry's Law constant for the specific system; x_1 is the mol fraction of the solute in the solvent.

Different values of H are applicable for each gas–liquid system.

For a gas which is only slightly soluble in a liquid, e.g. oxygen in water, a much higher partial pressure of the gas will be in equilibrium with a solution of given concentration than with a highly soluble gas, e.g. ammonia in water.

Generally the solubility of a gas decreases with increasing temperature. This is illustrated in Fig. 3.2 for ammonia–water mixtures of different concentrations.

> On one particularly hot afternoon a Winchester quart bottle of 0.88 ammonia solution was opened in a laboratory. It expelled ammonia gas so rapidly and

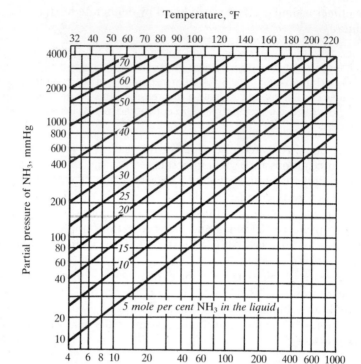

Fig. 3.2 Reference–substance plot for gas solubility. System ammonia–water

violently that the contents of the bottle were ejected on to the ceiling of the laboratory.[19] [The solubility of ammonia gas in water being temperature dependent, as in Fig. 3.2, would have been the major factor here.]

The solute in a solution formed out of contact with air will in time equilibrate with any atmosphere to which it is eventually exposed.

In oil refining foul waters may be produced containing dissolved hydrogen sulphide and other malodorous compounds. If these effluents are not treated for H_2S removal (by stripping in columns followed by incineration of the H_2S-rich gas) dangerous concentrations may form in the air above drainage sumps. Tests in one instance revealed concentrations of 200 to 400 parts per million.[3]

A confined explosion of methane gas in an underground pumping station at Abbeystead, UK, in May 1984 resulted in the deaths of 16 people and injuries to 26 others.[20] It was caused by ignition of a mixture of methane and air; the methane was displaced from a void formed in the end of a tunnel during a period of 17 days prior to the explosion when no water was pumped through the system.

Almost all the methane was of ancient geological origin and had percolated through the concrete walls of the tunnel between 2 and 2.5 km from Abbeystead either in gaseous form or in solution in water under pressure. It was concluded that the fact that significant quantities of methane might be dissolved in water had not apparently been recognised.[20] The relevant Henry's law constants[21] at 10 °C are:

Oxygen	3.27×10^4
Nitrogen	6.68×10^4
Methane	2.97×10^4

showing that methane and oxygen are twice as soluble in water as nitrogen. Hence contact of air containing methane with water at elevated pressure followed by release of dissolved gas from the supersaturated solution at atmospheric pressure will result in a vapour phase richer in methane and oxygen than the original air.

Phase changes resulting in volume changes

A very great increase in volume accompanies evaporation of a liquid to form vapour. For example, at 1 atmosphere pressure one volume of water will generate 1,600 volumes of vapour. In so-called 'steam' explosions resulting from flash formation from water the volume increase is even greater due to thermal expansion.

Similarly 4.54 litres (1 UK gallon) of gasoline on complete vaporisation under ambient conditions would yield 0.93 m^3 of neat vapour, sufficient if uniformly dispersed to fill a room or tank with a volume of 66.4 m^3 of vapour–air mixture at the Lower Explosive Limit of 1.4% in air (see Ch. 4). (In practice vaporisation would not in fact result in a homogeneous vapour–air mixture due to 'layering-out' effects of the heavy vapour, discussed below, but this does illustrate the hazards associated with spillages of relatively small volumes of volatile flammable liquids.)

This enormous increase in volume is calculable from basic chemistry, e.g. 1 mole of gas occupies 22.4 litres at 1 atmosphere pressure and 273 K. Thus 160 g bromine with a molecular weight of 160 and a density of 3.12 (as a liquid) would occupy $160/3.12 \simeq 51$ ml in the liquid state at 20 °C. The same weight of gas, however, would occupy $22.4 \times 293/273 \simeq 24$ litres.

Similarly assuming a liquid of specific gravity 0.8 and molecular weight 78, 1 gallon = 4.5 litres ≡ $4.5 \times 0.8 = 3.6$ kg

$$3.6 \text{ kg} \equiv \frac{22.4 \times 3.6}{78} \equiv 1.03 \text{ m}^3 \text{ of vapour.}$$

84   *The Safe Handling of Chemicals in Industry*

The calculation is equally applicable to toxic hazards, e.g. assuming homogeneous dispersion to achieve a concentration of 100 ppm the volume occupied by 1.03 m^3 would be 10,300 m^3 equivalent to a single-storey building of approximately 92 m × 19 m × 6 m high.

Conversely a condensation process results in a great decrease in volume. Similarly change in phase from solid to liquid, or the reverse, also involves volume changes. It is when these volume changes occur accidentally or due to an oversight, so that they are not allowed for in process design or operation, that accidents arise. Typical examples are given below (items (a)–(h)).

(a) Steam or vapour 'explosions' causing ejection of process materials due to rapid vapour generation

This is a particular hazard with the operation of 'hot oil' tanks, e.g. bitumen or heavy-fuel oil tanks operating above 100 °C; any water present results in steam generation and the formation of oil froth, often with sufficient violence to rupture the tank, which may 'boil-over'.[22]

> A road tanker barrel was being filled with hot tar, at a temperature just below 100 °C, when the pump failed. The barrel was about half-full at the time. In accordance with normal practice the tar line to the barrel was blown out with live steam.
>
> The driver went to another site to have the barrel topped up. At this site the tar was at a temperature in excess of 100 °C. A few seconds after filling commenced, hot tar erupted from the manhole on the barrel, due to flashing of steam condensate from the surface of the first batch of tar as the hotter tar was pumped on to it.[1]

This phenomenon may also arise from admixture of a hot liquid with a small amount of volatile liquid in a sealed container.

> An operator drained hot residue into a used 55-gallon drum from a reactor. He placed the plastic bung in its hole. Suddenly there was an eruption of hot residue from the bung hole and he was splashed with it. (The drum contained a residual amount of volatile chemical and capping allowed an increase in internal pressure. The plastic bung softened due to temperature and was hence easily displaced.)

The importance of a proper procedure with re-usable chemical containers is discussed in Chapter 13.

The following incidents are similar in terms of the mechanism involved.

> Heavy crude oil was being pumped into a 55,000 barrel tank when there was an internal 'explosion'; the tank roof was ruptured around the circumference and oil spilled on to the surrounding area.[23] Normally this crude was cooled to 132 °C by passage through a cooling box but, because it was fouled, this was by-passed. The oil vaporised water present in the inlet line and tank.

Hot oil, at a temperature in excess of 100 °C, was added to a large vertical storage tank which contained water. This caused the water to boil with explosive violence resulting in the top of the tank being displaced and oil being blown out. Some of the oil reached the top of a structure at 30 m elevation.[24]

A simplified theoretical analysis enables some estimates to be made of the minimum temperature excess, the maximum pressure produced in the vessel and the time to reach the critical flashing condition.[25]

The minimum temperature excess is given by,

$$(T_o - T_{WB}) = \frac{f \rho_w \Delta H_{vw}}{S_o} (1-f) \rho_o$$

The maximum pressure will be:

$$P = RT \rho_w \frac{f}{[f_H + f]}$$

The boiling time will be,

$$\tau = \frac{-Z \rho_w S_w}{h} \ln \left[1 - \frac{h \Delta H_{vw}}{a_w S_o S} Z (1-f) \rho_o \right]$$

where
T_o = oil temperature, K
T_{WB} = boiling point of water under conditions in the vessel, K
f = volumetric fraction, aqueous layer
ρ_w = water density, kg m^{-3}
ΔH_{vw} = heat of vaporisation of water, J.kg^{-1}
ρ_o = oil density, kg m^{-3}
S_o = heat capacity of oil, J.kg^{-1} K^{-1}
P = pressure, Pa
R = gas constant, J.kg^{-1} K^{-1}
T = temperature, K
f_H = volumetric fraction, headspace
τ = time, s
S_w = heat capacity of water, J.kg^{-1} K^{-1}
h = heat transfer coefficient across oil–water interface, W m^{-2} K^{-1}
z = depth of aqueous layer, m
Z = depth of tank contents, m

Water may enter tanks via open vents, through condensation as the tank is pumped out, or due to leakage from steam coils or steam tracing.

The solution of the problem of this type of transfer where the tank might contain water, even if unexpectedly, e.g. due to leakage from a steam coil, provides an interesting insight into how a potential hazard

may be reduced by some combination of process design measures and operating procedures, e.g. by, if feasible,
- Keeping the oil temperature <100 °C, and,
- Installing a high-temperature alarm on the inlet oil line if it is possible to exceed 100 °C, or alternatively, a combination of,
- Regularly draining-off water,
- Keeping the tank contents >100 °C (so that water is lost by vaporisation).
- Circulation of the tank contents.
- Adding <10% of the tank contents in any one transfer.
- Commencing all transfers slowly under close supervision, before building up to the full rate.

However simple, calculation shows that draining by itself is unlikely to provide a satisfactory solution with a flat-bottomed tank, e.g. a residual layer of only 3 mm of water on the bottom will displace 4.8 m above it on conversion to steam.

Boil-overs may also occur under some circumstances during oil fires in large fixed roof storage tanks.[26] Some burning crude or unrefined oils develop a 'heat wave' which travels down from the surface; the oil in this wave is at 250 °C to 300 °C so that when it reaches the bottom oil in which sufficient water is suspended, or bottom water, a violent boil-over will occur. Unless the froth produced in large volumes, e.g. 1,700 gallons per gallon of water, can break out as large bubbles on the surface a wave of burning oil may be thrown out possibly beyond the bund walls.

'Steam explosions' often occur through accidental admixture of moisture and molten metals, i.e. through charging damp metal pieces to a furnace or pouring of molten metal over, or on to, damp surfaces. Molten metal can be projected considerable distances from furnace openings or ladles in such incidents resulting in injuries to personnel, or the possibility of causing fires. Thus the refractory linings of launders through which molten metals, e.g. at 1,000 °C are allowed to flow, and which are replaced at intervals, must be thoroughly dried out before use. Similarly, moulds must be rendered moisture free; bars or tubes used for rodding-out, e.g. to remove slag from the periphery of burner ports or tapping holes, must also be free of surface moisture.

> Five men were killed and 13 injured as a result of a devastating steam explosion at a UK steelworks when water leaked into a 4.6 m high, 190 ton ladle full of molten steel in 1975. This ladle was one of two attached to a locomotive and filled from a blast furnace. Water was noticed entering the ladle; the explosion occurred after the ladle had been moved a few yards.[27]

The potential for steam explosions also exists with molten salt baths used in metal heat-treatment processes, e.g. carburising or nitriding. Wet work should never be introduced into a salt bath otherwise salt may spurt out. This is probably the most frequent cause of accidents with nitrate salt baths, the commonest source of water being the quenching tank which has to be situated nearby, but care is also needed to avoid explosions due to chemical reactions, e.g. between molten nitrate and carbon or metals.[28] (So far as moisture is concerned the established practice is to pre-heat components in an adjacent empty pot and it is clearly vital to avoid accidental ingress of water from water pipes, sprinkler systems, steam pipes, or off leaking roofs and to prohibit the use of water for firefighting.) Because of the magnitude of the volume change in passing from water to steam it is even necessary to exercise care in the selection of 'apparently dry' materials for covers or heat shields, e.g. asbestos-cement boards contain chemically combined water.

A fatal accident arose when a sheet of dry asbestos board which a man was using as a heat shield fell into a nitrate salt bath. After a short delay there was an explosion with violent disintegration of the board and a ejection of molten salt.[28]

(b) Vaporisation of a relatively small volume of liquid in a confined space leading to a flammable, or toxic, vapour hazard during cleaning or welding

An explosion at Sheffield Gas Works in 1973 wrecked a 1 million gallon capacity underground storage tank; 6 persons died and 29 others, including workers in nearby factories and members of the public, were injured.[29] This was caused by the explosion pressure arising from the ignition of a large volume of flammable vapour–air mixture in the tank. (Ignition was deduced to have occurred in one of the open pump shafts due to hot metal from flame-cutting operations, the flame propagating down the shaft into the main flammable mixture in the tank.)

The most probable source of flammable vapour in the tank was from the vaporisation of primary flash distillate, trapped either in the roof cavities or shallow domed areas of the roof-plates or drained from any hollow roof-supports having perforations in their welded seams. A residue of 70 to 80 gallons of primary flash distillate would, when completely vaporised, have been sufficient to fill the tank with vapour equivalent to the lower flammable limit. This quantity, spread uniformly over the surface of water in the tank, would result in a layer about 0.45 mm deep.

For flammable material an environmental concentration above the Lower Flammable Limit, generally in excess of 1%, is required to produce a flammable atmosphere, whereas for toxic chemicals, health risks can result from exposure to contaminant at the ppm level. Thus, the effect of vaporising small volumes of toxic liquids is even more striking as illustrated by reference to the small selection of examples given in Table 3.1.

88   The Safe Handling of Chemicals in Industry

Table 3.1   Volume changes on vaporisation of toxic liquids

Compound	Volume of liquid	Equiv. volume of 100% vapour	Equiv. volume in air at a concentration equivalent to the current OEL*
Ammonia	1	1.2×10^3	4.9×10^6
Sulphur dioxide	1	5×10^2	1×10^8
Chlorine	1	4.7×10^2	4.7×10^8
Carbon tetrachloride	1	2.3×10^2	2.3×10^7
Bromide	1	4.4×10^2	4.4×10^9

* Based on Guidance Note EH/40 1984

A stainless-steel tank at a perfume factory was normally cleaned with a detergent. However, one particular worker chose to use carbon tetrachloride instead, despite being warned of the risk of exposure to the vapours. He entered the tank and began to clean it with a mop impregnated with carbon tetrachloride but was soon overcome and required hospitalisation.

(c) The potential for leakage or spillage of material maintained as a liquid above its atmospheric boiling point by pressure, or as a liquid by means of refrigeration, to result in a large vapour cloud

The accidental release of 3 tons of liquid carbon dioxide at a brewery[30] resulted in a visible spread of carbon dioxide gas, denoted by condensed water vapour, of over 28 m². [It is of relevance to Ch. 7 that this incident, with a chemical generally considered to be of low toxicity, resulted in a fatality.]

This type of hazard is discussed later in Chapters 4 and 15 but so far as common flammable gases are concerned the extent is clear from the fundamental relationships:
1 volume of liquid petroleum gas (LPG) ≡ 225 volumes of neat vapour
1 volume of liquefied natural gas (LNG) ≡ 620 volumes of neat vapour

(d) Implosion of sealed, or inadequately 'vented', vessels due to vapour condensation on cooling

A 1.8 m diameter lead-lined still was not equipped with a vacuum relief valve. A spray of cold water was mistakenly injected into it while it was full of steam. The steam condensed rapidly to produce an internal vacuum and the still was crushed by the external atmospheric pressure.[31]

A vertical storage tank was pressure-tested by filling it with steam. While it was under pressure, and with all the valves still closed, there was a tropical storm. This caused rapid condensation of the steam and the tank sides were sucked in. [In this particular incident the vessel was subsequently restored to its original shape by the application by hydraulic pressure.]

(e) Implosion of sealed, or inadequately 'vented', vessels due to absorption of vapour in another liquid

A large tank used for aqueous ammonia was emptied and water was fed into it from overhead sprinklers. The tank had a 5 cm diameter vent which was closed and a 2.5 cm diameter vent which was open. Immediately after the water flow began, and before anhydrous ammonia was fed in via pipes in the bottom, there was an implosion; the tank collapsed to half its original volume.

Ammonia gas is highly soluble in water and the cause of this accident was its rapid absorption by the inflow of water producing a rapid reduction in internal pressure with which the 2.5 cm vent could not cope.[1]

(f) Sucking back of process materials, or ingress of air, following cooling and condensation of vapour in a vessel

This is relevant to distillation column operation; the column will contain a large volume of vapour and safeguards must be incorporated to protect it from drawing in air via the vent if there is a failure of heat supply to the reboiler, which is the normal vapour source, leading to excessive condensation. Provision for the controlled admission of inert gas, e.g. via a pressure-operated solenoid valve, may be required on normal column shut-down.

(g) Ejection of liquid from open pipelines when 'solid' blockages are released by 'steaming'

The phenomenon has followed 'freezing' of concentrated sulphuric acid, phenol, phosphorus and other potentially hazardous chemicals. In freeing such blockages by thawing the hazard has sometimes been increased by the application of process fluid, or steam, or pneumatic pressure upstream of the constriction; as a result the leakage due to expansion becomes insignificant compared with the jet and spray dispersion accompanying pressure release.

(h) Rupture of water lines following freezing

The relationship between force and pressure

Pressure is defined as force per unit area. It may be expressed in a variety of units, e.g.

$$N/m^2 \times (0.145 \times 10^{-3}) \equiv lbf/in^2$$
$$1 \text{ bar} \equiv 1 \times 10^5 \text{ N/m}^2$$

Also pressures are often measured in terms of the height of a column of liquid that the pressure will balance. Thus in consistent units

$$p = h \times \rho \qquad [3.9]$$

where p is the pressure, h the head of liquid and ρ the density.

It follows that, pressure × area = force. Consequently what may be considered a relatively small pressure can result in a very large force if it is applied over a large area.

For example:

If a pressure of 15 cm water gauge is applied to a 1,000 m³ vertical cylindrical storage tank with a diameter of 9.1 m then the roof is subjected to a force of 10 tonnes.[24]

Following displacement of plastic pellets from a road tanker using compressed air at 0.7 bar the practice was for it to be vented and the manhole cover on the top was then removed to check that the tanker was empty. On one occasion a driver commenced opening the manhole without releasing the pressure and when two of the quick-release couplings had been opened the cover blew open. The man was killed by falling off the tanker.[32] (Subsequently the vent valve was moved from the side of the tanker to the foot of the access ladder and the manhole cover replaced by a type which can be opened 1.3 cm while still capable of carrying full pressure, with a separate operation being necessary to release it fully.)

The local effects of a vessel rupturing under pressure can be devastating but, until consideration is given to the product of pressure × area, a pressure expressed in familiar units may seem moderate.

A tank, design pressure 0.35 bar, was in use for storage of a product of melting point 97 °C as a liquid; it was heated by an internal 7 bar steam coil. While it was empty and being prepared to receive product the inlet line was blown through with compressed air to check that it was clear. The air was not apparently getting through so the operator suspected a blockage in the line. In fact the vent on the tank was blocked and the applied air pressure of 5.2 bar ultimately blew the end off the tank resulting in serious injury to three men. (Initially the tank was apparently provided with a 15 cm diameter vent but this was blanked off and a 7.6 cm diameter, 7.6 cm deep dip branch used as a vent.)[33]

The results of a high-pressure release are even more serious as the following case history of what can happen to a compressed gas cylinder if the cylinder valve is knocked off illustrates.[34] (It also shows the need for such cylinders, which typically contain compressed gases at a pressure of 140 bar (2,000 psig), to be supported and secured while in use as described in Ch. 11. See also Fig. 3.3.)

> One of six 6.2 m^3 carbon dioxide cylinders, containing a pressure of about 60 bar was being moved back to its wall support when it was found to be leaking. A painter had this cylinder against his shoulder attempting to slide it along the floor when the valve separated from it and was projected backwards to hit a steel cabinet. The man wrestled the cylinder to the floor but could not hold it and it slid across and hit another cylinder which it knocked over, bending its valve. The rogue cylinder then turned 90° to the right and travelled 6 m before striking a scaffold, causing a painter to fall off. It spun around and then travelled back to near its starting position to strike a wall. It next turned 90° to the left, travelled the length of a room, with an electrician fleeing in front of it, and crashed into a wall 12 m away breaking 4 concrete blocks. It then turned 90° to the right through a doorway and travelled 18 m in a straight line before falling into a truck-well where it spun harmlessly around as the balance of the pressure was released.

On large-scale chemical plants the loss associated with an over-pressurisation incident may be out of all proportion to the error, or ommission, which was the prime cause.

> During commissioning of an olefin plant in 1973 the low pressure side of the refrigeration section of the gas separation plant was over-pressurised by a closed 30 cm valve on the exit from the low pressure side of the cold box. (This valve was installed to assist in drying the plant during pre-commissioning but the normal requirement for pressure relief protection on the low pressure steam was overlooked.) A valve starting line was opened allowing high-pressure gas to enter the low-pressure system; this system, designed for 3.5 bar, was subjected to 27.5 bar. Although no one was seriously injured the ensuing fire, confined to the refrigeration section, continued for 15 hours. The delay in commissioning the complex and the direct and consequential losses were considerable.[35]

The effects of over-pressures experienced at various distances from explosions are exemplified in Chapters 12 and 15.

Kinetics of chemical reactions

Chemical reaction rate is generally a function of reactants' concentration and temperature. For example with the reaction

$$A + B \rightarrow \text{product}$$

The reaction rate $= ka^\alpha b^\beta$ [3.10]

92 The Safe Handling of Chemicals in Industry

(a)

(b)

Fig. 3.3 Illustrations of forces created by sudden release of pressure
(a) Steam autoclave, one of three for curing bricks, ejected from building when steam pressure applied without the door being fully secured
(b) Vulcanising vessel propelled through wall of factory by release of steam suddenly via improperly secured vessel door (Courtesy Health and Safety Executive)

where: a, b = concentration of reactants
            k = rate of constant
            α, β = constants dependent on reaction
Usually $k = A \exp(-E/RT)$
where: E = activation energy, specific to the reaction
            T = absolute temperature
            A = integration constant

Clearly then the rate constant increases exponentially with temperature. Thus unless in the case of an exothermic reaction the heat of reaction is removed an increase in temperature may lead to 'run-away' conditions. (For most homogeneous reactions the rate is increased by factor of 2 or 3 for every 10 °C rise in temperature.)

For eqn [3.10], for a specific reaction

$$\text{Rate of heat generation} \propto e^{RT}$$

However, if the sole means of heat removal comprises an external jacket and/or internal coil through which coolant flows at temperature T_c

$$\text{Rate of heat removal} \propto (T_r - T_c) \quad [3.11]$$

Hence, while the generation rate is exponential with temperature T_r the removal rate is only linear; thus a critical value of T_r will exist at which control is lost.

> A compound was produced from an N-substituted aniline and epichlorohydrin, the reaction being carried out so that an agitated mixture of these two reagents was heated to 60 °C by means of an internal steam coil. When the exothermic reaction started a switch was made to cooling water to maintain the temperature at 60 °C. On one occasion an operating error allowed the temperature to exceed 70 °C and, even with full water flow, the reaction continued to accelerate. The temperature increased slowly over about 10 minutes so that a full evacuation was practicable before the ensuing explosion at 120 °C.[36] (While a pressure-relief system may have avoided an explosion, because of the relatively slow rate of pressure increase, this example clearly illustrates how the controllable range of an exothermic reactor is a function of T_r.)

Equation [3.10] also illustrates why reactions which are immeasurably slow at ambient temperature become increasingly rapid as the temperature is raised. The reaction of hydrogen and oxygen is an example.[37]

The rate of heat generation from an exothermic reaction is directly related to the mass of reactants involved. This, in conjunction with the heat transfer characteristics discussed later, is a very important consideration in scale-up, i.e. the performance of a reaction in a larger reactor. (Normal practice is to develop processes on a laboratory or pilot-plant scale and, if they appear viable, to increase the scale to a commercial-sized plant.)

In a conventional vertical, cylindrical reaction vessel of diameter D,

Volume ($\equiv$ mass) of reactants α $D^2.L$

where L is the height.

If reliance is placed upon convective heat transfer to a coolant in an external jacket

Heat transfer area α $D.L$

$$\frac{\text{Heat transfer area}}{\text{Potential heat release}} \; \alpha \; \frac{1}{D} \qquad [3.12]$$

and becomes increasingly unfavourable with size so that a critical value exists for D. Clearly, therefore, a reaction, such an as exothermic polymerisation, which is easily controlled on a laboratory scale, may become hazardous on a pilot-plant or commercial scale.

Incidentally, if heat is removed from an agitated reactor via an internal coil or external jacket, then if the process side offers the major resistance, which is commonly the case, the overall heat transfer coefficient U (eqn 3.16) depends upon speed of rotation of the agitator, e.g.

$$U \; \alpha \; (N)^{2/3}$$

Thus failure of agitation may result in hot spots and/or 'run-away'.

Exothermic reactions frequently occur under 'unexpected' uncontrolled conditions and result in ejection of the reactants or explosions.

Typical problems may arise from exothermic reactions due to inadvertant admixture of materials. The contacting of oxidising acids (e.g. chromic acid, nitric acid, permanganic acid or perchloric acid) with organic matter results in a violent reaction.

> A student disposed of a quantity of concentrated nitric acid by pouring it into a Winchester quart bottle normally reserved for organic waste. Two people were injured in an explosion which followed replacement of the cap.[38]

> A man was engaged in emptying a liquid chlorine pipe by passing chlorine as vapour via a flexible tube into an open container filled with an aqueous solution of 20% caustic. He was requested to fill about 10 cm^3 into a Dewar flask and, since residual chlorine was not needed in this flask, he discarded it into the caustic container. The ensuing reaction resulted in splashing of caustic on to the man's right foot.[39]

This is not an uncommon hazard with cleaning materials, e.g. vigorous sprays having been ejected when acids have been mixed with alkalis; explosions have also occurred.[29] (As a general rule the mixing of proprietary cleaning agents should be avoided.)

A list of the incompatible common chemical mixtures is given in Table 10.3(b) (p. 495).

Unexpected reactions may, of course, be inherently hazardous because toxic products are generated; common reactions in this category are summarised in Chapter 10. Alternatively, the generation of a gas phase may result in a pressure build-up in a confined space.

> An operator was killed by an explosion involving a 55-gallon drum. This was apparently caused by a delayed reaction between approximately 0.75 gallon of phosphorous oxychloride, left in the drum, and water. This quantity of $POCl_3$ would form 2.1 m^3 of hydrogen chloride gas on reaction with water; this, and the steam produced by the exothermic reaction, blew the bottom head off the drum and projected it 100 m.[1]

Differences in density between gases and vapours

As an approximation for most gases and vapours,

$$\text{Density of gas/vapour} \propto \frac{\text{molecular weight}}{T} \quad [3.13]$$

(at constant P_{total}).

Since few materials possess a molecular weight less than that of air (approximately 29), most gases and vapours are *under normal conditions* heavier than air. Common exceptions are hydrogen, methane and ammonia.

Therefore the higher the molecular weight of a gas/vapour the greater its density at constant temperature. Also at constant pressure this density is inversely proportional to temperature.

These phenomena are extremely important and arise in:

(a) The tendency of vapours heavier than air to spread and accumulate at low level giving rise to a fire/explosion hazard (e.g. propane, butane, 'heavy' hydrocarbons) or a toxic hazard (e.g. chlorine), or oxygen deficiency in confined areas.

> A leak developed in the connection between a torch and a propane gas cylinder; the hose was therefore bound with insulating tape. This cylinder and torch were left in an open trench with the valve on the torch closed, but not the valve on the cylinder; evidently propane was able to leak out for some time. Subsequently a workman entered the trench to find a nut which had dropped into it. He struck a match, as a source of light, and an explosion occurred.[40]
>
> Propane has a vapour density of 1.6 (air = 1.0) and therefore accumulated in the trench, as it always tends to do at low level.

The related hazards of bulk toxic releases to the atmosphere are described in Chapter 15. The density characteristics of the common bulk toxics are given in Table 3.2, which shows that all but two are denser than air at ambient temperature.

Table 3.2 Density characteristics of selected toxic substances with the potential to create a vapour cloud

Substance	Molecular weight	Density ρ_o, at 20 °C (kg/m^3)	ρ_o/ρ air* at 20 °C
Chlorine	71	2.95	2.46
Ammonia	17	0.71	0.59
Sulphur dioxide	64	2.66	2.22
Phosgene	99	4.12	3.43
HF vapour	20	0.83	0.69
HCN vapour	27	1.12	0.94
Bromine vapour	160	6.65	5.54
Acrylonitrile vapour	53	2.20	1.84

* ρ_{air} is taken to be 1.20 kg/m^3 at 20 °C

While it would be expected that an ammonia vapour cloud, resulting from a sudden catastrophic failure of a pressurised tank, would disperse safely by rising due to buoyancy, there have been many cases where this has not occurred; the composition, and physical and chemical changes within the cloud, affect its form and density.[41]

> When ammonia vapour escapes to atmosphere as the result of the accidental failure of a pressurised container, the formation of a cold ammonia–air mixture denser than the surrounding air is likely. Such a cloud 'slumps' and becomes very broad.[42] In 1976 a road tanker carrying 19 tonnes of anhydrous ammonia crashed through a barrier on an elevated section of motorway. The pressurised tank burst on hitting the roadway below and rocketed. The initial cloud-height was 35 m and after about one minute the visible cloud was approximately 300 m wide by 600 m in length; this could not be attained under the influence of atmospheric dispersion only.[42]

Moreover, although the molecular weight of hydrogen fluoride (HF) is 20, vapour density measurements indicate that the saturated vapour released from liquid is highly associated so that the molecular weight at 1 atm and 20 °C appears to be 70 to 80; therefore, a vapour cloud of pure HF undiluted with air is likely initially to be denser than the surrounding air. It is unclear whether it would remain so, or become neutral or buoyant as dispersion proceeds.[43]

(b) Vapours which are less dense than air at ambient temperature nevertheless spread at low level when cold, e.g. vapours from liquid ammonia or LNG spillages.

> A major release of anhydrous ammonia occurred in Sweden in 1976 due to the rupture of a rubber hose from a tanker to a quay-side sphere. About 180 tonnes of ammonia were released and the tanker was completely enveloped in a cloud of ammonia gas just a few minutes after rupture; the cloud dispersed after about an hour.[44]

(c) Accumulation of gases less dense than air at high points in equipment or unvented buildings (e.g. hydrogen or methane).

Thus leaks from natural gas pipes or appliances in sealed basements, etc., will tend to diffuse up through buildings. Similarly in battery-charging rooms ventilation is desirable at high level.

(d) Hot gases rising by 'thermal lift' and, therefore, in general dispersing to atmosphere.

In this respect it is important not to overlook the possibility of workmen being occupied on tasks above the source of gases, e.g. on maintenance or painting work or for prolonged periods, as in the case of overhead crane drivers in foundries and steelworks.

Once a gas or vapour has diffused into and mixed with the surrounding air, however, it is the mean density of the mixture which is important. For example, in the design of exhaust ventilation, comparison of the relative density of a gas with air may suggest that down-draught extraction is needed to exhaust heavy gas; however, as illustrated in Table 3.3, the relative density of saturated air at ambient temperature may indicate that down-draught would offer no special advantage.[45]

When a petroleum vapour of density about 3 (with respect to air) is mixed with air in the ratio of 5% by volume, that is near the middle of the flammable range, the average density of the mixture is 1.1. Thus the 'flammable mixture' is only a little denser than air.

Table 3.3 Relative densities of saturated air for selected chemicals[45]

Substance	Relative density of vapour (air = 1) 25 °C	Saturated air concentration, ppm, 25 °C, 760 mmHg	Relative density of saturated air (air = 1) 25 °C
Benzene	2.7	125,000	1.21
Bromochloromethane	4.47	194,000	1.67
Carbon tetrachloride	5.3	152,000	1.65
Diisobutyl ketone	4.9	2,100	1.01
Nitroethane	2.58	27,500	1.04
Parathion	10.0	0,04	1.0

Heterogeneous liquid–liquid systems

As explained earlier, if two immiscible liquids are present in a mixture, each exerts its own characteristic vapour pressure independent of the other; thus the total pressure is the sum of the two vapour pressures and the boiling point of the mixture can be significantly lower than that of either component in the pure state.

> The boiling point of cyclohexane at atmospheric pressure is 80 °C and that of water 100 °C whereas, because of the above phenomena, a mixture of the

two will boil at 69 °C. Therefore if water and cyclohexane were separately heated to 75 °C and then mixed the mixture would boil violently and the liquid temperature would fall to 69 °C.[46]

If a solution of a liquid A in liquid C is mixed with a second liquid B, where C and B are immiscible (as for example in a solvent extraction operation as discussed in Ch. 9), then some transfer will occur of A to B. Given sufficient time, depending on the area of contact and the degree of turbulence, equilibrium will be reached between the concentration of A in each phase. For many dilute solutions these concentrations are related by

$$\frac{y}{x} = m = \text{a constant} \qquad [3.14]$$

where: x is the mass (or mole) fraction of A in C; y is the mass (or mole) fraction of A in B; m is the partition coefficient.

In concentrated solutions the equilibria are better represented by a distribution curve.

The relevance of eqn [3.14] to safe handling of chemicals is that following separation, e.g in a gravity settling chamber, both liquids will contain a proportion of the third component (and indeed traces of each other since some small degree of mutual solubility is unavoidable).

> A team of labourers were employed to dig a drainage trench within a tetraethyl-lead manufacturing plant. An old drain carrying water contaminated with tetraethyl lead ran parallel to the new one. Contaminated water escaped into the new trench and on to the labourers' feet; as a result some of the men suffered from tetraethyl-lead poisoning.[47]

On exposure to the atmosphere a different equilibrium relationship is established according to a form of eqn [3.8]. Thus toxic or flammable vapours may accidentally be released from aqueous effluent streams in sumps or sewers, or from 'water washings'.

Density difference of liquids

Liquid densities may vary significantly and specific gravity is defined as the ratio of the density of a material to that of water. Thus just as ρ_a/ρ air gives an indication as to whether a vapour will rise or fall in air, the specific gravity of a liquid indicates whether it is 'heavier' or 'lighter' than water. Materials such as natural oils and fats have specific gravities greater than one; materials such as gasoline and kerosene have specific gravities less than one. Density is also generally reduced by an increase in temperature. Hazards associated with these effects include:

(a) Stratification of immiscible liquids in process and storage vessels.

Layering of reactants may occur in the absence of agitation during charging, with violent reaction then being initiated when agitation commences.

> A chemical operative was involved in the preparation of a mixture of phenol and liquid caustic soda. The phenol was added without agitation resulting in layering in the mixing vessel; when the agitator was operated a violent reaction occurred and sufficient heat was generated to cause explosive, local boiling. About half the contents of the reactor were projected through the hinged lid on to the operator who suffered extensive phenol burns.[48]

> A bucket containing 25% sodium hydroxide solution was used to catch and neutralise bromine dripping from a leak. In the absence of stirring a layer of unreacted bromine formed beneath the alkali solution When the layers were disturbed some hours later during disposal operations there was a violent eruption.[49]

Several interesting case histories have been published relating to layering phenomena involving either immiscible liquids or miscible liquids of different density and viscosity.[50]

(b) Rupture of closed pipelines and equipment due to thermal expansion exerting very great hydraulic pressures (e.g. ammonnia).

> A centrifugal pump containing liquid ethylene glycol was left running with inlet and outlet valves closed. Eventually the pump housing ruptured with explosive force. The cause was thermal expansion of the liquid due to heating by mechanical energy from the pump.[1]

A similar incident confirms that once liquid is trapped between two closed ends any source of heating can result in failure.

> A cast-iron centrifugal pump ruptured when it was left full of stock with valves on the inlet and outlet closed but with steam tracing left on.[51] [This arose following a pipework modification to allow the pump to be by-passed for normal operation; control of modifications is discussed in Ch. 17.]

The ability of liquids to exert pressures which may rupture a container, e.g. a road-tanker barrel, due to thermal expansion is a very important consideration in chemicals storage and transportation.

> Initial failure of the road tanker involved in the Spanish campsite disaster [Ch. 13 page 816] was apparently due to overfilling. The tanker held 23 tonnes of propylene, whereas the maximum should have been approximately 19 tonnes to leave a substantial voidage for thermal expansion.

Hence the reason for specific filling ratios.
A formula used generally is:

$$\text{Degree of filling} = \frac{U}{1 + \alpha\,(t_R - t_F)} \text{ \% of capacity}$$

Where U = the % filling permitted at the reference temperature
t_R = reference temperature (°C)
t_F = mean filling temperature of liquid (°C)
$\alpha = \dfrac{d_{15} - d_R}{(t_R - t_F) d_R}$

and d_R = density of liquid at reference temperature
d_{15} = density of liquid at 15 °C

With increasing hazard U may be reduced but increasing ullage leads to the contents 'slopping about' more.

(c) The potential for lighter liquids to spread, or accumulate on top of, denser liquids.

A fitter removed a large slip plate and a quantity of light oil ran into a flooded pipe trench. The oil spread along the water surface and its vapour was ignited by a welding torch 20 m away.[4] In this incident it was the spreading of the less dense oil on top of the water which created the hazard; normally vapour from a small spillage of such oil would not have spread such a distance.

In the event of water being added to concentrated sulphuric acid, which has a specific gravity of 1.84 (cf. water = 1.0) there is a tendency for it to remain on the surface. Mixing is, therefore, restricted to the interface and violent boiling of the water will result in the ejection of a mixture of acid and hot water.[19] This is a typical example of an uncontrolled enthalpy release on mixing of the type referred to later. (Dilution should be by controlled addition of acid to water with the temperature restricted to 80 to 90 °C.) Local hot-spots should be avoided by careful stirring throughout.

Gasoline, kerosene and many organic liquids are less dense than water and hence on accidental discharge into drainage systems may create a flammable or toxic hazard, because they accumulate on the surfaces of water in sumps, pits or sewers. This is also an important consideration if water is used in fire-fighting.

(d) Spillage or discharge of liquids from heated storage tanks or process vessels. For example 'slop-overs' due to thermal expansion may occur with almost full, fixed-roof oil tanks in fire (or due to the presence of water, or due to surface boiling).[26]

Vaporisation associated with pressure reduction

When a liquid near its boiling point at one pressure is let down to a reduced pressure vapour 'flashing' occurs. This proceeds until the liquid temperature is reduced to below the saturation temperature at the new pressure.

Examples arise in:

(a) 'Flashing' of steam or vapour from vessels or liquids on venting to atmosphere. (A dispersion of fine droplets, i.e. a mist, may also be entrained in the vapour.)

> A stoppered flask containing formic acid was placed in a laboratory oven at 105 °C by a research student. The temperature of the oven hence exceeded the boiling point of the liquid at atmospheric pressure. When the flask was removed from the oven and opened the contents evaporated rapidly and splashed the student.[52]

(b) Rapid vaporisation of LPG escapes or spillages to produce a vapour cloud. Similar effects occur with ammonia and chlorine.[53]

While the amount of vapour that will flash off can be calculated for a given pressure reduction, the amount and fineness of spray is unpredictable; under some conditions all the liquid may form a mixture of vapour and spray.[54] If the vapour is flammable a fine spray of the corresponding liquid will also be flammable and remain so below the flash point.[55] Alternatively the aerosol can constitute a health hazard, e.g. in the case of corrosive liquids.

Enthalpy changes on mixing of liquids

Mixing of two or more chemicals of dissimilar molecular structure may be exothermic or endothermic. The enthalpy release when some liquids are mixed has been a frequent cause of accidents due to ejection of the liquids from equipment or, in extreme cases, explosions.

> The dissolution of flake sodium hydroxide in water takes place with a considerable evolution of heat. The potential hazards are:[19] if too small a quantity of water is used there may be violent local boiling resulting in ejection of solution from the container. During dissolution a fine mist of sodium hydroxide solution is evolved; this may present a hazard to the eyes, respiratory tract or skin as discussed in Chapter 5.

> A road tanker of hydrochloric acid was inadvertantly unloaded into a large storage tank used for concentrated sulphuric acid. About half of the 13.6 metres^3 load had been discharged when there was a violent explosion. The tank was raised several centimetres off the ground, and badly buckled, and the inlet and outlet lines broken.[1]

Corrosion

Corrosive substances are those which by chemical action may cause damage to living tissue or materials of construction. This section is restricted to the latter since health effects are the subject of Chapter 5.

Corrosion is a common cause of hazards due either to the selection of wrong, or unsatisfactory, materials of construction at the design stage or changes in process conditions. The effects of corrosion resulting in weakness of equipment, pipework, bolts or supporting steelwork due to thinning and subsequent failure under pressure, impact or load are well documented.

Technically, corrosion is the gradual chemical or electrochemical destructive attack of metal by atmosphere, moisture or other chemical environments. The dual action of stress and a corrodent can give rise to stress corrosion cracking or corrosion fatigue. Serious conseqences may result from corrosion of supporting structures or plant items such as bursting discs, centrifuges, cooling towers, diaphragms, electrical components, gaskets, non-return valves, pipes, reactors, storage or transport vessels, etc. Ramifications include unscheduled shut-down, financial loss from plant replacement, safety hazards to operatives, loss of efficiency, loss of product, product contamination and over-design.

Corrosion may be uniform (i.e. the metal is attacked evenly) or intensely localised (termed pitting). Mechanisms can be direct oxidation or electrochemical. The former occurs when metals are heated in oxidising environments. Most galvanic corrosion occurs in gaseous or aqueous media and involves oxidation and reduction reactions occurring at metal interfaces. The rate of reaction is controlled by flow of electrons through the metal which constitutes the corrosion current. When electron flow ceases, corrosion stops.

By convention, the anodic reaction causes the corroding metal (M) to go into solution via an oxidation reaction since most metals are more stable as the oxide or other compounds.

$$M \rightarrow M^{n+} + ne$$

At the cathode an electrochemically equivalent reduction reaction takes place such as one of the following:

(i) $2H^+ + 2e \rightarrow H_2$
(ii) $O_2 + 4H^+ + 4e \rightarrow 2H_2O$ (oxygen reduction in acid solutions)
(iii) $O_2 + H_2O + 4e \rightarrow 4OH^-$ (oxygen reduction in neutral or alkaline solutions)
(iv) $M^{3+} + e \rightarrow M^{2+}$ (metal ion reduction)
(v) $M^+ + e \rightarrow M$ (metal plating)

The evolution of hydrogen in (i), e.g. in drums of acid can be hazardous for various reasons.[56] These include the following:

(a) Build-up of flammable atmosphere

> An explosion occurred in a large lead and acid resisting brickwork-lined, mild-steel vessel. This was part of a plant for sulphuric acid concentration. It

had been emptied for cleaning and inspection, and after washing-out with water the top cover was replaced. While two men were welding pipe connections to this cover there was a violent explosion.[9] An explosive hydrogen–air mixture had been formed in the vessel due to leakage of diluted residual sulphuric acid through a fault in the lead lining and its subsequent reaction with steel to generate hydrogen.

A 25% solution of sodium hydroxide was filtered into a tank trailer thought to be constructed of mild steel. Copious volumes of hydrogen were given off before it was discovered that the tank was made of aluminium.[57]

(b) *Build-up of internal pressure in an ever-weakening container*

A 200 litre galvanised steel drum was filled with a mixture of water, methanol and butyric acid which had been neutralised to pH 9.8 with ammonia. During the night it burst spontaneously; the bottom was blown off and the drum was projected on to the loading ramp.[58] The cause was dissolution of the zinc coating with evolution of hydrogen which occurs under alkaline conditions, or under neutral conditions in the presence of ammonium salts.

(c) *Production of atomic hydrogen as a species.* Penetration of this species into metal produces blistering (causing bulging) or embrittlement (reducing ductile strength).

(d) *The consumption of oxygen by atmospheric corrosion in sealed tanks may result in*:
- 'Oxygen-deficiency' a hazardous condition for entry (see Ch. 6).
- In rare cases, tanks collapse.

Under certain conditions, promoting accelerated rusting of steel, the rate of oxygen depletion may be surprisingly rapid.[59]

A further feature of corrosion is the potential for stresses to develop as a result of the increased volume of corrosion products, e.g. rust occupies about 7 times the volume of steel from which it is formed.

Contrary to common belief, dilute solutions can be more corrosive than concentrated versions. Thus, 36% sulphuric acid will attack carbon-steel or stainless-steel type 316L whereas 96% sulphuric acid can be handled in either. Similarly, carbon-steel storage tanks coated with neoprene latex are suitable for sodium hydroxide solution of 50% strength but not 20%.[60] Some corrosion-resistant materials for concentrated aqueous solutions and acids are given in Tables 3.4 and 3.5, respectively.[60]

Many salts are corrosive to common materials of construction[60] as exemplified by Tables 3.6 and 3.7. Traces of contaminants may result in, or accelerate, corrosion.

Table 3.4 Some corrosion-resistant materials for concentrated aqueous solutions

Material	Acids — Oxidising			Acids — Reducing			Acids — Organic		Alkalis		Salts — Aqueous solutions		
	Oleum	70–100% H_2SO_4	HNO_3 (conc.)	0–70% H_2SO_4	0–37% HCl	0.80% HF	Acetic acid	Formic acid	NaOH (caustic)	NH_4OH (ammonia)	NaCl	NaOCl	$FeCl_3$
Alloy C	L	L	L	R	L	L	R	R	R	R	R	R	L
Tantalum	V	R	R	L	R	N	R	R	N	N	R	R	R
Glass/silicates	R	R	R	R	R	N	R	R	L	L	R	R	R
Carbon, impreg. with furan	N	L	L	—	—	L	—	—	R	R	R	R	R
Carbon, impreg. with phenolic	N	L	L	R	R	L	R	R	N	R	R	R	R
FEP/TFE	R	R	R	R	R	R	R	R	R	R	R	R	R
Furan resin	N	N	V	R	L	N	L	L	R	R	R	V	R
Phenolic resin	N	L	V	R	R	L	L	R	N	N	R	N	L
Epoxy resin†	N	N	V	L	R	—	L	N	R	L	R	N	R

Key: R = Recommended for full range of concentrations up to boiling or to temperature limit of (non-metallic) product form.
L = Generally good service but limited in concentration and/or temperature.
V = Very limited in concentration and/or temperature for service.
N = Not recommended.

* See Table 3.8 for detailed polymer resistance to organic solvents.
† Epoxy hardener will strongly affect chemical resistance.

Organic solvents		Not recommended for
Aliphatics	Aromatics	
R	R	Acid services > 65 °C, especially hydrochloric acid or acid solutions with high chloride contents.
R	R	Hot oleum (> 50 °C), strong alkalis, fluoride solutions, sulphur trioxide.
R	R	Strong alkalis, especially > 54 °C, distilled water > 82 °C, hydrofluoric acid, acid fluorides, hot, concentrated phosphoric acid, lithium compounds > 177 °C, severe shock or impact applications.
—	—	Strong oxidisers, very strong solvents.
R	R	Strong alkalis, very strong oxidisers.
R	R*	Molten alkali metals, elemental fluorine, strong fluorinating agents.
R	L*	Strongly oxidising solutions, liquid bromine, pyridine.
R	L*	Strong alkalis or alkali salts, very strong oxidisers.
L	L*	Strong oxidising conditions, very strong organic solvents.

Table 3.5 Common materials of construction for strong acids

Acid	Construction material	Important safety considerations
Acetic	316 L stainless steel	Excess acetic anhydride in glacial acetic acid can accelerate corrosion; chloride impurities (ppm levels) can cause pitting and stress-corrosion cracking.
	Copper/copper alloys	Not for highly oxidising conditions.
	Aluminium alloys	Sensitive to contaminants; requires very clean welding; attacked very rapidly in concentrations near 100% or with excess acetic anhydride.
Formic	304 L stainless steel	For ambient temperature only.
	Copper/copper alloys	Not for oxidising conditions, including air.
Hydrochloric	Rubber-lined steel (natural rubber)	Low tolerance for organic solvent impurities; temperature limited according to hardness of rubber; steel fabrication must be properly done.
	Alloy C	Not for hot concentrated HCl.
	Alloy B	Not for oxidising conditions (test if reducing conditions are in doubt).
	Tantalum	Not for fluoride impurities.
	Impervious graphite	Fragile.
	Reinforced plastic	HCl may attack or permeate laminate; temperature limited (< 65.5 °C); requires excellent engineering design and fabrication quality.
Hydrofluoric	Alloy 400	Not for oxidising conditions (test if in doubt).
	Copper	Not for oxidising conditions (test if in doubt); < 65.5 °C only.
	Cupro-nickel	Not for oxidising conditions (test if in doubt); concentration and temperature slightly limited.
	Carbon steel	Not below 60% concentration, depending on impurities.
	Impervious graphite	Fragile; limit to below 60% concentration.
	Polyvinylidene fluoride	—
Nitric	304 L stainless steel	Must use low-carbon (or stabilised grade) if welded; not for fuming acid concentrations above 65.5 °C.
	High-silicon iron	Castings only; limited shock resistance; only for concentrations above 45% if temperatures over 71 °C.
	Aluminium (e.g. 3003, 5052)	Mostly for over 95% concentration, not for below 85% concentration; requires very clean welding.
	Titanium	May ignite in red fuming nitric acid if water is below 1.5% and nitrogen dioxide is above 2.5%.

Table 3.5 (continued)

Acid	Construction material	Important safety considerations
Oleum	Carbon steel	Not for 100–101% H_2SO_4 concentration; limited in temperature.
	Glass-lined steel	Limited shock resistance.
Phosphoric	316 L stainless steel	Must use low-carbon or stabilised grade if welded, up to 85% concentration and 93 °C.
Sulphuric	Carbon steel	Not for below 70% concentration; ambient temperatures only; flow velocities below 0.6 to 1.2 m/s.
	Alloy 20 variations	Limited temperature at 65–75% concentration.
	High-silicon iron	Castings only; limited shock resistance.
	Chemical lead	Soft and suffers from erosion; creeps at room temperature; limit to below 90% concentration.
	Glass	Limited shock resistance.
	Alloy C	Better for reducing acid strengths (< 60% concentration).
	Rubber-lined steel	For dilute not concentrated (oxidising) strengths; temperature limited according to rubber hardness and acid concentration; steel fabrication must be properly done.
	Brick linings with silicate mortar	Absorption of the corrosive by the masonry (use membrane substrate); poor properties in tension or shear (use in compression); many brick linings 'grow' in service but if used in archlike contours, growth merely increases compression.

Rapid corrosion may occur when oil and water meet because the water may react with impurities in the oil to produce corrosive products. If water is added to an oil line via a branch, rapid corrosion may occur at the welds as shown in Fig. 3.4(a); a better arrangement for water addition is shown in Fig. 3.4(b).[61]

Possibly the most troublesome impurity in plant liquors and cooling waters is the chloride ion. Its presence can result in severe pitting conditions for a wide range of materials and, under certain conditions of temperature and pH, stress-corrosion cracking. The most common stress-corrosion cracking environments for a number of metals and alloys are listed in Table 3.10.[62]

Besides metals, other materials are vulnerable to chemical attack. Organic solvents may degrade polymers causing swelling, reduced strength or even failure. Table 3.8 indicates the resistance of some common polymers to organic solvents (further examples are given in Ch. 11). The attack process is accelerated by increasing temperature.

Table 3.6 Comparison of corrosion rates by solutions of salts

Salts (Type and examples)	Corrosion rates for listed alloys				
	Carbon steel	304 SS	316 SS	Alloy 400 (65Ni-32Cu)	Nickel 200
Non-oxidising non-halides					
Alkaline (pH > 10)					
e.g. sodium carbonate	L	L	L	L	L
Neutral					
e.g. sodium sulphate	M	L-M; SCC	L-M	L	L
sodium nitrate	L-M; SCC	L; pits	L, pits	L	L
Acid					
e.g. nickel sulphate	S	L-M	L-M	M	M; pits
Non-oxidising halides					
Neutral					
e.g. sodium chloride	M; pits	M; SCC; pits	M; SCC; pits	M	M
Acid					
e.g zinc chloride	S	S; SCC; pits	S; SCC; pits	M	M
ammonium chloride	S	S; SCC; pits	M; SCC; pits	M	M-S
Oxidising non-halides					
Neutral					
e.g. sodium chromate	L*	L	L	L	L
sodium nitrite	L*	L	L	L	L
potassium permanganate	M	M	M	M	M
Acid					
e.g. ferric sulphate	S	L	L	S	—
silver nitrate	S; SCC	M	M	S	S

Oxidising halides					
Alkaline					
e.g. sodium hypochlorite	S		S; pits	S; pits	M-S; pits
Acid					
e.g. Ferric chloride	S; SCC	S; pits	S; pits	S	S
Cupric chloride	S	S; pits	S; pits	S	S
Mercuric chloride	S	S; SCC; pits	S; SCC; pits	S; SCC	S

Key: L = Low – less than 5 mpy; for all concentrations and temperatures below boiling.
 M = Moderate – less than 20 mpy; perhaps limited to lower concentrations and/or temperatures.
 S = Severe – more than 50 mpy.

* Chemical acts as corrosion inhibitor if present in sufficient amounts, but may cause pitting if in lower amounts.

Table 3.7 Some inorganic salts which are highly corrosive to carbon steel

(Corrosion rate >50 mpy)	
Aluminium sulphate	Magnesium fluorosilicate
Ammonium bifluoride	Mercuric chloride
Ammonium bisulphite	Nickel chloride
Ammonium bromide	Nickel sulphate
Ammonium persulphate	Potassium bisulphate
Antimony trichloride	Potassium bisulphite
Beryllium chloride	Potassium sulphite
Cadmium chloride	Silver nitrate
Calcium hypochlorite	Sodium aluminium sulphate
Copper nitrate	Sodium bisulphate
Copper sulphate	Sodium hypochlorite
Cupric chloride	Sodium perchlorate
Cuprous chloride	Sodium thiocyanate
Ferric chloride	Stannic ammonium chloride
Ferric nitrate	Stannic chloride
Ferrous ammonium sulphate	Stannous chloride
Ferrous chloride	Uranyl nitrate
Ferrous sulphate	Zinc chloride
Lead nitrate	Zincfluorosilicate
Lithium chloride	

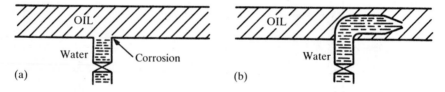

Fig. 3.4 Water to oil addition

Glass or masonry is susceptible to attack by acid, fluoride or caustic. Concrete may be degraded by acids and other chemicals. Table 3.9 identifies concentration and temperature limits for some non-metals.[60]

Corrosion prevention

Prevention is usually achieved by careful choice of material of construction, by physical means (such as paints, metallic, porcelain, plastic or enamel linings or coatings), and by chemical means by alloying or coating or by changing the process. Some metals such as aluminium are rendered passive by the formation of an inert protective film. A metal to be protected may be linked electrically to a more easily corroded metal such as magnesium which serves as a sacrificial anode.

Physico-chemical principles and safety 111

Table 3.8 Organic solvent resistance of common polymers

Solvent	Example	Epoxy (120–148 °C)	Furan (120 °C)	Phenolic (93 °C)	Polyester FRP (Bisphenol-A fumarate) (104 °C)	Vinyl ester FRP (Low-temperature variety) (93 °C)	FEP/TFE (204/260 °C)	Nylon 6/6 (93–120 °C)	Polyethylene (65 °C)	Polypropylene (107 °C)	Polyvinyl chloride (60 °C)	Polyvinylidene fluoride (135 °C)	Butyl rubber (93 °C)	Natural rubber (65 °C)	Neoprene (93 °C)
		Thermosetting resins					*Thermoplastics*						*Elastomers*		
Alcohols	Methanol	G	G	G	G	G	E	E	E	E	E	E	E	G	G
Aldehydes	Formaldehyde	G	E	G	G	G	E	E	G	E	E	G	E	G	G
Aliphatics	Heptane	G	E	E	G	E	E	E	P	G	G	E	P	P	E
Aliphatic amines	Diethylamine	G	E	P	P	P	E	P	G	G	G	G	—	—	G
Aromatics (and derivatives)	Benzene	E	G	G	P	P	E	G	P	G	P	G	P	P	P
	Aniline	P	G	G	F	P	E	P	G	E	P	G	G	P	P
	Phenol	G	G	G	P	P	E	P	G	G	P	G	P	P	P
	Xylene	G	E	E	G	G	E	G	P	G	P	G	P	P	P
Chlorinated aliphatics	Trichloroethylene	G	G	E	G	P	E	—	P	F	P	E	P	P	P
	Ethylene chloride	G	G	G	P	—	E	G	P	F	P	E	P	P	P
Ketones	Acetone	G	G	P	G	G	E	G	P	G	P	P	G	P	G
Miscellaneous:															
Pyridine		P	P	P	P	P	E	F	F	E	E	F	P	P	P
Tetrahydrofuran	Aromatics	—	P	—	P	P	E	F	P	F	F	F	P	P	P
Furfural		G	E	P	G	P	E	—	P	P	P	F	—	—	—

E = Recommended to maximum temperature of product form
G = Recommendation limited to somewhat lower temperature, or restricted in product form
F = Very limited recommendation; for ambient temperature only
P = Not recommended. Severe attack
(Temperatures are approximate maxima for each polymer)

Table 3.9 Concentration and temperature limits for some non-metals in acids

Non-metallic	Hydrochloric acid (reducing)	Hydrofluoric acid (reducing)	Nitric acid (oxidising)	Sulphuric acid (oxidising above 60%)
Polypropylene	20% at 79 °C 38% at 52 °C	60% at 38 °C Not recommended over 60%	30% at 52 °C 60% at 25 °C	60% at 93 °C 93% at 79 °C 98% at 52 °C
Polyvinylidene fluoride	38% at 134 °C	70% at 100 °C 100% at 66 °C	10% at 93 °C 50% at 79 °C 100% at 52 °C	60% at 110 °C 85% at 66 °C 98% at 49 °C
PVC*	38% at 60 °C	60% at 49 °C Not recommended over 60%	40% at 60 °C 60% at 25 °C	70% at 60 °C 93% at 25 °C
Impervious graphite	38% at 100 °C	48% at boiling 60% at 85 °C Not recommended over 60%	10% at 85 °C 20% at 60 °C 30% at 25 °C	93% at 71 °C 96% at 25 °C
Hard natural rubber	38% at 85 °C	20% at 49 °C Not recommended over 20%	Not recommended	30% at 85 °C 50% at 66 °C Not recommended over 50%
FEP/TFE	100% at temperature limit for material	100% at temperature limit for material	100% at temperature limit for material	100% at temperature limit for material

* Resistance will vary with plasticiser additions.

Table 3.10 Some causes of stress corrosion

Metal	Environment
Aluminium alloys (high strength)	atmosphere tap water sodium chloride solutions
Copper alloys (brass)	ammoniacal solutions amines mercury salt solutions
Magnesium alloys	atmosphere tap water sodium chloride solutions
Monel 400	fluosilic acid chromic acid hydrofluoric acid vapour mercury salt solutions
Nickel 200	fused sodium and potassium hydroxides strong sodium and potassium hydroxide solutions
Steel carbon	sodium hydroxide solutions nitrates of calcium, sodium, ammonia mixed nitric and sulphuric acids calcium chloride brines
Steel-stainless-austenitic (300 series)	chloride solutions sea water caustic solutions steam
Steel-stainless (400 series)	chloride solutions sea water nitric acid caustic solutions mixed nitric and sulphuric acids

Particular attention should be given to gasket materials and the materials used for any flexible piping. The selection of the correct materials of construction for any particular duty is equally important for temporary apparatus such as hoses, small containers, or sample bottles.

> A postgraduate student made up 10 litres of a mixture of concentrated sulphuric acid and potassium dichromate in a polythene bucket, to use for cleaning glassware, but found the following morning that only the bucket handle remained.

The selection of materials of construction is outside the scope of this text but advice on the prevention of corrosion for specific plant items is to be found in standard references[63] and in a range of British Standards and Codes of Practice, etc., e.g. for pipelines [64] and pressure vessels.[65] General precautions to be followed when handling corrosive chemicals are discussed in Chapter 6. Inspection of plant is critical where corrosion occurs. This should include removal of specimen floor tiles, concrete encasing of steelwork, etc. Mechanical and non-destructive testing may also be required. Once corrosion has started on structures it is generally easiest to remove and replace the damaged components.

External corrosion of steel beneath thermal insulation has resulted in a number of serious incidents from sudden failure of high pressure containment.[66] External corrosion has been found on carbon, chrome–molybdenum, Al-killed carbon and 3.5% Ni steels due mainly to water ingress. Inadequate weatherproofing and cladding may be a factor, and corrosion is more severe when the operating temperature is between −4 °C and 120 °C or where equipment is operated intermittently between ambient and much higher temperatures. The rate of corrosion may be accelerated by process materials, the local environment or the presence of salts from the insulation. A sudden massive failure, as distinct from a pinhole leak, may result from the extensive external wastage. Detailed design and inspection strategies are required to guard against this hazard.

Surface area effects in mass transfer processes (and heterogeneous reactions)

In many operations involving chemicals mass is transferred across a phase boundary under the influence of a concentration gradient. This may be by intent (as in the separation processes–absorption, distillation, solvent extraction, etc., discussed in Ch. 11) or incidentally. In any event the rate of mass transfer can be expressed in an equation of the general form

$$N = K.A.(\Delta C)_m \qquad [3.15]$$

where: N is mass transferred/unit time
K is an overall mass transfer coefficient
A is the interfacial area
$(\Delta C)_m$ is the mean concentration gradient, representing the deviation from equilibrium

It is self-evident from eqn [3.15] that the rate is directly related to the concentration difference whether it is a pressure differential or a solubility relationship. Put simply, the greater the initial concentration at a source of transferable solute the more rapidly it will diffuse. The rate is also proportional to a coefficient which in general will increase with the degree of turbulence, e.g. agitation, or velocity in either phase (or in the phase offering the greatest resistance to the transfer).

The rate is also a function of the interfacial area A. Hence the rate of flammable or toxic vapour evolution from a liquid, for example from an open vessel or from a spillage, is directly related to the exposed surface area. (Such processes involve in fact a combination of heat and mass transfer, since evaporation can only continue at a rate commensurate with

the supply of latent heat of vaporisation to the liquid, but in the present context the dependence on surface area is similar.) The rates at which vapour may be evolved can also create hazards in situations involving:
- Dispersion of liquids as sprays or mists. These may be produced intentionally (e.g. from aerosol sprays used in agriculture or horticulture or in fuel injection systems) or unintentionally (e.g. during turbulent overhead filling of tanks or from leaks in pressurised pipework systems).
- Disposal of solvent-impregnated tissues or waste rags.
- Spreading of solvent-based formulations as a film, e.g. adhesives, over significant areas.
- Foam formation (e.g. due to gas bubbles leaving the surface because of de-entrainment or due to vigorous agitation of the liquid).

In such cases the area exposed, and hence the rate of vapour evolution may be many times greater than involved with the bulk liquid surface. (With toxic liquids dispersed as mists there is of course also a potential hazard due to the inhalation of fine droplets.)

The surface-area effect can be equally useful in the control of hazards, e.g. a water fog is an extremely effective means of dealing with a leakage of anhydrous ammonia since it is highly water soluble.

The rates of gas–solid reactions are also surface-area dependent. Thus dust which is finely milled may create an unexpected hazard. Accumulations of fine combustible dust in extraction ducts from grinding or polishing machinery may hence be prone to smouldering and spontaneous combustion.

Spontaneous ignition of thermal insulation on hot pipes and which has become impregnated with oil is a well-known phenomenon. Oxidation of oil is promoted within the insulation since the open structure and large surface area increase the interfacial area of the oil film with air. Furthermore, following the mechanism described under exothermic reactions, since the insulation minimises the dissipation of heat, the temperature rises until the oil ignites spontaneously. Coal is similarly subject to spontaneous heating and ignition when fresh, reactive surfaces are exposed. (The hazards and control of spontaneous combustion are discussed further in Ch. 4.)

Normally when a solid, flammable material is ignited it burns layer by layer because the exposed surface area is limited; the energy release is hence controllable. However, when a flammable solid is dispersed as a finely divided dust the magnitude of the surface area is increased enormously; on ignition rapid burning occurs with accelerated energy release. This can result in dust explosions of the type discussed in Chapter 4.

Analogous surface-area effects arise with chemicals which are toxic by absorption via the skin (refer to Ch. 5). The time of exposure *and* the contact area are then important parameters.

A pipefitter was engaged on removing a gasket from a flange when he was splashed on the face, neck, shoulders and arms with phenol. He lost consciousness within a few minutes. Hospital treatment which involved thorough washing probably saved his life but he was severely poisoned.[68] [In any event a burn to one eye resulted in partial loss or vision – refer to Ch. 6 for the essential use of eye protection with such materials.]

Heat transfer by forced convection

Many processes involving chemicals rely upon the transfer of heat, either for heating or cooling, by forced convection, i.e. depending upon flow of fluids. In such cases the rate of heat transfer may be expressed by an equation (analogous to eqn 3.15).

$$Q = U.A.(\Delta t)_m \qquad [3.16]$$

where: Q is the heat transferred/unit time
$(\Delta t)_m$ is the mean temperature difference between process fluid and the heating/cooling medium
U is the overall heat transfer coefficient
A is the area for heat transfer, e.g. of a coil, tube bank, or jacket.

The overall resistance to heat transfer $1/U$ is made up of a series of resistances.

$$\frac{1}{U} = \frac{1}{h_i} + \frac{1}{h_o} + \frac{x_w}{k_w} + R_{d_i} + R_{d_o} \qquad [3.17]$$

where: h_i is the heat transfer film coefficient on the inside
h_o is the heat transfer film coefficient on the outside
x_w/k_w is the resistance of the wall between the fluids
(x_w = thickness, k_w = thermal conductivity)
R_{d_i}, R_{d_o} are resistances due to dirt or scale deposits

Clearly anything which causes a significant increase in resistance, for example an accumulation of sludge increasing R_d; or reductions in flow reducing h_o or h_i, will result in a decrease in the rate of heat transfer. This will cause the process fluid to reach a higher (or lower) temperature than expected and cause a hazard. The example of an exothermic reaction has already been mentioned.

Similarly the removal of a resistance to heat transfer will result in adjustment of interface temperatures.

A secondary ammonia reformer was provided with a 30.6 cm refractory lining. When this lining cracked the vessel contents at 1,100 °C came into contact with the shell plate. The latter had a design temperature limit of 175 °C and the overheating caused it to rupture.[69]

[The safeguards subsequently recommended included more frequent inspections of the refractory lining and daily checks for hot spots indicative of refractory deterioration.]

The effects of particle size

Airborne particulate matter can be composed of liquid (mists) or solids (dusts, fumes). Two striking effects of spraying liquid as an aerosol or fragmenting solid to produce dust are to increase both the surface area of the material and the volume it occupies.

Thus, crushing 1 cm^3 of solid into cubes of 1 μm produces 10^{12} particles and increases the total surface area from 6 cm^2 to 12 m^2. If released into the air to produce a concentration of 100,000,000 particles per m^3, the volume occupied will be 10,000 m^3.

Increasing the surface area can increase the reactivity of the material and may render certain substances pyrophoric or explosive as discussed in Chapter 4.

Particles of metal arising from condensed sodium vapour, or sodium in the 1 μm to 10 μm range from a dispersion prepared in an organic liquid, have a very high specific surface area and are pyrophoric. An escape of liquid droplets from a closed system is also likely to ignite immediately on contact with the atmosphere. Conversely a solid stream of liquid of comparatively low surface area discharged from a fractured pipe or vessel may not ignite unless it is > 200 °C, and a pool of metal may not commence burning until its temperature reaches 400 °C or more.[70]

An example involving a dust explosion is given in Chapter 9, page 472. Increase in interfacial area as mentioned earlier also increases the ease with which a substance may enter the body; this is discussed further in Chapter 5.

Once airborne the behaviour pattern depends upon the size, shape and density of the particle. In a vacuum particles fall under gravity according to the equation,

$$v = \sqrt{2gS} \qquad [3.18]$$

where: v = velocity (m/s)
S = distance of fall (m)
g = 9.81 m/s^2

In air the resistance of air soon prevents the dust from accelerating until the resistance equals the force of gravity. The particle then continues to fall at a constant rate known as the 'terminal velocity'. Table 3.11 shows the terminal velocity of particles as a function of particle diameter. Table 3.12 gives common examples of particles of various diameters.

Table 3.11 Terminal velocity v. particle size

Diameter in microns	Rate of fall	
	Ft/min	Metres/second
5,000	1,750	9
1,000	790	4
500	555	3
100	59.2	3×10^{-1}
50	14.8	75×10^{-3}
10	0.592	3×10^{-3}
5	0.148	75×10^{-5}
1	0.007	36×10^{-6}
0.5	0.002	10×10^{-6}
0.1	0.00007	36×10^{-8}

Table 3.12 Typical particle-size ranges

Particulate matter	Particle size in microns
Viruses	0.02–0.1
Bacteria (Cocci)	0.8–1
Industrial airborne dust	0.1–10
Spores	3–10
Face powder	1–10
Water droplets in fog	Up to 40
Pollens	10–80

The time it takes particles to reach their terminal velocity depends upon the density, size and shape of the particle. For example, the terminal velocity of a 1 μm silica particle is approximately 1 mm in 30 seconds whereas that of a 100 μm particle will be 10,000 times faster, of about 0.3 m s^{-1}. Thus a dust cloud composed of particles of various sizes soon 'fractionates', the visible matter settling to the ground in a few minutes, while smaller particles take several hours to settle, if at all, and are sensitive to slight air currents. Clearly the size distribution of the airborne particles can be vastly different from that of the source material.

The implication of particle size for its ease of entry into the body and how it 'settles' in different parts of the respiratory tract is covered in Chapter 5.

References

(*All places of publication are London unless otherwise stated*)
1. Jenkins, A. J. D., *Chem. Engr*, 290 (Oct. 1974), 637.
2. Henley, E. J. & Rosen, E. M., *Material and Energy Balance Computations*. Wiley, Chichester 1969, 284.
3. Hughes, J. R., *Storage and Handling of Petroleum Liquids; Practice and Law*. Griffith 1967.
4. *Fires and Explosions*, Hazard Workshop Module 003 Institution of Chemical Engineers, 13.
5. Watts, H. E., *Report on Explosion and Fire at Region Oil Co. Ltd Premises Royal Edward Dock, Avonmouth, Bristol, on 7th September 1951*. HMSO 1952.
6. Kletz, T. A., *Loss Prevention Bulletin 052*. Institution of Chemical Engineers, Aug. 1983.
7. Matheson, E., Symposium on Chemical Process Hazards with Special Reference to Plant Design. Institution of Chemical Engineers 1960.
8. *Public Inquiry into a Fire at Dudgeons Wharf, UK*. HMSO Sept. 1970.
9. 'Repair of drums and small tanks: explosion and fire risk', *Health & Safety at Work* No. 32, 20.
10. Tennant, R., Johnston, H. J. and Wells, J. B., *Conn. Med.*, **25** (1961) 106.
11. Watkin, R. P. St. J., 'Mercury – a hazard to health', *B.S.S.G. Jl*, Vol. 9, No. 1, 6.
12. Universities Safety Association, 'Acute mercury poisoning', *Safety News*, No. 4, Nov 1973.
13. Department of Employment, 'Precautions in the handling, storage and use of chlorine', *Health & Safety at Work Booklet 37* 1970.
14. *Annual Report of H. M. Chief Inspector of Factories, 1968*, HMSO, 37.
15. *Fire Journal*, May 1979, 20.
16. Strehlow, R. A., referred to in G. Jones & R. L. Sands, *Great Balls of Fire*. Jones & Sands, Coventry, 1981.
17. Manufacturing Chemists Assoc., *Case History 1880*.
18. Anon., Universities Safety Association, *Safety News*, **8** (Feb. 1977), 13.
19. Gaston, P. J., *The Care, Handling and Disposal of Dangerous Chemicals*. Institute of Science Technology, 1965, 32.
20. Health and Safety Executive, *The Abbeystead Explosion*. HMSO 1985.
21. Anon., News Review, *Chem. Engr.*, **409** (Dec. 1984), 7.
22. Keey, R. B., *Instn Chem. Engrs Symp. Series*, No. 49, 1977, 147–51.
23. Risinger, J. L. & Vervalin, C. H., in C. H. Vervalin (ed.), *Fire Protection Manual for Hydrocarbon Processing Plants* (2nd edn). Gulf, Houston, Texas, 1973, pp 125–37.
24. *Hazards of Over- and Under-Pressuring of Vessels*, Hazard Workshop Module 001. Institution of Chemical Engineers, 1980.
25. Keey, R. B., *Instn Chem. Engrs. Symp. Series*, No. 49, 1977, 151–5.
26. Home Office (Fire Department), *Manual of Firemanship, 6b – Practical Firemanship – II*. HMSO 1973, 320.
27. Health and Safety Executive *The Explosion at B.S.C. Scunthorpe, 4th November 1975*. HMSO.

120   *The Safe Handling of Chemicals in Industry*

28. 'Precautions in the use of nitrate salt baths', *Safety, Health and Welfare*, No. 27. HMSO.
29. *Annual Report of H. M. Chief Inspector of Factories, 1974*. HMSO.
30. *Annual Report of H.M. Chief Inspector of Factories, 1972*. HMSO.
31. Austin, G. T., in H. W. Fawcett, & W. S. Wood, (eds) *Safety and Accident Prevention in Chemical Operations*. Wiley, New York, 1965, 101.
32. Kletz, T. A., *Organisations Have No Memory*.
33. Henderson, J. M. & Kletz, T. A., *Instn Chem. Engrs Symp. Series*, No. 47, 1976.
34. Anon. *Industrial Safety*, Jan. 1974, 35.
35. Heron, P. M., *Chem. Eng Prog.*, Sept. 1976, 72–5.
36. Schierwater, F. W., *Instn Chem. Engrs Symp. Series*, No. 34. 1971, 47.
37. Glasstone, S. & Lewis, D., *The Elements of Physical Chemistry*. Macmillan 1961.
38. Buttolph, M. A., 'People in laboratories' Symposium on Safety in University Chemical Engineering Departments. Instn Chem. Engrs, 18 Dec. 1979.
39. Manufacturing Chemists Assoc., *Case History 1880*.
40. Universities Safety Association, *Safety News*, No. 6 (Feb. 1975), 16
41. Kaloutas, G. J. 'The prediction of hazards from bulk toxic releases'. M. Sc. thesis, Univ. of Aston 1980.
42. Kaiser, G. D. & Walker, B. D., *Atmospheric Environment*, 1978, 12, 2289–300.
43. Beattie, J. E., Abbey, F., Haddock, S. R., & Kaiser, G. D., *The Toxic and Airborne Dispersal Characteristics of Hydrogen Fluoride; Canvey Island Report 1978*. HMSO
44. Hakansson, R., 'Ammonia line rupture', *Ammonia Plant Safety*, 1977 Vol 19. American Institute of Chemical Engineers, New York
45. Halley, P. D., *Chem. Eng Progress*, **61**. No. 2 (Feb. 1965), 59.
46. King, R., 'Major fire and explosion hazards in hydrocarbon processing plants,' 11.
47. Foulger, J. H., 'Effects of toxic agents', in H. W. Fawcett & W. S. Wood, *Safety and Accident Prevention in Chemical Operations*, Wiley 1965, 270.
48. *Annual Report of H.M. Chief Inspector of Factories, 1967*. HMSO.
49. *Manufacturing Chemists Assoc., Case History 1636*.
50. Anon. *Loss Prevention Bulletin*. Institution of Chemical Engineers, 1979 (029), 124–33.
51. Lloyd, A. W., & Roberts, K. W. J., *Instn Chem. Engrs Symp. Series*, No. 34, 1971, 243–50.
52. Universities Safety Association, *Safety News*, No. 9 (May 1978), 30.
53. Westbrook, G. W., 'Safety assessment of the carriage of liquid chlorine', in Buchmann (ed.). *Proceedings of the 1st International Loss Prevention Symposium*, 1974.
54. Kletz, T. A., *Journal of Hazardous Materials*, 2 (1977/78), 1–10.
55. Burgoyne, J. H., *Instn Chem. Engrs Symp. Series*, No. 15, 1963, 1.
56. Anon. *Loss Prevention Bulletin*. Institution of Chemical Engineers, 1975 (006), 1–12.
57. *Manufacturing Chemists Assoc., Case History 1115*, 1965.
58. Anon., *Loss Prevention Bulletin*. Institution of Chemical Engineers, 1979 (027), 74.

59. Anon. *Loss Prevention Bulletin*. Institution of Chemical Engineers, Apr. 1984 (056), 26.
60. Kirkby, G. H., *Chem. Engng*, 1980 (Nov.), 86.
61. Imperial Chemical Industries, *Safety Newsletter* **27**, 1.
62. Maylor, J. B., *Chem. Engr*, Mar. 1969, No. 226, CE74–79.
63. Perry, J. H. & Chilton, C. H., (eds), *Chemical Engineers Handbook*, 5th edn. McGraw Hill 1973.
64. Institute of Petroleum, *Model Code of Safe Practice: Pt 6. Petroleum Pipelines Safety Code*. 1976.
65. British Standards Institution, *BS 5500: 1976 Unfired Fusion Welded Pressure Vessels*.
66. Anon. *Loss Prevention Bulletin*. Institution of Chemical Engineers, Apr. 1984 (056), 1–8.
67. *Manufacturing Chemists Assoc., Case History 618*, 1966.
68. Hamilton, A. & Hardy, H. L., *Industrial Toxicology*. Hoeber, New York, 1949.
69. *Snow, M. S., Chem. Engng*, 1979, Vol **86** (1), 109–112.
70. Bulmer, G. H., *Instn. Chem. Engrs Symp. Series*, No. 33, 1972, 79.

CHAPTER 4

Flammable materials handling

The minimum requirements for a flame, or the propagation of a flame front, are generally:
1. A fuel (gas or vapour) within certain limits of concentration.
2. A supply of oxygen above a certain minimum concentration (generally from air).
3. An ignition source of minimum temperature, energy and duration.

All three, represented as the corners of a 'fire triangle' as in Fig. 4.1, must be present. This provides the basis for safe handling of flammable materials by a combination of atmosphere control (i.e. adjustment of 1 and 2 above to provide a non-flammable mixture) and elimination of ignition sources. However, this concept must be applied cautiously, e.g. no additional ignition source is necessary if the flammable material is itself at a high temperature (in excess of its spontaneous ignition temperature) and no additional oxygen is required if an oxidising agent is already present. Furthermore, some vapours may spread so that an ignition source may be reached at a considerable distance from the source.

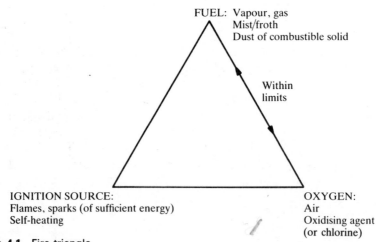

Fig. 4.1 Fire triangle

Liquids

Liquids and solids do not burn as such; they give off vapours and gases which can mix with air and be ignited within certain limits. (Metal fires are an exception; combustion is then solely a surface phenomenon with very high temperatures, e.g. 1,100 °C being attained, or in exceptional cases with magnesium, aluminium, zirconium or uranium >2,760 °C.)[1] Therefore, for liquids volatility is a major factor. The 'flash point' is an empirically determined method of obtaining the minimum temperature at which an ignitable mixture exists above a liquid surface. Typical values for petroleum fractions are given in Table 4.1 and for paraffins in Table 4.2.

Clearly 'flash point' is, by definition, inapplicable to gases but some solids, e.g. naphthalene and camphor, can produce flammable vapours and therefore have flash points. While an ignitable mixture of this type can spread flame from an ignition source, it will not support combustion until the temperature reaches the 'fire point', generally a few degrees higher.

Caution is required in relying too heavily upon flash point as a criterion of safety since:

1. An atomised mist or fog of liquid droplets can be ignited well below the flash point of the material. (This is the cause of crankcase explosions.) Thus if a liquid is splash filled so that a liquid mist is formed the flash point may be reduced by up to 50 °C. Froths may behave similarly.
2. A small amount of volatile material in a relatively high-boiling mixture may drastically lower the flash point, as illustrated in Fig. 4.2. This renders the whole mass flammable due to the phenomena discussed in Chapter 3. (For example, it is necessary to avoid ingress of proprietary kerosene-based solvents or paraffin into crankcases when cleaning engines.)
3. Combustibles absorbed in porous media (e.g. insulation) may ignite at mild temperatures.
4. 'Safe' material may be accidentally heated above its flash point during processing or maintenance operations.[3] Application of heat may also result in 'cracking', i.e. thermal breakdown, of liquids or solid deposits resulting in volatile compounds with reduced flash points.
5. On some occasions confusion has arisen over the identity of flammable liquids, with the result that a more volatile liquid has been transferred into the wrong storage tanks or processes, or marketed.

However, flash point is used as a primary measure of flammable hazards by authorities dealing with the storage, transportation or use of flammable liquids. In the UK the Institution of Petroleum classification is:

- Petroleum, Class 'A'; flash point less than 73 °F (22.8 °C), e.g. motor and aviation gasolines, most crude oils.

Table 4.1 Properties of (typical) petroleum fractions (after Meidl, 1978; see ref. 2)

	Flash point (°C)	Ignition temp (°C)	Flammable limits (% by volume in air)		Specific gravity (water = 1.0)	Vapour density (air = 1.0)	Boiling point (°C)
			Lower	Upper			
Asphalt typical	204*	485			1.0–1.1	—	371
Crude petroleum	−7 to 32				1.0		
Fuel oil 1 (kerosene)	38†	229	0.7	5.0	≤1.0	—	151 to 301
Fuel oil 2	38†	257			≤1.0	—	
Fuel oil 4	54†	263			≤1.0	—	
Fuel oil 5	54†				≤1.0	—	
Fuel oil 6	66†	407			1.0	—	
Gasoline (60 octane) (100 octane)	−43 −38	280 } 456 }	1.4	7.6	0.8	3.0–4.0	38 to 204
Lubricating oil (motor oil)	149 to 232	260 to 371			≤1.0	—	360
Mineral oil	193				0.8–0.9	—	360
Naphtha (Stoddard solvent)	38 to 60	149 to 232	1.1	6.0	0.8	—	181

Naphtha							
V M & P (50 flash)	10†	232‡	0.9	6.7	1.0	4.1	116 to 143
(High flash)	30‡	232‡	1.0	6.0	1.0	4.3	138 to 177
(Regular)	−2‡	232‡	0.9	6.0	1.0	—	100 to 160
Petroleum ether (Petroleum naphtha)	−46 to −18	228	1.1	5.9	0.6	2.5	35 to 60
Benzene	−11	562	1.3	7.1	0.9	2.8	80
Naphtha (coal tar)§	38 to 44	277			0.8	—	149 to 204
Naphthalene	79	526	0.9	5.9	1.1	4.4	218
Toluene	4	536	1.2	7.1	0.9	3.1	111
Xylene: *ortho*	17	464	1.0	6.0	0.9	3.7	144
meta	25	528	1.1	7.0	0.9	3.7	139
para	25	529	1.1	7.0	0.9	3.7	138

* Depending on the curing qualities, the flash point of asphalt varies
Cutback asphalt has a flash point of 10 °C.
† Minimum or legal flash point, which varies in different States in USA.
‡ Flash point and ignition temperature will vary depending on the manufacturer.
§ Coal tar naphtha, mainly toluene and xylene.

Table 4.2 Properties of paraffins[2]

	Flash-point (°C)	Ignition temp. (°C)	Lower flammable limit (% by vol. in air)	Upper flammable limit (% by vol. in air)	Specific gravity (water = 1.0)	Vapour density (air = 1.0)	Boiling point (°C)
Methane (CH_4) (Marsh gas)	Gas	537	5.3	14.0		0.6	−162
Ethane (C_2H_6)	Gas	515	3.0	12.5		1.0	−89
Propane (C_3H_8)	Gas	466	2.2	9.5		1.6	−42
Butane (C_4H_{10})	Gas	405	1.9	8.5		2.0	−0.6
Petane (C_5H_{12})	−40	309	1.5	7.8	0.6	2.5	37
Hexane (C_6H_{14})	−28	234	1.2	7.5	0.7	3.0	69
Heptane (C_7H_{16})	−4	223	1.2	6.7	0.7	3.5	98
Octane (C_8H_{18})	13	220	1.0	3.2	0.7	3.9	126
Nonane (C_9H_{20})	31	206	0.8	2.9	0.7	4.4	151
Decane ($C_{10}H_{22}$)	46	208	0.8	5.4	0.7	4.9	174

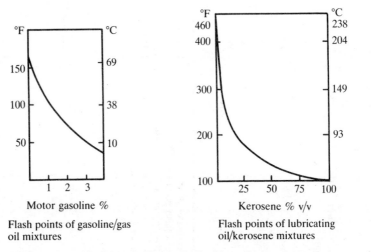

Flash points of gasoline/gas oil mixtures

Flash points of lubricating oil/kerosene mixtures

Fig. 4.2 Effect of small additions of volatile liquid on flash points of mixtures

- Petroleum, Class 'B'; flash point 73 °F to 150 °F (22.8–65.6 °C), e.g. kerosenes, white spirits, tractor vaporising oil.
- Petroleum, Class 'C'; flash point more than 150 °F (65.6 °C), e.g. gas oils, diesel oils, heavy fuel oils.

Ambient temperature variation with geographic location is clearly important since a solvent which would always be below its flash point under ambient conditions in the UK will be more hazardous in hotter climates. A list of common flammable liquids with flash points less than 90 °F (32 °C) is given in Table 4.3.

United Kingdom legislation concerning the delivery, storage and handling of flammable liquids includes:

The Petroleum (Consolidation) Act 1928 applies to the transport and storage of all hydrocarbon liquids having a flash point of below 23 °C. *SI 1929 No. 993 (The Petroleum Mixtures Order 1929)* applies the whole of the Petroleum (Consolidation) Act 1928 to mixtures containing petroleum (not just petroleum spirit) which have flash points below 23 °C. Transport of these substances by road is restricted to certain approved types of vehicle operating within specified limits. Their storage is only permitted under a licence issued by a petroleum officer authorised by a local authority, county council or similar body.

The Dangerous Substances (Conveyance by Road in Road Tankers and Tank Containers) Regulations 1981 control the conveyance by road of dangerous substances in road tankers of any capacity and tank containers having a capacity of more than 3 m^3. Dangerous substances are defined, including flammable liquid, flammable gas, flammable solid (i.e. a solid which is combustible under conditions encountered in conveyance by road or which may cause or contribute to fire through friction), spontaneously combustible substance and oxidising substance. Regulations apply, for example, to the construction, testing and examination of road tankers and tank containers; information to be obtained by the operator and information to be available during conveyance; precautions against fire and explosion; supervision of vehicles; the carrying of hazard warning panels and labels; instruction and training of drivers, etc. Detailed rules are given for the unloading of petroleum spirit at petroleum filling stations and certain other licenced premises.

The Factories Act 1961: The Highly Flammable Liquids and Liquefied Petroleum Gases Regulations 1972 (SI 1972 No 917) impose restrictions on the storage, and handling, of all solvents having a flash point below 32 °C (which also when heated under the conditions of a specific test and exposed to an external source of flame, applied in a standard manner, support combustion). Buildings were highly flammable liquids are used or stored must be fire-resistant structures. The quantity allowed in any workroom is limited to practical needs. The movement or transfer of such liquids within the factory should be in pipelines or closed containers. The provision of adequate signs warning of the presence of highly flammable liquids is stipulated. (NB The Factories Act and the Petroleum (Consolidation) Act cover similar aspects in some cases and it is necessary to comply with the more stringent requirement in any particular case.)

Table 4.3 Selection of highly flammable liquids, i.e. flash-point < 32 °C

Acetaldehyde	1,2-Dichloropropane	Methyl formate
Acetone	1,1-Diethoxyethane	2-Methyl furan
Acetonitrile	Diethylamine	Methyl hydrazine
Acetyl chloride	Diethylcarbonate	Methyl isobutyl ketone
Acrolein	Diethyl ketone	Methyl isobutyrate
Acrylonitrile	3,4-Dihydro-2H-pyran	Methyl methacrylate
Allyl acetate	Diisopropyl amine	Methyl-n-propyl ketone
Allyl alcohol	Diisopropyl ether	1-Methylpyrrole
Allyl amine	Dimethoxymethane	
Allyl bromide	Dimethoxypropane	Naphtha (50°) (Reg)
Allyl chloride	Dimethylcarbonate	Nitroethane
Allyl chloroformate	Dimethyl sulphide	
Amyl acetate	1,4-Dioxane	Octane
t-Amyl alcohol		Phosgene solutions in benzene
Amyl amine	Endrin	Piperidine
Amylene	Ethanethiol	1,3-Propanediamine
iso-Amyl formate	Ethanol	Propargyl bromide
Amyl mercaptan	Ethoxy acetylene	iso-Propenyl acetate
Amyl nitrite	Ethyl acetate	Propionaldehyde
Aziridine	Ethyl acrylate	Propionitrile
	Ethyl benzene	Propionyl chloride
Benzene	Ethyl butyrate	iso-Propyl acetate
Benzotrifluoride	Ethyl chloroformate	n-Propyl acetate
Butyl acetate	Ethyl crotonate	Propyl alcohol
Butyl alcohol	Ethylene dichloride	Propyl amine
Butyl amine	Ethyl formate	Propyl benzene
Butyl bromide	N-ethyl morpholine	Propyl chloride
Butyl chloride	Ethyl vinyl ether	iso-Propyl ether
Butyl formate		n-Propyl formate
Butyl hydroperoxide	Furan	iso-Propyl formate
Butyl mercaptan		Propyl nitrate
Isobutyl methyl ketone	Gasoline	Pyridine
tert-Butyl peracetate	Heptane	Pyrrolidine
Butyl vinyl ether	Hexane	Pyruvic acid
n-Butyraldehyde	Hexene	
n-Butyronitrile	Hexyl amine	Styrene
Butyryl-chloride	Hydrocyanic acid	
		Tetrahydrofuran
Carbon disulphide	Kerosene	Thiophene
Cellulose nitrate		Toluene
Chlorobenzene	Ligroin	Triethyl o-formate
p-Chloro-m-cresol		Triisobutyl aluminium
Chloroprene	Mesityl oxide	Trimethyl borate
Collodion	Methyl acetate	2,2,4-Trimethyl pentane
Crotonaldehyde	Methyl acrylate	2,4,4-Trimethyl-2-pentane
Cyclohexane	Methyl alcohol	
Cyclohexene	2-Methyl-1-butene	Unsymmetrical dimethyl hydrazine
Cyclopentane	2-Methyl-2-butene	
Cyclopentanone	N-Methyl butyl amine	Valeraldehyde
	Methyl butyrate	Vinyl acetate
Di-n-butyl ether	Methyl chloroformate	Vinyl ether
Dibutyl peroxide	Methyl cyclohexane	Vinylidene chloride
1,3-Dichloro-2-butene	4-Methyl cyclohexene	
1,1-Dichloroethane	Methyl ethyl ketone	Xylene
1,2-Dichloroethylene		

Flammable materials handling

In addition, most factory premises require a fire certificate issued under the *Fire Precautions Act 1971*. This specifies:
- The particular use(s) of the premises which it covers.
- The means of escape in case of fire.
- The means for securing that the means of escape can be safely and effectively used, e.g. measures to restrict the spread of fire, smoke and fumes, emergency lighting and direction signs.
- The means for fighting fire for use by occupants.
- The means for giving warning in case of fire.
- Particulars of any explosive or highly flammable materials which may be stored/used in the premises.

It may also impose requirements as to:
- The maintenance of the means of escape and keeping them free from obstruction.
- The maintenance of other specified fire precautions.
- The training of employees on the action to be taken in the event of fire and the keeping of appropriate records.
- The limitation of the number of people on the premises.
- Any other relevant fire precautions.

The Fire Certificates (Special Premises) Regulations 1976 administered by the Health and Safety Executive deal with 'Special Premises', as listed in Schedule 1, Part 1, as in Table 4.4. These premises include those where the process hazards may be on such a scale, or be of such a character, or have such a direct bearing on general fire precautions

Table 4.4 Fire Certificates (Special Premises) Regulations 1976–Schedule 1, Part 1, Premises for which a fire certificate is required

---

1. Any premises at which are carried on any manufacturing processes in which the total quantity of any highly flammable liquid under pressure greater than atmospheric pressure and above its boiling point at atmospheric pressure may exceed 50 tonnes.

2. Any premises at which is carried on the manufacture of expanded cellular plastics and at which the quantities manufactured are normally of, or in excess of, 50 tonnes per week.

3. Any premises at which there is stored, or there are facilities provided for the storage of, liquefied petroleum gas in quantities of, or in excess of, 100 tonnes except where the liquefied petroleum gas is kept for use at the premises either as fuel, or for the production of an atmosphere for the heat-treatment of metals.

4. Any premises at which there is stored, or there are facilities provided for the storage of, liquefied natural gas in quantities of, or in excess of, 100 tonnes except where the liquefied natural gas is kept solely for use at the premises as a fuel.

5. Any premises at which there is stored, or there are facilities provided for the storage of, any liquefied flammable gas consisting predominantly of methyl acetylene in quantities of, or in excess of, 100 tonnes except where the liquefied flammable gas is kept solely for use at the premises as a fuel.

Table 4.4 (continued)

6. Any premises at which oxygen is manufactured and at which there are stored, or there are facilities provided for the storage of, quantities of liquid oxygen of, or in excess of, 135 tonnes.

7. Any premises at which there are stored, or there are facilities provided for the storage of, quantities of chlorine of, or in excess of, 50 tonnes except where the chlorine is kept solely for the purpose of water purification.

8. Any premises at which artificial fertilisers are manufactured and at which there are stored, or there are facilities provided for the storage of, quantities of ammonia of, or in excess of, 250 tonnes.

9. Any premises at which there are in process, manufacture, use or storage at any one time, or there are facilities provided for such processing, manufacture, use or storage of, quantities of any of the materials listed below in, or in excess of, the quantities specified:

Phosgene	5 tonnes
Ethylene oxide	20 tonnes
Carbon disulphide	50 tonnes
Acrylonitrile	50 tonnes
Hydrogen cyanide	50 tonnes
Ethylene	100 tonnes
Propylene	100 tonnes
Any highly flammable liquid not otherwise specified	4,000 tonnes

10. Explosives factories or magazines which are required to be licensed under the Explosives Act 1875.

11. Any building on the surface at any mine within the meaning of the Mines and Quarries Act 1954.

12. Any premises in which there is comprised:
    (a) any undertaking on a site for which a licence is required in accordance with Section 1 of the Nuclear Installations Act 1965 or for which a permit is required in accordance with Section 2 of that Act; or
    (b) any undertaking which would, except for the fact that it is carried on by the United Kingdom Atomic Energy Authority, or by, or on behalf of, the Crown, be required to have a licence or permit in accordance with the provisions mentioned in sub-paragraph (a) above.

13. Any premises containing any machine or apparatus in which charged particles can be accelerated by the equivalent of a voltage of not less than 50 megavolts except where the premises are used as a hospital.

14. Any premises at which there are in process, manufacture, use or storage at any one time, or there are facilities provided for such processing, manufacture, use or storage of, quantities of unsealed radioactive substances classified according to Schedule 3 of the Ionising Radiations (Unsealed Radioactive Substances) Regulations 1968 in, or in excess of the quantities specified:

Class I radionuclides	—	10 curies
Class II and III radionuclides	—	100 curies
Class IV radionuclides	—	1000 curies

15. Any building, or part of a building, which either:
    (a) is constructed for temporary occupation for the purposes of building operations or works of engineering construction; or
    (b) is in existence at the first commencement there of any further such operations or works

and which is used for any process or work ancillary to any such operations or works.

that the two aspects cannot be considered apart, or where highly specialised processes are carried on. The certificates specify broadly similar provisions, and are able to impose similar requirements, as under the Fire Precautions Act.

Control of Pollution Act 1974. The deposit of solvents on any land is prohibited unless the land is occupied by the holder of a disposal licence authorising the deposit. Advice on proposed deposits and applications for a disposal licence should be sought from the relevant disposal authority, which will be the county council for the area. Solvents must also be prevented from leaking into drains, watercourses, etc.

Customs & Excise Act 1952. This concerns the receipt, storage and use of those solvents on which duty is, or could be, payable. Ethanol, including industrial methylated or denatured spirits, and hydrocarbons come within this category. HM Customs & Excise must be advised of any change in the storage arrangements.

Most situations involving flammable-material safety occur in the presence of normal air (approximately 21% oxygen). In this context the use of air for blowing through vessels or lines containing flammable liquids or vapours should be avoided; inert gas should be provided instead. This precaution also applies to flammable liquids/solids generally considered 'safe' at ambient temperatures, e.g. phenol, since these readily produce flammable vapour on heating.

Ignition is more easily accomplished, and combustion is more rapid, in pure oxygen or in oxygen enriched air, as discussed later. Alternatively, dilution with inert gas (nitrogen, carbon dioxide or flue gases) may be used to inhibit combustion. A concentration of vapour can be reached below which a flame will not propagate; this concentration is the Lower Explosive Limit (LEL). Conversely, the vapour concentration can be made so 'rich' that there is insufficient oxygen for combustion; this is the Upper Explosive Limit (UEL). The intermediate range is the Flammable Range; within these limits any mixture will burn. Typical values are given in Tables 4.1 and 4.2.

Clearly, a vapour or gas with wide limits of flammability is more hazardous. For example, the range for hydrogen is 4% to 75%, for hydrogen sulphide 4.3% to 45%, and for acetylene 2.5% to 81% compared with 16% to 25% for ammonia.

Limits of flammability are not affected appreciably by normal variations in atmospheric pressure. The effect of increase in pressure above atmospheric is complex and specific for each flammable material. The range of flammability is generally narrowed by reduction in pressure appreciably below atmospheric and at a specific low pressure the limits coincide. In applying quoted flammable limits to atmosphere control it should be remembered that:

1. Increases in temperature increase the flammable range by decreasing the lower and increasing the upper flammable limit.
2. 'Pockets' or 'fringe areas' of mixtures in the flammable range may be present in the vicinity of any 'over-rich' mixtures; the turbulence and expansion following ignition of such a pocket may enhance admixture of air and vapour and immediately bring the whole mass into the flammable range.

Dilution with inert gas serves to depress the flammable limits. The effect is more pronounced with carbon dioxide than with nitrogen, for a given percentage of added inert gas. The relationships are illustrated for carbon monoxide and hydrogen in the diagrams in Figs 4.3 and 4.4. The area inside each triangle represents flammable compositions; the area outside represents non-flammable compositions. Thus the effect of added inert is to reduce the percentage of oxygen to support combustion until,

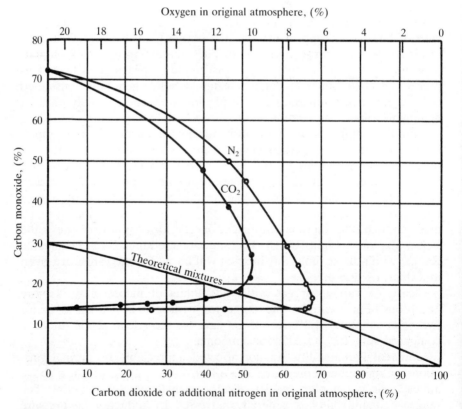

Fig. 4.3 Limits of flammability – carbon monoxide in air and carbon dioxide (or nitrogen)

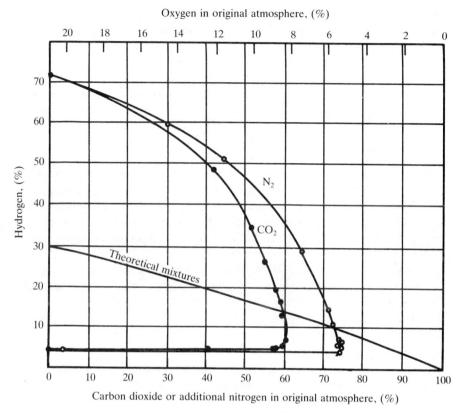

Fig. 4.4 Limits of flammability – hydrogen in air and carbon dioxide (or nitrogen)

below a critical value, there is insufficient present. This critical (minimum) oxygen level varies with the inert gas and with the flammable vapour involved. For example, the mixture of hexane–air and added nitrogen that propagates flame with a minimum concentration of oxygen, contains 22 volumes of added nitrogen per volume of hexane, this will propagate flame when the oxygen concentration is 11.9% or higher. All other mixtures require higher oxygen concentrations. Minimum oxygen values have been similarly determined for other hydrocarbons.[4] The effect of added inert, except for some extinguishing materials (e.g. chlorobromomethane) which have a specific inhibitory effect, is related directly to thermal capacity; hence carbon dioxide is more effective than nitrogen. Flue gas is also used as a blanketing medium, e.g. the inerting of cargo spaces of oil tankers during ballast voyages using engine exhaust gas.

For most petroleum products the critical oxygen content is 10–11% and levels may be kept below this by 'blanketing'. (This does not account for the stricter limitations necessary with some materials, e.g. diolefins to minimise peroxide formation.) In practice the oxygen concentration must be reduced further than quoted values to allow for variations in mixture composition; hence when purging air from plant using inert gas, purging should generally continue until the proportion of oxygen is reduced to well below 8%. If an ignition source is to be applied < 2% oxygen may be necessary.

Gases

The principles associated with the combustion of pressurised flammable gases are similar to those for vapours except that they have no flash point, i.e. they are ready to burn without preheating. Their flammable limits are generally also wider. One significant difference is that the vapours of nearly all flammable liquids, including LPGs are appreciably heavier than air; thus they tend to move downwards and outwards from any point of emission. Conversely natural gas (vapour density 0.584) and to an even greater extent hydrogen (vapour density 0.07) are lighter than air and will rise even when cold. Acetylene (vapour density 0.91) also tends to rise following leakage.

> A natural gas explosion blew out the shop window and the rear of the premises of a men's outfitters and 38 people were injured, mainly by flying glass. The source of the gas was a 10 cm grey cast-iron main beneath the footway outside. (This was point-supported on the remains of a brick wall in a cellar and settlement of the ground outside had occurred due to traffic loading; this had caused the pipe to hog over the brickwork and fracture.) Gas entered the premises via the path of the gas, electricity and water services and rubble infill allowed ready permeation of the gas.[5]

The lighter gases disperse readily whereas the heavier gases (e.g. butane 2.11, propane 1.55) may spread out over a considerable area and collect in trenches and basements; ignition sources, discussed later, are more common at low level so that, in general, leakage of a heavy gas is potentially more serious.

> At about 01.31 hr an explosion ruptured the 15 cm transfer line between a refinery at Gothenburg (Sweden) and a depot some 5 km away. About 15 tonnes of propane were released about 200 m outside the refinery boundary fence and adjacent to a public road. During the next 50 minutes or so there was confusion. At 02.00 hr the police and fire brigade set up road blocks and evacuated inhabitants of nearby houses; the road towards Torshamnen was however left open. At 02.15 hr the AC (part of the fire brigade) reported that the leaking pipe was shut at both ends but the emission continued, with undiminished force. (Once the rupture occurred there was no means of stopping the leakage.) Five minutes later a car approached from Torshamnen and developed engine trouble while passing through the

'fog'. At 02.25 hr vehicles from the fire brigade seen in the fog were ordered to leave the area the same way they had come. At that moment the unconfined vapour cloud ignited and burned as a flash fire, initiating several explosions. Three firemen received burns and one subsequently died. Houses, cars and nearby factories were also extensively damaged.[6]

The cause of pipe rupture was sabotage involving detonation of a small charge of plastic explosive. The ignition source of the propane vapour cloud was probably a petrol-driven salvage vehicle from the fire brigade which had been driven into it. [Following this incident consideration was given to the need for marking pipelines, especially those passing over public roads, so that the public/emergency services would know whom to contact in an emergency. Also, the necessity for incorporating isolating valves was raised, particularly in populated areas or vulnerable sections of pipeline. General precautions are summarised in Table 4.11]

Dusts

Solid particles less than about 100 μm in diameter settle slowly in air and comprise dust (larger particles are grit). Such finely divided particles behave in some ways similar to gas; hence with a combustible solid a flammable dust–air mixture is obtainable within certain limits.

Dust explosions are relatively rare but can involve an enormous energy release. A common phenomenon involves a primary explosion, involving a limited quantity of material, distributing dust in the atmosphere and initiating a severe secondary explosion.

> In a provender factory cotton-seed cake was being ground in a basement; the ground material was carried by a bucket elevator via several floors to bins at the top of the factory building. The ground material was ignited, possibly by friction at a bucket subsequently found to be loose inside the casing. No one was injured by the primary explosion in the elevator. However, accumulations of loose dust in the room through which the elevator passed were dislodged. Flame from the burst elevator casing ignited these dust clouds resulting in a secondary dust explosion; six people were killed, a large number of people injured and the building was wrecked.[7]

> An explosion occurred in a ship while the interior surfaces of tanks were being sprayed with zinc resulting in extensive damage to the structure. A primary explosion was believed to have been caused by ignition of gas leaking from a hose. This threw into suspension accumulated deposits of zinc dust from 'over-spray'.[8]

Small particles are required in order to provide a large surface area to mass ratio and for the solid to remain in suspension. Surface absorption of air (oxygen) by the solid or evolution of a combustible gas on heating (e.g. as in the case of bituminous coal) may be a predisposing factor; the presence of moisture reduces the tendency to ignite, and can in practice favour agglomeration.

Table 4.5 Explosion characteristics of dusts[9]

Type of dust (−200 mesh)	Ignition temperature of dust cloud (°C)	Minimum spark energy required for ignition of cloud (millijoules)
Metal powders		
Aluminium, atomised	640	15
Aluminium, milled	550	—
Magnesium, milled	520	40
Magnesium/aluminium alloy 50:50	535	80
Titanium	330	10
Zinc	600	650
Zirconium	*	5
Plastics		
Cellulose acetate	320	10
Methyl cellulose	360	20
Methyl methacrylate	440	15
Phenolic resin	460	10
Polyamide	500	20
Polystyrene	490	15
Urea resin	450	80
Agricultural products		
Cocoa	420	100
Coffee	410	160
Cornstarch	380	30
Cotton seed	470	80
Dextrin, corn	400	40
Grain dust	430	30
Sugar	350	30
Miscellaneous		
Adipic acid	550	70
Aluminium stearate	400	15
Coal, high volatile (Pgh. seam)	610	60
Cork	470	45
Phthalic anhydride	650	15
Pitch, coal tar (58% volatile)	710	20
Rubber, crude, hard	350	50
Sulphur	190	15
Wood flour	430	20

* Dust cloud ignited under some conditions when dispersed in air at ambient temperature.

The explosive range of dust in air can be very wide; the limits vary with both chemical composition and the size and shape of the particles. The lower limits are around 20 g/m^3 (0.02 oz/ft^3), equivalent to a dense fog in appearance; the upper limits are ill-defined but are generally of little practical significance. The explosion characteristics of the most common hazardous dusts are summarised in Table 4.5.[9] (See also Tables 4.13 and 11.24.) Apart from the differing ignition characteristics and lower explosive limits, the important data are the maximum explosion pressure and

Minimum explosive concentration oz. per 1,000 cu. ft.	Maximum explosion pressure (psi)	Rates of pressure rise (psi per sec.)	
		Average	Maximum
40	90	3,500	10,000+
45	70	2,000	4,250
20	95	3,000	10,000+
50	90	4,000	10,000+
45	80	3,400	10,000+
480	50	600	1,750
40	65	800	8,750
25	110	2,800	6,750
30	100	1,900	6,000
20	100	500	1,750
25	80	1,700	6,000
30	90	1,800	7,000
15	90	2,400	7,000
70	85	800	2,000
45	62	550	1,200
85	50	150	250
40	110	2,200	6,750
55	90	800	2,500
40	105	1,800	7,000
55	95	1,000	2,750
35	90	1,600	5,000
35	75	1,200	2,750
15	95	1,200	4,750
55	85	800	2,250
35	100	2,000	5,500
15	70	1,300	4,250
35	95	1,900	6,000
25	80	1,200	3,800
35	80	1,700	4,750
40	110	1,600	5,500

the rates of pressure rise. In general, organic or carbonaceous materials, or easily oxidisable metals like aluminium or magnesium, are more hazardous than nitrogenous organic materials, e.g. oil seeds or leather; the least hazardous materials are those containing an appreciable amount of mineral matter, e.g. coke or bone-dust.

As with vapours/gases, there is a critical oxygen content below which ignition of combustible dusts will not occur; these are summarised for common dusts in Table 4.6,[9] where it will be noted that hot surfaces are

Table 4.6 Minimum permissible oxygen content to prevent ignition of combustible dusts[9]

Type of dust	Using carbon dioxide for inerting		Spark ignition
	Hot surface (850 °C ignition)		
	O_2 percentage above which ignition can take place	Maximum recommended O_2 per cent	O_2 percentage above which ignition can take place
Metal powders			
Aluminium (atomised)			3
Magnesium			3
Magnesium/aluminium alloy			0
Titanium			0
Zinc			10
Zirconium			0
Resins			
Cellulose acetate	5	4	13
Phenolic	9	7	14½
Polystyrene	7	5½	14½
Urea	11	9	17
Moulding compositions			
Cellulose acetate	7	5½	11½
Methyl methacrylate	7	5½	14½
Phenolic	7	5½	14½
Polystyrene	9	7	14½
Urea	9	7	17
Resin ingredients			
Hexamethylene tetramine	11	9	14½
Pentaerythritol	7	5½	14½
Phthalic anhydride	11	9	14½
Miscellaneous			
Cornstarch	5	4	11
Sulphur	—	—	11
White dextrin	—	—	12
Coal dust, high volatile	10½	8½	15
Petroleum pitch			11

- The 'Maximum recommended O_2 per cent' applies only to maintaining an inert atmosphere for protection against unexpected or unlikely sources of ignition. Much higher factors of safety are required where sources of ignition are deliberately applied.

more efficient ignition sources than sparks – since a finite amount of material has to be heated up.

Ignition and combustion of flammable mixtures – gas or vapour

The ignition temperature is the temperature at which a small amount of material will spontaneously ignite in a given atmosphere and burn without

	Using nitrogen for inerting		
	Spark ignition		
Maximum recommended O_2 per cent	O_2 percentage above which ignition can take place	Maximum recommended O_2 per cent	
2½	9	7	
2	2	1½	
0	6	5	
0	6	5	
8	10	8	
0	4	3	
10½			
11½			
11½			
13½			
9			
11½			
11½			
11½			
13½			
11½			
11½			
11½			
9			
9			
9½			
12			
9			

● In the furnace test dust clouds of Zr ignited in CO_2. During heating for several minutes, undispersed layers of the following metals ignited (glowed) in CO_2: Mg, Zn, Mg/Al, Ti.

a further heat input. It is dependent upon conditions, e.g. mixture composition, temperature, pressure, shape of space or container, nature of ignition source (flame, sparks, etc.) and direction of flame propagation. However, a temperature range can be given for ordinary conditions. This is not the temperature of the igniting medium, which must obviously be higher, but is a useful guide to safe surface temperatures.

When a gas or vapour, or a dust cloud burns in a confined space, e.g. a vessel or building, the heat of combustion causes rapid expansion of the gaseous combustion products which are restrained by the walls of the confined space. The pressure in fact developed depends upon the composition of the flammable mixture. A mixture just within the limits of flammability may result in an over-pressure of several pounds per square inch, but a stoichiometric mixture, i.e. one with the correct quantity of air for complete combustion, may generate a pressure exceeding 100 psi (700 kN/m^2). The rapid pressure rise may reach the rupture pressure of the walls which burst with a loud noise so that the pressure is 'immediately' released. With a gas or vapour this is termed a confined vapour cloud explosion (CVCE).

An alternative type of explosion is a BLEVE (boiling liquid expanding vapour explosion) which occurs when a pressure vessel containing a liquefied gas fails under exposure to fire. Here the damage results from the blast wave due to the relief of internal pressure, thermal radiation from the fireball and the projection of large fragments. Figure 4.5 illus-

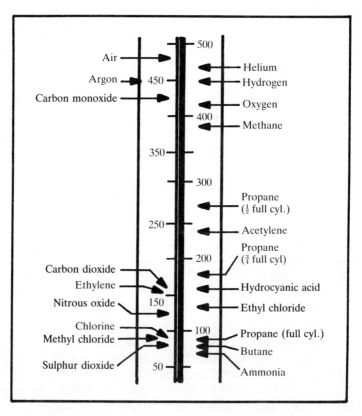

Fig. 4.5 Indication of temperature (°C) at which gas cylinders will be in danger of bursting[10]

trates the temperatures at which gas cylinders will be in danger of bursting. Hydrogen, carbon monoxide and methane are considerably less hazardous than the other flammable gases in this respect. While an acetylene cylinder is rather more vulnerable to bursting in a fire than a half-full propane cylinder, the figure shows that a full propane cylinder is in danger of bursting below 100 °C.

With a very large volume of petroleum vapour intermixed with air, e.g. from a massive LPG escape, ignition causes sudden expansion and turbulence accompanied by noise and blast. This is a classical unconfined vapour cloud explosion (UVCE) discussed in Chapter 15.

A major fire is the probable outcome of any large spillage of flammable materials. This may take the form of a running liquid fire, a pool fire or, rarely, a fireball. Since fuel is required to initiate and sustain the combustion process, the largest potential fire will correspond to where there is the greatest storage of flammable liquid or gases. With flammable liquids kept under normal pressures and temperatures the hazard is one of surface spread of flame. However, with liquefied flammable gases, or flammable materials stored and processed at high temperatures and pressures, there is the potential for the generation of a vapour cloud rich in fuel resulting in a fireball.

A running liquid fire involves the spread of flame across the surface of a flammable liquid spillage. From the ignition source the flame flashes back to the surface of a large spillage; this can result in a running flame which spreads due to gravity effects on the liquid. The speed and extent of the spread will depend upon local topography and will be increased by drains and trenches. However, tank storages are normally bunded to prevent the spread of liquid in the event of loss of containment.

Ignition sources

Generally a combustion process is initiated by the introduction of a finite amount of energy to raise a finite volume of the material to its ignition temperature. Potential ignition sources for vapour–air mixtures are discussed below.

Flame-cutting and welding, etc.

This includes oxy-acetylene or electric arc welding, not cutting operations, brazing or soldering and blow torches.[3] The situation in which even if the residues within a tank or pipeline are relatively non-volatile a flammable vapour–air mixture may be generated by the heat from the torch was discussed in Chapter 3.

Either the flame, or arc, itself or the shower of metal particles may ignite flammable vapours. Many fires have been started by hot slag and the globules of molten metal seen as sparks. There are reports of sparks rolling, bouncing, flying or simply falling into trouble areas.[11]

> Welding sparks set fire to drums of a dye-stuff standing about 10 m from the welding site. As a result several hundred kg of a very valuable product were lost. The welding was controlled by a hot-work permit [see Ch. 17] but the supervisor who signed it underestimated the distance welding sparks can fly.[12]

In the situation when an open flame is applied to a vessel or pipe containing a flammable liquid or gas, ignition can occur even when the flame has not penetrated the wall. This arises because the vapour or gas can reach its auto-ignition temperature, e.g. 205 °C to 540 °C, before the softening point of the metal is reached, for example with steel before it even becomes red hot. (Note that some vapours, e.g. carbon disulphide (100 °C) have even lower auto-ignition temperatures.)

Sparks produced from stray currents from arc welding sets have also caused explosions.

> Contractors' employees obtained an empty 45-gallon drum marked 'IPA' from within a factory and placed it on end for use as a working platform from which to do welding. About 2 weeks later another contract employee placed his arc-welding equipment on the drum which was earthed through contact with a process vessel. An arc was struck, the top of the drum was pierced and the worker was severely injured by an explosion. The drum had contained isopropyl alcohol.[13]

Blowlamps cause many fires during burning-off operations as a consequence of flame reaching concealed timber or other combustibles, resulting in their smouldering for hours before bursting into flame. Materials used for sheeting-up may contribute markedly to the rapid development and spread of fire.

Smoking materials

The flame from matches, the friction spark from lighters or a glowing cigarette end are all potential ignition sources. Their efficiency varies, e.g. the temperature of a cigarette end ranges between 290 °C and 730 °C whereas a match flame exceeds 1,100 °C.

Therefore it is standard practice to forbid smoking, except in certain defined areas, and the carrying of lighters or matches in factories, etc., where there are flammable materials. In some plants specific personnel are permitted to use matches solely for lighting furnaces.

> A man dropped a lighter while filling it with naphtha from a 1-gallon can. A spark was apparently generated from the wheel and an explosion resulted from ignition of vapour from a drum filling plant.[14]

Sparks, etc., from electrical equipment

Ignition sources from electrical equipment generally arise from sparks when an electric current jumps across a gap between conductors, or arcs, when current-carrying contacts are separated. Such an electrical short circuit or arc may involve a temperature of 3,870 °C (6,998 °F).[15]

Direct current motors, generators and slipring and synchronous motors, produce sparks at commutators and sliprings continuously; switches spark or arc when opening/closing. (Other types of electrical equipment are non-sparking in normal operation, e.g. well-designed squirrel-cage induction motors, and transistors operating as switches.)

There is adequate energy present in the majority of such systems to ignite a flammable mixture. Therefore in areas where flammable vapour is likely to be present all electrical equipment, including lighting or analytical apparatus, should be either flameproof, non-sparking, intrinsically safe or purged. Selection depends upon the severity of the hazard and local conditions.

> A fire started in a domestic refrigerator used for storage of a variety of chemicals considered unstable or likely to be dangerous at room temperature. The motor and refrigeration parts of the refrigerator were not involved but scorch marks indicated that the contents had been involved. The cause of the fire was considered to be a flammable vapour from the contents being ignited by a spark from the electrical contacts of the thermostatic control.[16]

(Further references to this hazard are to be found in the literature (refs 17, 18) and in Ch. 11. If flammable liquids are to be stored in a refrigerator care must be taken to remove electrical contacts, including a light and switch, from inside the body or near the door unless flameproof switches are fitted. Even if it is electrically safe a fire may be initiated by an external ignition source when the door is opened; this is one reason for prohibiting smoking in laboratories.)

The local heating effect of an electric current may also be sufficient to cause a surface temperature to exceed the spontaneous ignition temperature of a flammable atmosphere. The hot filament of an electric lamp can cause ignition, as can apparatus in which mechanical breakdown of insulation has occurred and resulted in sparking, over-heating, or arcing. Sparks are usually at a temperature above ordinary flame temperatures, but to cause ignition they must persist for a sufficient time; therefore a certain intensity must be available. (Operation below this minimum energy requirement forms the basis for 'intrinsically safe' electrical apparatus.)

Just as with fixed equipment, special precautions are necessary before portable apparatus (e.g. inspection lights, test apparatus or portable instruments) are introduced into any area where flammable gas or vapour may be present. Thus in some petroleum plants there is prohibition of transistor radios, bicycle lamps and non-flameproof torches.

> A short circuit in a mains-operated calculator ignited insulation and the fire spread to the plastic body and then to the timber desk on which it stood. The fire developed in the room, a study, for about 2 hours before spreading to an adjacent research laboratory where heat detectors actuated an auto-dialler to call the fire brigade. Extensive damage resulted in a loss of £1,730,000.[19]

Some discussion has centred on the use of electronic watches in areas restricted due to the likely presence of flammable mixtures. In general, however, modern, low-powered electronic watches are considered safe on Zone 1 or Zone 2 areas.[20]

> A motor fitter was fatally burned in a vehicle inspection pit while attempting to drain a car petrol tank. He accidentally knocked a low-voltage inspection light which fell into the pit, broke and the hot filament ignited the petrol vapour.[21] (Petrol tanks should be drained in good ventilation conditions, preferably in the open air, and not over pits.)

If non-flameproof apparatus has to be introduced into potentially hazardous areas, atmosphere monitoring and permits to work are necessary (see Ch. 17).

Static electricity[22, 23]

The generation of static electricity is a well-known phenomenon associated with the separation of two dissimilar materials. In contact the two surfaces are equally and oppositely charged; on separation charges of opposite polarity also separate and spark discharge may occur between the two surfaces or a third body. The charges may be transported/conducted some distance after separation before there is sufficient accumulation to produce a spark, e.g. in the flow of liquids or powders. While the size of the charge is generally small the potential difference may be very high so that the spark is of sufficient energy for ignition.

> A powder was normally emptied from a metal hopper down a metal chute into a metal vessel. When the chute was replaced by a plastic hose it developed an electrostatic charge due to flow of the powder. The hose was in fact an electrical conductor but polyethylene end pieces which were used for attachment prevented leakage to earth. Eventually a spark was generated which ignited the dust plug.[24] [This is also an example of a hazard being introduced by a simple maintenance modification; see Ch. 17.]

Examples of operations which can result in static charge generation are given in Table 4.7.

Table 4.7 Generation of static charges

System	Examples
Solid–solid	Persons walking Grit blasting Conveying of powders Belts and pulleys Fluidised beds
Solid–liquid	Flow of liquids in pipelines/filters Settling of particles in liquid (e.g. rust and sludge)
Gas–liquid	Released gas (air) bubbles rising in a large tank Mist formation from LPG evaporation[26] Splash filling Cleaning with wet steam[26] Mist formation from high pressure water jets
Liquid–liquid	Settling of water drops in oil
Solid–gas	Mixing of immiscible liquids Pneumatic conveying of solids Fluidised beds

Static generation is likely in any operation involving fluidisation of particles, because of the large surface area of the particles and the random collisions between them and with the container walls. Earthing will facilitate discharge of charge accumulated on the container but discharge of static from the fluidised particles is generally not controllable.[25] The hazard with combustible particles depends on numerous factors including the size distribution, the concentration of the suspension and the electrical conductivity and heat of combustion of the particles. Generally with air-fluidised beds the dust concentration is high and outside the explosive range; however, above the surface of the bed there is a sharp decrease in concentration and a zone within the explosive range is likely. Furthermore, this zone will contain an increased proportion of finer particles. Another hazardous zone is likely to occur where the fluidised material is delivered into a storage vessel or pipeline. A particularly hazardous situation arises if flammable gases or vapours are also present. The essential precautions include permanent bonding of the whole plant to earth, antistatic flooring and the provision and use of antistatic footwear,[26] as well as provision of explosion relief.

> Crude cresylic acid was being transferred into a tank via both a top and bottom connection. The tank was earthed by the pipelines which led to and from it.[27] Since the top connection had no downpipe the operation involved 'splash filling'. A spark due to an electrostatic discharge ignited the vapour in the tank resulting in a fire in which it was severely damaged and much of the contents lost. (Note that external earthing does not control such spark generation inside tanks.)

Electrostatic charges generated when a liquid is pumped through pipelines depend on the electrical conductivity of the liquid; with a liquid of high electrical conductivity the charge is easily generated but quickly dissipated. Hazardous liquids are generally those with conductivities in the range 0.1 to 1000 pS/m. (picosiemens/metre, 1 siemen = 1 mho.) The rate of charge generation increases with turbulence so that an increase is associated with high flowrates and constrictions in the pipeline, e.g. filters, valves, bends or orifice plates. A low flowrate reduces the rate of charge accumulation and also allows more time for dissipation to earth by conduction. Generally water-free products can be pumped at velocities of up to 7 m/s through pipelines up to 20 cm diameter without excessive static charge generation but, if traces of water, or particulate matter, may be present the velocity should be limited to 1 m/s. Earthing and bonding is normally an effective method of charge dissipation except with liquids of very low electrical conductivity. An antistatic additive may then be used to raise the electrical conductivity to a safe value, e.g. 200×10^{-12} to $1,000 \times 10^{-12}$ ohm m^{-1}.

Storage tanks, pumps, tanker filling arms, drum filling points and associated hoses and pipework handling flammable liquid should be electrically bonded and earthed. Thus drum filling points, road-tanker filling points, and plastic/polymer powder or granules filling devices associated with metal receptacles, require earthing clips and bonding cables; these are connected to earth and effectively ground the tanker, drum or receptacle when coupled up prior to filling.

> An explosion occurred while an operator was pumping toluene into a 250 litre drum at 0.4 litre/s (5 gal/min) through a 2-cm line. Indications were that a static discharge occurred because the drum was not grounded. Since the drum was lined with a thin non-conductive, pigmented, epoxy coating, experiments were performed to determine whether conventional bonding and grounding would have been effective; it was concluded that external grounding of this type of plastic-lined drum was equally effective as for an unlined drum.[28]

> A man poured gasoline from a rusty 5 (22.7 litres) gallon can into the tank of a car via a chamois leather stretched over the top of a metal funnel. The funnel had a wooden collar so that it stood vertically; this wood effectively insulated the funnel from the tank. Before all the gasoline had been poured, a static discharge jumped between the funnel and tank and ignited the vapour.[29]

Because of their relatively low electrical conductivities, detailed precautions and procedures are generally necessary with petroleum liquids.

> An operator died as a result of an explosion in a cone-roof gas–oil tank while he was taking a sample through the gauging hatch. The explosion

ruptured the weak roof-to-shell seam and projected the roof into the bunded area beside the tank; the ensuing fire was relatively minor.

The tank was in hydrofined gas–oil product rundown-service and the explosion apparently resulted from ignition of a hydrogen-rich, treat gas–air mixture in the vapour space. While the vapour space above gas-oil (flash point approx. 66 °C) was normally non-flammable, a flammable atmosphere had arisen by blow-through of treat gas from the hydrofiner stripper. The probable ignition source was electrostatic sparking between the charged sample can, suspended on a polypropylene rope, and the gauging hatch as the can was being withdrawn.[30]

Design and operating rules for the prevention of electrostatic ignitions during tankage operations are given in standard references 31 to 33. Procedures for the sampling and gauging of storage tanks now include the following[30]:

- A prohibition on the lowering of gauging tapes, sample containers, thermometers or other devices into a tank during, or for 30 minutes after, all pumping in or circulating within has ceased. This applies to all atmospheric product tanks excluding products which do not accumulate static (e.g. crude, bitumen and residual products); floating roof tanks provided the roof is floating; cone-roof tanks where portable gauging, sampling or other devices are lowered into a gauging well or pipe which extends to the base of the tank.
- Sampling, gauging and other devices lowered into tanks must be constructed entirely of non-conductive materials or (less preferably) have all exposed metal parts securely grounded to the tank shell while in use.
- To eliminate the possibility of electrostatic charge generation when sliding through an operator's gloved hand, synthetic fibre rope should not be used.

Spontaneous ignition

Liquids may ignite spontaneously when heated out of contact with air and then allowed to escape into open air *or* when sprayed on to hot surfaces, e.g. steam pipes or diesel engine exhaust pipes. Under these circumstances the whole body of material is heated to the spontaneous ignition temperature. Values of paraffins are given in Table 4.2 (method of test ASTMD286), and for petroleum fractions in Table 4.1.

> A fire occurred, resulting in the deaths of three men and extensive damage to a plant, when hot oil above its auto-ignition temperature came out of a pump and ignited. Fitters had been dismantling the pump and the leakage occurred when they removed the cover plate because the suction valve had been left open. [This is an instance of inadequate isolation of equipment the proper procedures for which are described in Ch. 17.][34]

While the auto-ignition temperature is not an exact value, being dependent on the experimental conditions during determination, it represents an indication of the maximum safe surface temperature for contact by a material in the presence of air and the maximum temperature for unignited venting into the air.

Spontaneous combustion

Oxidation processes are exothermic and, with readily oxidisable materials, the heat generated through reaction may be inadequately dissipated. The reaction is accelerated by increasing temperature and under certain conditions the material ignites spontaneously.

'Spontaneous combustion' refers to ignition and combustion in a body of material without an external source of ignition. It is a phenomenon which occurs with certain materials generally considered stable at ordinary temperatures and excludes those materials proven to be instantaneously flammable in air (see page 150), explosives, or unstable chemicals. It involves self-heating resulting from an exothermic process within the mass of material.[35] Some materials known to self-heat to ignition under certain conditions are listed in Table 4.8[35]; this list is not exhaustive but does indicate the types of material likely to be involved.

Table 4.8 Materials prone to self-heating

Carbon	Fertilisers	Rags
Celluloid	Iron pyrites	Sawdust
Coal	Lagging (contaminated)	Foam and plastic
Copper ore concentrates	Monomers for polymerisation	Iron filings/wood
Cotton	Oils	Soap powder
Distillers dried grains	Plastics, powdered	Zinc powder
	Hay	

Self-heating requires that heat is generated in a mass of material more rapidly than it can be removed, e.g. by conduction to the surface, or by natural or forced convection. The rate of an exothermic reaction, generally in this case oxidation by air, typically increases exponentially with temperature – curve H in Fig. 4.6. Lines A, B, C and D represent different rates of heat loss. With a mass initially at temperature T_A, the temperature of the surroundings, and the cooling rate following line A then clearly, since the generation rate exceeds the removal rate, the temperature of the mass will increase. A stable situation will be at X. If the initial temperature was T_B and the cooling rate followed line B the cooling is inadequate and a thermal runaway occurs which can lead to a fire. In this situation T_C represents the critical temperature of the surroundings.

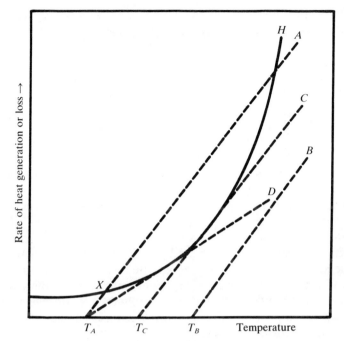

Fig. 4.6 Heat balance relationship: H = rate of heat generation; A,B,C,D = rates of heat loss

The rate of cooling commencing at T_A might fall to follow line D, e.g. due to insulation or an increase in size of the mass (due to the relationship between volume and heat-transfer area discussed on page 94). Line D then represents a critical minimum rate to avoid a thermal runaway.

General criteria for the possibility of spontaneous combustion of susceptible materials (after ref. 35) are:
- Porous material soaked with a reactive liquid.
- Accumulations, e.g. as layers, in heated process equipment (i.e. driers), on hot surfaces (i.e. electric motors or hot bearings).
- Large piles of materials with restricted cooling.
- Over-dried material when moisture absorption may contribute to the rate of heat release.
- Finely divided material – exposing a large surface area (see Ch. 3), e.g. dust accumulations in extract ducts.

The precautionary measures in any specific case can hence be deduced, aided by small-scale tests.[35]

A particular hazard arises from oil contamination of fibrous lagging (insulation) materials operating at high temperatures. Ignition occurs because the oil-wetted lagging exposes a high surface area of oxidisable material to atmospheric oxidation under conditions where the heat of reaction cannot be dissipated at a sufficient rate.

Spillages of oil frequently contaminate insulation because of leaks from pipe joints, pump glands, holed pipes or gearboxes. If this oil-soaked insulation is in contact with hot equipment (e.g. very hot process lines, still kettles or reboilers, or heat exchangers), then ideal conditions are created for spontaneous ignition. If no action is taken then, based on experience, fire may occur in the lagging after about 3 hours following the spillage; the greatest hazard arises when the insulation is broken.

Oil-soaked cotton, or similar rags can also create a fire hazard although petroleum oils are not as liable to 'self-heating' as animal or vegetable oils.

Best practice is to avoid placing oil-contaminated overalls, sacks, textile waste or rags near hot equipment or radiators. Waste rags should be placed in closed metal bins.

No external heat source is required with unsaturated (i.e. natural) oils, (nor indeed with a large pile of rags saturated with petroleum-based oils). Thus in the storage of turpentine at paper mills it is recommended that air should not be blown through transfer lines into storage tanks and monitoring of temperature, peroxide levels and pH should be carried out to avoid hazards from air oxidation.[36]

Vegetable oils with 'iodine values' above 100 are more susceptible to self-heating and the higher the iodine value the greater the hazard. Thus in descending order of hazard, commonly used oils include perilla, linseed (used for preserving leather and a constituent of some polishes), stillingia tung, cic, hemp-seed, cod-liver, poppy-seed, soyabean, seal, whale (sperm), walrus, maize, olive, cottonseed, sesame, rape and castor.

Spontaneous combustion without any external heating is a problem with some other materials with a high specific surface, e.g. soft grades of coal, porous charcoal. With coal the liability to spontaneous heating is greater the softer and newer the coal, i.e. anthracite is least susceptible to it. High stacks of powdered coal, where there is little heat escape, are most hazardous. Charcoal is porous as freshly made; absorption of air generates heat which may induce slow oxidation and eventual ignition. The risk is removed by moisture absorption but is re-created by drying out.

Pyrophoric materials

A range of chemicals exists which are so reactive that on contact with air, and its moisture, oxidation/hydrolysis occurs at such a rate as to cause ignition. Some of these reactions liberate flammable gases. Examples of these pyrophoric chemicals are listed in Table 4.9.

Table 4.9 Pyrophoric chemicals

(a) Pyrophoric alkyl metals and derivatives

Groups
Dialkylzincs
Diplumbanes
Trialkylaluminiums
Trialkylbismuths

Compounds
Bis-dimethylstibinyl oxide
Bis (dimethylthallium) acetylide
Butyllithium
Diethylberyllium
Diethylcadmium
Diethylmagnesium
Diethylzinc
Diisopropylberyllium
Dimethylberyllium
Dimethylbismuth chloride
Dimethylcadmium
Dimethylmagnesium
Dimethylmercury
Dimethyl-phenylethynylthallium
Dimethyl-l-propynylthallium
Dimethylzinc
Ethoxydiethylaluminium
Methylbismuth oxide
Methylcopper
Methyllithium
Methylpotassium
Methylsilver
Methylsodium
Poly (methylenemagnesium)
Propylcopper
Tetramethyldistibine
Tetramethyllead
Tetramethylplatinum
Tetramethyltin
Tetravinyllead
Triethylantimony
Triethyl bismuth
Triethylgallium
Trimethylantimony
Trimethylgallium
Trimethylthallium
Trivinylbismuth
Vinyllithium

(b) Pyrophoric carbonyl metals
Carbonyllithium
Carbonylpotassium
Carbonylsodium
Dodecacarbonyldivanadium
Dodecacarbonyltetracobalt
Dodecacarbonyltriiron
Hexacarbonylchromium
Hexacarbonylmolybdenum
Hexacarbonyltungsten
Nonacarbonyldiiron
Octacarbonyldicobalt
Pentacarbonyliron
Tetracarbonylnickel

(c) Pyrophoric metals (in finely divided state)

Caesium	Rubidium
Calcium	Sodium
Cerium	Tantalum
Chromium	Thorium
Cobalt	Titanium
Hafnium	Uranium
Iridium	Zirconium
Iron	
Lead	*Alloys*
Lithium	Aluminium – Mercury
Manganese	Bismuth – Plutonium
Nickel	Copper – Zirconium
Palladium	Nickel – Titanium
Palladium	
Platinum	
Plutonium	
Potassium	

(d) Pyrophoric metal sulphides
(Ammonium sulphide)
Barium sulphide
Calcium sulphide
Chromium(II) sulphide
Copper(II) sulphide
Diantimony trisulphide
Dibismuth trisulphide
Dicaesium selenide
Dicerium trisulphide
Digold trisulphide
Europium(II) sulphide
Germanium(II) sulphide
Iron disulphide
Iron(II) sulphide
Manganese(II) sulphide
Mercury(II) sulphide
Molybdenum(IV) sulphide
Potassium sulphide
Rhenium(VII) sulphide
Silver sulphide
Sodium disulphide
Sodium polysulphide
Sodium sulphide
Tin(II) sulphide
Tin(IV) sulphide
Titanium(IV) sulphide
Uranium(IV) sulphide

Table 4.9 (continued)

(e) Pyrophoric alkyl non-metals	(f) Pyrophoric alkyl non-metal halides
Bis-(dibutylborino) acetylene	Butyldichloroborane
Bis-dimethylarsinyl oxide	Dichlorodiethylsilane
Bis-dimethylarsinyl sulphide	Dichlorodimethylsilane
Bis-trimethylsilyl oxide	Dichloro(ethyl)silane
Dibutyl-3-methyl-3-buten-1-Ynlborane	Dichloro(methyl)silane
Diethoxydimethylsilane	Iododimethylarsine
Diethylmethylphosphine	Trichloro(ethyl)silane
Ethyldimethylphosphine	Trichloro(methyl)silane
Tetraethyldiarsine	Trichloro(vinyl)silane
Tetramethyldiarsine	
Tetramethylsilane	(g) Pyrophoric alkyl non-metal hydrides
Tribenzylarsine	Diethylarsine
mixo-Tributylborane	Diethylphosphine
Tributylphosphine	Dimethylarsine
Triethylarsine	1,1-Dimethyldiborane
Triethylborane	1,2-Dimethyldiborane
Triethylphosphine	Dimethylphosphine
Triisopropylphosphine	Ethylphosphine
Trimethylarsine	Methylphosphine
Trimethylborane	Methylsilane
Trimethylphosphine	

A well-known example is pyrophoric iron sulphide which can form a scale on the inside surfaces of carbon-steel equipment containing H_2S-bearing hydrocarbons (e.g. sour crude or sour naphtha). Oxidation of this scale when air is introduced is highly exothermic and the temperature of this scale may act as a local source of ignition.

> A 45,000 barrel noded spheroid collapsed completely due to an internal explosion during gas-freeing. The explosion, which occurred while draining water and filling and overflowing to an established procedure, was caused by a remaining gas pocket of air and iron sulphide.

Precautionary measures to avoid the risk of ignition from pyrophoric iron sulphide are well establised in petroleum refinery operations and have to be applied, for example whenever equipment handling sour streams is taken out of service.[37] These comprise:
- Purging free of hydrocarbon before allowing air to enter (normal procedure in any event).
- Immediately on opening, the interior surfaces are kept wetted using water hoses until pyrophoric deposits have been removed.
- Safe disposal of pyrophoric sludge, e.g. by removal, while wet, to an open area where ignition and burning will not cause damage.

Reactions leading to high local temperatures may also occur on traces of catalyst; hence Raney nickel glows when exposed to the air.

A few seconds after opening a half-used, 500 ml bottle of sodamide it began to emit sparks and generate heat and fumes. (Sodamide reacts violently with water and is also attacked by moist air.) The fumes necessitated evacuation of the laboratory and the bottle became so hot that it collapsed, the contents scattered with sparks and explosions.[38]

On a plant scale the hazard associated with opening equipment containing pyrophoric chemicals or residues may be increased by the presence of flammable materials. Hence the inside surfaces of opened equipment containing any pyrophoric chemical should be damped down periodically. Pyrophoric chemical actually removed from equipment or pipelines has to be soaked with water to avoid ignition; this must be repeated until controlled oxidation renders the chemical non-pyrophoric.

Frictional ignition

Three kinds of frictional contact can result in ignition, namely[39]:
- Impact by one material striking another either in free fall or through a blow.
- Rubbing friction, e.g. when moving parts of a machine are fouled on a stationary surface.
- Frictional smearing, when a steel surface coated with a softer light metal is subjected to a high specific bearing pressure with sliding or grazing.

Because of the hazard of frictional ignition there are severe restrictions on the use of aluminium, magnesium, titanium and alloys of these metals in coal mines in the UK. These metals have an affinity for oxygen and in a 'thermite' reaction with rust produce temperatures up to 3,000 °C. Therefore a 'thermite' flash can be produced by striking a smear, or thin coating, of alloy on rusty steel with a hammer. The smear and the blow can be simultaneous, e.g. when a piece of light-alloy equipment is dropped on to rusty steel. The hazard may also arise when a glancing blow is struck on silver paper (i.e. aluminium foil) on a rusty surface or on a rusty surface painted with a formulation containing light alloys. Hence aluminium paints may be prohibited, or restricted to certain areas, or only applied to steel surfaces which have been blasted clean and primed with zinc-rich paint.

The glancing impact of stainless steel, mild steel, brass, copper–beryllium bronze, aluminium copper and zinc on to aluminium smears on rusty mild steel can initiate a thermite reaction of sufficient thermal energy to ignite flammable gas–air and solvent–air atmospheres and dust clouds typical of those found in the chemical industry.[40]

In general, therefore, light alloys, particularly those containing magnesium, aluminium paint and grit, or rock, are important when considering the possibility of incendive sparks.

A fire occurred in an open tank while a mechanic was tightening a screwed fitting which was leaking. The tank had been emptied but still contained flammable vapour. The tank was constructed of aluminium but the fitting, a valve, was of steel. Either friction between the aluminium and steel caused oxidation to aluminium oxide – an exothermic reaction or a thermite reaction occurred between aluminium and rust.

When hand tools are used in an area where flammable vapours may be present the following precautions are advisable:
- Primarily, the surrounding atmosphere should be rendered free from flammable vapour concentrations wherever possible.
- Light alloy hand tools should only be used under gas-free conditions.
- Operations such as the chipping, or striking, of concrete or stone should only be permitted with thoroughly wetted surfaces.
- All hand tools should be kept in good condition and free from rust or smears of light alloy.

Petroleum vapour is very unlikely to be ignited by impact of steel on steel produced by hand, and power operation is required to produce incendive sparks.

An explosion occurred in a fixed roof storage tank. Since the liquid inside had a high electrical conductivity, static discharge was ruled out. However, frictional heating was suspected between a taut vibrating wire, which supported a swing arm, rubbing against a pulley which had seized on to its bearing. Experiments have shown that steel wires subject to friction can produce glowing filaments of thin wire: these filaments cannot ignite methane but may ignite other gases.[41]

It is possible to ignite hydrogen and perhaps ethylene, acetylene or carbon disulphide by the impact of steel on steel using hand tools.[14] When 'non-sparking' tools are used, e.g. hammers, care must be taken to avoid embedded grit particles since impact of steel on 'rock' poses a greater hazard.

90 tons of solvent burned in 9 minutes when a stock of drums of flammable solvent caught fire. The average heat output was 4,500 megawatts; buildings were damaged or destroyed at 75 m, burns requiring medical treatment occurred at 120 m, timber was charred or ignited at 120 m. The likely source of ignition was friction caused by a drum falling off a fork-lift truck.

Flint, rock or grit can produce incendive sparks irrespective of the striking material; hence the need for copious wetting of hard surfaces when they are chipped in potentially hazardous areas.

The first apparatus for producing light in a coal mine was Spedding's steel mill patented in 1760. It comprised a spur wheel and steel disc in a small steel frame. The operator held a piece of flint against the rotating disc and the continuous succession of sparks gave warning of danger by indicating the presence of firedamp (i.e. methane). However, faith in the mill was rudely shattered by a serious explosion at Wallsend Colliery in 1785.[42]

Friction in bearings has been a common ignition source; the spread of fire in such circumstances has frequently been aggravated by the presence of oil or flammable materials on nearby surfaces. A proper system of lubrication, inspection and routine maintenance clearly reduces the hazard.

Engines

Vehicles with petrol engines are all potential ignition sources, due to the spark–ignition system, dynamo and battery. Therefore, their use is confined to roads in 'safe' areas. 'Non-flameproof' diesel engines, e.g. fork-lift trucks, electric generators or cranes are potential ignition sources due to hot carbonaceous particles or flames from the exhaust system or the temperature attained by the exhaust pipe.

> A fire and a series of explosions occurred at a warehouse in a factory. The warehouse contained about 49 tonnes of LPG in cartridges and aerosol containers and about 1 tonne of petroleum mixtures in small containers, raw materials and packaging materials. The fire brigade arrived within 3 minutes of being called but the fire spread rapidly totally destroying the warehouse and its contents. In the intense heat pressurised containers ruptured violently and were ejected from the building. Minor damage was caused to nearby premises.[43]
>
> A leakage of aerosol or fuel cartridge contents was deduced to have been ignited by the electrical system of a battery-operated fork-lift truck.
>
> It was concluded that the warehouse did not comply in several respects with the appropriate Code of Practice for LPG storage[44] and that it should have been classified as a Zone 2 area and only apparatus suitably explosion-protected for such a classification should have been installed or used in it. Hence the 48 volt battery-powered truck was unsuitable.[43]

Lightning

Earthing for lightning discharge needs to be of low resistance, e.g. $<7\ \Omega$, and must be short and direct. The recommended value for protection of plant is $<10\ \Omega$.[45]

Open flames

Small fires, furnaces, pilot lights and lighted tapers are all efficient ignition sources. For example, furnaces afford a ready ignition source for drifting flammable vapours. Not only the combustion chamber itself but overheated flues or stacks, or incandescent material from the stack, may cause ignition.

An explosion occurred in the base of a flare stack, used in a chemical plant for burning off combustible waste gases. The source of ignition was the pilot flame on top of the stack and the fuel was combustible gas. The air entered via a leak in a large bolted joint.[46] Similar explosions have arisen because of air diffusing down the stack or entering via a broken joint not rapidly blanked off.

Explosions within direct-fired ovens and furnaces have occurred in the past due to delayed ignition, sometimes due to abortive lighting attempts, or because inadequate purging has been allowed after flame failure (sometimes because the automatic controls have been tampered with). Indeed two-thirds of the accidents reported in the UK with gas- and oil-fired appliances occur during the lighting-up sequence.

The furnace illustrated in Fig. 4.7 tripped out on flame failure due to a reduction in fuel-oil pressure. The operator closed the two isolation valves and opened the bleed valve.

Subsequently the oil pressure was restored and a supervisor tested the interior of the furnace with a combustible gas detector. Since he got no response he inserted a lighted poker; there was a bang, the supervisor was slightly injured and the furnace brickwork was damaged.[47] (It took a few seconds for the solenoid valve to close on flame failure during which oil entered the furnace; the line between the valve and burner may also have drained into the furnace. The oil had a flash point of 65 °C and was not detectable by the combustible gas detector due to condensation in the sample tube. The correct procedure is to sweep out the furnace for a sufficient period of time to ensure that any unburned oil has evaporated.)

Portable heaters constitute an obvious potential source of ignition. Therefore, the manufacturers' instructions should be followed precisely with regard to clearance distances and electrical cables should not be trailed across access ways.

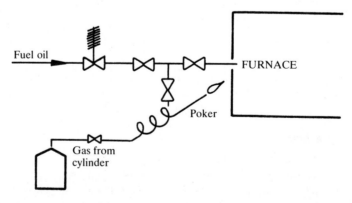

Fig. 4.7 Furnace oil supply and lighting arrangement

Adherence to instructions is vital with heaters dependent upon a cylinder-gas supply, e.g. with regard to the correct location of the cylinder, maintenance of hose connections and the proper procedure for cylinder change-over. The supply should be isolated at the cylinder when not in use.

The potential hazards associated with fixed heaters generally arise from lack of adequate guarding and/or their being surrounded by combustible waste material.

Exothermic reactions and admixture of incompatible chemicals

Special flammable hazards may be created from uncontrolled exothermic reactions, e.g. oxidation, halogenation or polymerisation.

A hazard may also arise from the admixture of any incompatible chemicals, for example inadvertant addition of any oxidising acid to organic matter. The combination of nitrates or chlorates with carbonaceous materials, or the violent reaction of peroxides with organic materials, are other examples. Further details are given in Chapter 10.

While these are the most common ignition sources there are other less foreseeable possibilities, e.g. the inadvertant formation of pyrophoric compounds or, in one particular incident, the resonance of traces of mercury vapour in a mixture with the energy emitted from nearby fluorescent tubes.

Ignition sources for dust–air mixtures

Given a dust concentration within explosive limits, ignition is not so easy as with air–vapour mixtures. A larger source of heat is required, since the combustible material is widely scattered in denser particles and a larger volume of mixture must be heated to the ignition point.

In general, the same range of potential ignition sources applies as for air–vapour mixtures. However, in view of the most common operations in which combustible dusts are processed, summarised in Table 4.10, certain sources are particularly significant, e.g. friction, overheated bearings.

The determination of dust explosion characteristics, together with characteristics of vessels, sizing and design of vents, and additional measures and precautions are described in reference[48].

Oxygen enrichment

The manner in which common materials burn in air is well known. However, if the oxygen content of the atmosphere is accidentally increased above 21%, the flammable range is widened and the ease of

Table 4.10 Typical dust-producing opertions

Operation	Examples
Pulverisation	Coal dust (foundries, mines, power stations)
Grinding – fine dust as a by-product	Bone dust, sugar, metals
Chemical action	Carbon-black production, zinc distillation
Transportation	Elevating (e.g. bucket-elevators) conveying, bagging-up
Spray drying	Cereals, coffee, dyestuffs
Fluidised bed drying	
Grading, screening, blending, packing	All materials
Polishing (dry)	Metal articles
Dust collection plant	Cyclones, filters

ignition, combustion rates and combustion temperatures will all be increased. For example, materials which do not normally burn readily in air, such as clothing, will burn much more actively in an oxygen-enriched atmosphere. In a room enrichment of the atmosphere to around 25% oxygen creates a hazardous situation; this level is particularly dangerous in 'confined spaces' as discussed in Chapter 17.

> An explosion occurred in a ship's forepeak. Propane gas had leaked from an appliance during the night and a worker 'sweetened' the atmosphere with oxygen. When a cutting torch was lit there was an explosion and fire, injuring four men.[40]

> A man at a shipyard was asked to couple a half-inch airline for operation of two drills to a one-inch air outlet by means of a threaded adaptor. However, the nearest one-inch outlet to the confined space in which work was to be carried out was for oxygen. This resulted in a wrong connection being made. Some hours later one man lit a cigarette which burned to his lips in seconds; the remains of the cigarette ignited the clothing of several men resulting in five fatalities. Enrichment had occurred due to dispersal of oxygen from the drill outlets into the confined space.[50]

There is a spontaneous combustion hazard with some combustible materials. For example oil or grease can burn or explode if their ignition temperatures are reached in the presence of pure, and especially pressurised, oxygen. These temperatures can be reached when a cylinder valve is opened quickly; hence oil or grease should not be used on valves in high-pressure oxygen service.

> A cylinder was being filled with a mixture of oxygen, methane and nitrogen. It had been charged with the methane and nitrogen and was

being charged with oxygen when an ignition occurred in the valve, attributed to the presence of grease. (A more serious incident was avoided by the operator rapidly shutting off the oxygen flow.)[51]

Similarly there are real dangers in the misuse of oxygen as a 'replacement' for compressed air. For example this is known to have caused an explosion in a diesel engine when it was admitted into the cylinders for starting, due to compression ignition of oily residues in associated piping in the presence of oxygen at high pressure, and in a petrol engine when used to clear a blockage.[16]

Control with flammable materials

Control principles

In theory elimination of one corner of the 'fire triangle' renders fire or explosion impossible. Practical experience is that this is misleading, i.e. if flammable gases or vapours are mixed with air in flammable concentrations it must be anticipated that sooner or later the mixture will catch fire or explode. This is because of the difficulty of eliminating every source of ignition. Therefore, while approved practice is to avoid the introduction of any of the ignition sources described earlier (except in combustion process applications, e.g. furnaces), the first defence is generally (again excluding certain processes, e.g. oxidation) to prevent the formation of a mixture within the flammable limits.

This entails measures to prevent flammable liquid or gas escaping from the plant into the open air, to prevent air entering equipment, and the use of inert gas blanketing where appropriate. The elimination of ignition sources is a second defence. In common practice for reliable control with flammable materials, including combustible dusts, the aim is to remove two corners from the 'fire triangle'.

Escalation of fires

Fire spreads by means of:
- Heat transfer, by conduction/convection or radiation – or generally a combination.
- Direct burning.
- The release of flammable liquids or vapours.

Within buildings fire spreads principally via direct burning and convective heat transfer (see Fig. 4.8). Hot gases spread laterally under ceilings and other horizontal surfaces and rise by natural convection up staircases, lift shafts, vertical ducts or through floor openings. As more materials are ignited the volume of smoke and hot gases will increase; flue effects may

Fig. 4.8 Escalation of a factory fire. Extensive fire protection was not provided since the products were non-flammable. However, a small fire, which developed on a Sunday morning, escalated rapidly via the PVC plastic-clad, metal sheet construction (dripping, inflamed molten plastic is clearly illustrated) (Courtesy E. Davies)

be caused by openings to atmosphere so that the burning is intensified by the drawing in of extra air. Flying sparks, carried by convection currents, and radiant heat, may assist fire spread by igniting combustible materials. When a sufficiently high heat-release rate occurs from solid materials within a confined space, flammable vapour emitted into the space results in a 'flash-over' in which the complete volume may be filled with flame.

In process plants fire escalates predominantly by radiative heat transfer and by the release of flammable vapour or liquids. The release may follow, for example, rupture of a process vessel or flanges being forced apart due to thermal expansion. As described in Chapter 3 flammable vapours heavier than air, namely most hydrocarbons and industrial solvents, may drift a significant distance from a source of release, e.g. downhill or through trenches and ducts. Ignition at some point then results in a 'flash-back'.

The phenomenon of a 'boil-over', explained in Chapter 3, may also cause escalation of fire from oil storage tanks.

A fire on the floating roof of a crude-oil storage tank containing 60,000 m^3 of light crude oils was believed to have been caused by ignition of vapour from oil seepage on to the roof by incandescent carbon particles dislodged from a flare stack 108 m from the tank.[52] During the course of fire-fighting a boil-over occurred with flames rising 60 m to 90 m into the air and oil slopped into the bund; this caught fire covering 4 acres. Hose lines, used to provide water for a cooling curtain around adjacent tanks, were melted. A second boil-over followed about 2½ hours later, again disrupting the cooling provisions.

The fire eventually burned for about 50 hours with the loss of some 25,000 tonnes of oil; over 680,000 litres of foam compound were used in the fire-fighting operation.

Flammable material can also be spread over a wide area if a pressurised vessel ruptures and rockets into the air, i.e. a BLEVE (see p. 140).

Normally measures such as firewalls, bunds or dykes should restrict the spread of liquids. However, unless precautions are taken, firewater which becomes contaminated with flammable liquid may spread the fire to a wider area via drains, sewers, trenches or ducts.

Fire characteristics of materials

The fire characteristics of solid products and of construction materials may be tested in small scale, or simulated, conditions. These tests must be carefully formulated for the specific hazard and the results should be presented to indicate the extent to which the item is capable of promoting danger.[53,54]

The main properties of relevance are[53]:

Ignitability – a measure of the ease of ignition, e.g. flash point test.
Flammability – the property determining the rate at which fire develops, which is clearly dependent on temperature and availability of oxygen.
Surface spread of flame – allied to flammability but concerned only with the surface.
Heat release – a measure of how much the burning material will contribute to a fire in progress.
Smoke/gas release – a measure of the tendency to result in a toxic hazard (see Table 15.9) or impede escape.
Fire resistance – an indication of the length of time for which a structure will resist the effect of a standardised fire before failure, e.g. time to failure under load; insulation failure – allowing ignition on the non-fire side; integrity failure – allowing passage of combustion products.
Flame penetration – penetration of flame through the material, cf. insulation failure.
Smoke/gas penetration – failure of integrity of an enclosure.

Detailed advice has been published on fire testing of building materials and the preparation of specifications.[54,55]

Fire prevention in factories generally

The four main causes of fires in UK factories were found in one study to be 'deliberate', 'electrical', 'smoking' and (the majority) 'unknown'. Poor housekeeping was probably a major contributory factor in many cases.

Arson

Arson, i.e. the wilful damage and destruction of property by fire, is often an underestimated consideration in loss control. In 1976 at least 250 large fires in the UK were attributed to arson with an estimated cost of £32.7 m. Furthermore, since the cause of 50% of all large fires is 'unknown' statistics relating to arson[56,57] may underestimate the problem.

In 1982 two youths started a fire with flammable liquids outside a chemicals warehouse. About 2,000 tonnes of chemicals were stored in the warehouse, including sodium chlorate, 29 tonnes (in polyethylene bags in 50 kilogram steel drums, stacked on wooden pallets); paradichlorobenzene, 262 tonnes; orthodichlorobenzene, 110 tonnes; 1,2,4-trichlorobenzene, 8 tonnes; sodium nitrate, 4.2 tonnes.[58,59]

The fire spread into the warehouse, probably under the eaves (asbestos cement sheet roof on steel purlins and trusses) and spread rapidly throughout the building, possibly because of the presence of flammable vapours. Eleven tonnes of sodium chlorate became involved in the fire and exploded violently less than 1 minute later. The night watchman was knocked off his seat by the explosion; a few minutes later a second major explosion occurred involving another stack of sodium chlorate and further minor explosions occurred at intervals. Surrounding property was extensively damaged and several hundred residents were evacuated.

Neither the HSE nor the fire authorities were aware that chemicals were being stored in this building.

The management of the premises were unaware of guidance on the storage of sodium chlorate issued by HSE and they had received no information from the manufacturers.

A fire at a plastics components premises was started deliberately in the raw-materials stores and caused £850,000 damage: 170 tonnes of PVC granules and powder, 30 tonnes of other chemicals including acrylic copolymers and hydrogenated castor oil, and 800 litres engine oil were destroyed. Seven firemen were affected by toxic fumes and residents from nearby homes were evacuated while fire fighting was in progress.[58]

The detailed precautions required to minimise loss by arson, or sabotage (see page 135) are obviously dictated by the circumstances. However, a summary of general technical and administrative measures is given in Table 4.11[60] and in reference 61. Strategies and scientific techniques involved in the investigation of fires where arson is suspected are described in reference 62.

Children playing with fire may pose a similar problem to arsonists.

A fire started in a den which children had made from cardboard cartons outside a transport warehouse.[58,65] It spread to wooden pallets carrying folded cardboard cartons and then into the warehouse through a hole in a door. The bitumen coating on the sheet steel roof caught fire and flaming

Table 4.11 Some precautions against loss due to arson[58]

Keep out intruders
Automatic intruder alarms linked to a central monitoring point will reduce opportunities for fire raising, as well as ordinary break-ins; if intruders are disturbed within five minutes of access the damage is considerably reduced.

Advice on intruder alarm systems, methods of surveillance, perimeter fencing, security lighting, doors, locks and window protection are provided by the FPA.[63]

Control of access at all times
- Supervise entrance into the premises by day.
- Lock all gates that cannot be supervised.
- Secure the whole premises at the end of the day, and when it is unoccupied.
- Check that all doors, windows and gates are locked.
- Keep unauthorised persons out of storage areas at all times.
- Keep fences, walls and gates in good repair; seal gaps beneath doors, mend broken windows and cut back vegetation.
- Keep security staff well informed of their duties.

Kill fires quickly
- An automatic sprinkler or fire detection system designed to call the fire brigade immediately will detect fires before they become serious.
- Whether there is an automatic system or not, there should also be regular fire safety patrols outside working hours. Guidance for fire safety patrols is published by the FPA.[64]

Deny fuel to the fire-raiser
- Keep waste and rubbish out of reach, particularly away from doors and windows where fire-raisers can easily use it to start a fire.
- Check that secluded areas, e.g. beneath stairways, are not depositories for rubbish.
- Clear packaging and waste from storage areas at the end of the day.
- Lock flammable substances away when not in use.

Obtain employees' collaboration
- Screen new employees and follow up references.
- Strictly supervise casual labour, maintenance men and outside contractors working after normal hours.
- Call for work re-appraisal reports on all staff from time to time.
- Allocate security duties either to full-time security staff or to reliable supervisors and other employees.
- Enlist the co-operation of all staff in reporting suspicious characters and keeping intruders off the site.

Consider layout of premises
- Avoid employees having to pass through storage areas to and from their work areas; when this is unavoidable, special attention should be paid to security of the storage, and similar unmanned areas.
- Easy access points such as loading bays, doors and windows should, if possible, be positioned away from site access points; this precaution is particularly desirable with fire exit doors.

---

molten bitumen dripped on to the stored goods: 150 tonnes of milk powder in paper sacks caught fire and heated up 68 tonnes of sodium chlorate kept in polyethylene bags within sealed steel drums (25 and 50 kilogram capacity) and stacked on timber pallets. Within 10 minutes of the discovery of the fire a series of major explosions occurred involving the sodium chlorate. The

Fig. 4.9 General view of the seat of the explosion at the Braehead container depot. Damaged drums of sodium chlorate are in the foreground (Courtesy Health and Safety Executive)

blast destroyed the warehouse and contents (Fig. 4.9) scattered debris (open drums, steel roofing sheets, burning timber) and damaged about 200 properties within a radius of 1.5 km. Three similar nearby storage sheds were damaged; in one a large quantity of whisky and other spirits were consumed by fire. A security guard and twelve people outside the site were treated for shock and minor injuries. Injuries were minimised because the premises were closed for the New Year holiday and the attendance of the fire brigade was delayed because they were attending another incident when the fire call came. (Particular care had to be taken during clearing up operations because of the presence of asbestos fibres from the burnt bitumen coating.)

Both in this incident in 1977 and that described on page 162 in 1982, explosions occurred involving sodium chlorate stacked in drums; in the UK guidance on storage has been issued by HSE and any inventory $\geqslant 25$ tonnes is notifiable and subject to control under major hazards legislation – see Chapter 15.

Preventative measures

Good housekeeping is a primary consideration in fire prevention. Even in cases where fires were started maliciously, better housekeeping would have made it more difficult or limited the spread. Therefore, attention is generally recommended for the following areas.

Electrical equipment Fires are frequently started by badly maintained, or 'temporary', wiring with loose connections, inattention to lighting equipment, overheating of motors and overloaded circuits, or misuse of substations and switchrooms.

Regular inspection and maintenance by qualified electricians and the prohibition of makeshift installations by other employees, are important preventive measures.

Waste disposal Combustible waste should not be allowed to accumulate in corners, passageways, or other convenient 'storage' areas. Otherwise it may serve as a ready fuel for ignition by a discarded match or cigarette-end, or to feed a fire initiated in some other area or manner. Special provision should be made for the storage of flammable liquid wastes.

Storage If combustible material is stored, care is necessary to avoid congestion, narrow gangways, or inadequate breaks between storage racks or areas. Even when a material is itself relatively 'inert', its plastic or cardboard packaging may be combustible so that similar rules are applicable. Material stored in the open should not be stacked near windows; this avoids the potential for fire to spread into adjacent buildings.

Contractors Contractors should generally be supervised by means of a Clearance Certificate of the type described in Chapter 17. Close control should be exercised over any temporary heating, lighting or cooking facilities introduced on to the site. In addition the condition in which the site is left at nights and at weekends should be checked.

Escape and access Escape doors and routes must be kept free from obstructions. Frequent inspections are especially necessary if the doors or routes are not regular means of access or egress.[66]

Access for fire fighting, for example to enable the fire brigade to enter the rear of a building, is also important.

Fire equipment The continuing efficiency of fire-fighting equipment should be ensured by regular inspection and maintenance. Such servicing should be recorded.

Extinguishers should have designated locations; they should not be hidden or be used for propping open doors. Fire points should be identified by clear notices and be checked regularly. The extinguishers should be of the correct type, as discussed later. However, this needs to be supplemented by the provision of instructions as to where, and how, to use them. Practice in setting them off is also necessary.

Fire and smoke stop doors Heavy fireproof doors and shutters should be self-closing. It is essential to ensure that they will, in fact, close automatically; this involves regular checks on fusible links, counter-weights and the lubrication of tracks and bearings. The doorway itself must also be kept clear of obstructions. Fire check doors also require inspection, e.g. to ensure that they are not wedged or propped open.

Flues Passages for services and other ducts might act as flues for the transmission of fire and smoke; they should be adequately fire-stopped. If they are not stopped, or their duty necessitates they be left open, they should be kept free of combustible materials. Such materials should also be cleared away from the entrances. Any hatches fitted, e.g. for access, should be kept closed.

Sprinklers Providing they are properly installed and maintained automatic sprinklers are the most simple and effective alarm and extinguishing device for the majority of factories and warehouses. Structural alterations, change of use, or internal modifications within the building require alterations to the sprinkler system.

The specification of sprinkler systems is covered comprehensively in references 67 and 68.

A fire in a polyethylene extruders warehouse in 1981 resulted in damage valued at £3.6 m.[58] The warehouse contained 400 tonnes of plastics products, including reels of polyethylene film, and 1,600 tonnes of polyethylene granules, stacked partly on timber pallets and partly in metal cages. Goods were stacked at heights up to 5 metres. The building was provided with a sprinkler system (installed to 29th edition FOC Rules, Extra high-hazard–high-piled storage, Category 11).

The fire was started maliciously at 02.02 hours. A production worker was alerted to it by the sprinkler alarm bell; he saw a fire in one corner of the warehouse and called the fire brigade. The fire brigade were given an incomplete address which delayed their arrival by about 4 minutes. The water from the sprinkler system was failing to extinguish the fire because it was cascading off the top of the stacks and not reaching the seat of the fire. Firemen deployed one jet but the hydrant water supply was poor and the fire continued to develop. The nearest static supply was a pond 500 metres away. The sprinkler system was turned off at 02.35 hours to allow an attack with protein foam (180 litres of concentrate) but this too was unsuccessful. The system was turned on again at 03.10 hours but by then about 440 heads had opened and no water reached the heads over the seat of the fire.

By 03.55 hours most of the polyethylene film and granules were burning fiercely and priority was given to protecting adjacent property. About 100 nearby residents were evacuated at 05.00 hours, because of the huge quantities of dense black smoke evolved and the danger from flying brands of burning plastics. They were allowed to return home at 16.40 hours. The fire was brought under control by 08.00 hours.

Fire detection

Where flammable gases are present a pre-fire condition can be identified by a flammable gas detector of the type described in Chapter 8. Actual fire detection is commonly achieved by:
1. Heat sensing, as actual temperature or rate of temperature rise depending upon:
    (a) melting of a metal (fusion);
    (b) expansion of a solid, liquid or gas;
    (c) electrical sensing.
2. Smoke detection depending upon:
    (a) absorption of ionising radiations by smoke particles;
    (b) light scattering by smoke particles;
    (c) light obscuration.
3. Flame detection depending upon:
    UV or IR radiation sensing.

A combination of these detectors, or duplication, or a voting system, may be used to minimise spurious trips and to improve reliability. The detection system may actuate an alarm or an alarm/fire extinguishment system.

Fire extinguishment

The characteristics of common fire extinguishing media are reviewed below. In assessing these it has been customary to classify fires as:

Class	Type
A	Fire involving solid materials, usually of an organic nature, in which combustion normally takes place with the formation of glowing embers.
B	Fire involving a liquid or a liquefiable solid. (The miscibility or otherwise with water is an important characteristic.)
C	Fire involving a gas.
D	Fire involving a burning metal, e.g. magnesium, aluminium, sodium, potassium, calcium or zirconium.

There is a significant difference between the two most common types of fire, namely A and B, and hence in the appropriate means for extinguishment. With a Class A fire (typically involving wood, refuse, cloth, paper, etc.) the fire burns deep and may continue to smoulder after the flames are out and eventually reflash; the penetration and cooling action of water is therefore required. Conversely, with a Class B fire only the vapours burn at the surface; the smothering action of dry chemical, carbon dioxide or foam will thus extinguish the flames and put the fire out. (Water is not suitable since most flammable liquids will float on it and continue to burn.)

As a *general* guide:
Water should not be used on fires involving electrical equipment or on materials with which will react.
Normal extinguishing agents should not be used on Class D fires.

Once ignited vapours or other combustible gas will continue to burn as long as there is a continuous supply of sufficient fuel and air to provide a flammable mixture. Fire extinction, therefore, involves depriving the combustion zone of either vapour (e.g. by cooling with water fog) or air (e.g. using a layer of dry powder) or both (e.g. by applying a continuous foam blanket on the surface of burning liquid to exclude air and to cut off the supply of vapour by shielding the liquid from flame). Propagation of flame can also be interrupted by inhibiting the chain reactions involved using, e.g. organic-halogen vaporising liquids or small particles of dry chemicals, generally sodium or potassium bicarbonate.

Water

Water is the most widely used extinguishing agent, applied either in a spray or a jet. Its high specific and latent heats are advantageous when cooling combustibles. However, a jet is completely unsuitable for fires in flammable liquids because its force is liable to disperse the liquid and hence spread the area of fire; moreover most flammable liquids will float and continue to burn. If there is a significant depth of liquid, i.e. in a tank, or with light oils, water in a jet will not be evaporated by the heat of the fire and will sink to the bottom hence raising the liquid level and possibly causing an overflow. In fires involving liquids with high boiling points and flash points bulk liquid may be heated in a tank to a temperature above the boiling point of water; evaporation to steam beneath the liquid will cause severe foaming and possibly a 'boil-over' (see page 160).

Sprinkler systems are suitable for Class A risks, e.g. offices, storages of specific products (boxed products, polyurethane blocks, high racked/stacked storages), canteens. Detection is by sprinklers each with a heat-sensitive element and discharge densities are generally between 2.25 and 30 dm^3 m^{-2} min^{-1} with specified minimum areas of coverage at design density of 72 to 300 m^2.[69] Systems may be wet, dry, alternate, pre-action or cycling depending upon the design requirements.

Water fog is effective in cooling some burning fuels to below the flash point and hence extinguishing fire by reducing the vapour concentration to below the LEL. Simultaneously the steam formed by vaporisation of the water droplets displaces air and assists extinguishment. Hence fixed systems employing water sprays are often found on petroleum processing plant. Oils with very low flash points, e.g. gasoline, cannot be cooled sufficiently for extinguishment although some diminution is achieved in intensity of the flames.

Water is also valuable as protection for tanks exposed to heat, i.e. radiant heat and/or direct flame lick. The method of application may be either fixed, e.g. spray nozzles, drencher heads and monitors, or mobile.

Dry powder

Powders exert little cooling action but act by forming an air-excluding layer. They are effective on flammable liquids and electrical fires.

A system comprises a powder container, pressurising gas – usually dry carbon dioxide, an actuating valve, discharge piping and nozzles. Powders may be selected for flammable liquids or gases – BC powders, or for solids, flammable liquids and gases – ABC powders: special powders, are available for use on metals. 'Packaged' systems have an individual capacity of 1.5 to 15 kg powder; 'engineered' systems have capacities up to 1,000 kg with multi-point discharge for effective coverage.[69] The rate of discharge is generally twice that satisfactory in tests and must be sufficient to extinguish any fire completely. They are considered particularly useful for dealing economically with running fuel or gaseous fires from a jet.[69] They are suitable for indoor or outdoor use, and are compatible with water sprays and inerting gas, but not normally with foams.

Foam

Foam is produced when foam solution, comprising a proportioned mixture of water and foam concentrate, is aspirated with air causing expansion. The expansion varies depending upon foam quality, the water temperature and quality, and the efficiency of the foam maker; typically expansion ratios are in the range 6 to 10 (low expansion foam), 20 to 100 (medium expansion foam) or >100 (high expansion foam). With the majority of foams the foam solution is simply a vehicle to transport the water to the surface of, e.g. a flammable liquid and cause it to float; it is the water which extinguishes the fire.

With regard to the application of foam, flammable liquids may be divided into two groups – the insoluble hydrocarbons and the water-miscible polar solvents. While the hydrocarbons, including gasoline, diesel fuel and naphtha, may be extinguished with standard protein-based or synthetic foams, the water-miscible fuels such as alcohols, esters and ketones require special alcohol-resistant foams. The effectiveness of foam is due to its ability:

- To cool the fuel surface and surrounding metal shells.
- To suppress flammable vapour emitted by the fuel.
- To smother the flammable vapour and prevent it coming into contact with air.
- To separate the fuel surface from the flames.

Foam is particularly effective against fires in heavy oils in tanks since it will remain in position long enough for the oil and tank to cool and hence prevents re-ignition. However, it is unsuitable for 'running fires' because the flowing liquid transports foam away and a continuous blanket cannot form.

> A fire started when lightning struck a 58 m diameter cone-roof tank containing kerosene and fragments of that tank struck two other tanks, a 55 m diameter open-top floating-roof tank containing gasoline and a 34 m diameter covered floating-roof tank. This was a difficult fire to attack because of the exposure hazard presented by a butane sphere in the area. The extreme radiant heat of the burning tanks prevented the use of ground-level monitors. Therefore, it was determined to first extinguish the kerosene tank by subsurface injection.[70]
>
> The cone-roof kerosene tank was cooled by use of a 16 m articulated squirt device. Fire in the dyke area between the two burning tanks was attacked with monitors. A site for injection was selected 488 m away from the tank and a manifold was fabricated containing ten high back-pressure foam makers, each foam maker capable of 300 gallons per minute at 150 psi inlet pressure. Three foam trucks were used for proportioning. However, foam injection was delayed when gasoline spilled into the dyke area causing a tremendous ground fire. Foam injection had to be abandoned until the gasoline burned off. Once the subsurface injection began, a crack was discovered in the product line leading into the tank. Foam escaped from this crack and extinguished the dyke fire near the valve. The fire near the injection point was extinguished next and the majority of the fire in the tank was also extinguished except the rear quadrant opposite the point of injection. The next day another effort to extinguish the tank fire with ground-level monitors proved successful. (Subsequent examination showed that two tears in the tank shell had folded over, so that fuel trapped between the two layers of shell plating had continued to burn since foam was unable to enter the holes.)[70]

Foams are not suitable for, and must not be used on, flammable metals or material which may react with water, or on high-voltage electrical equipment. Their usual applications are where flammable liquids are involved, e.g. boiler-rooms, engine-rooms, oil-rig modules and chemical process plants.

Low expansion foams, usually a 4% to 6% concentration, are suitable for the extinguishment of fires in flammable liquids with fire points <100 °C. There is a critical rate of application, e.g. approximately 12 m^2 min^{-1} of foaming solution relative to the free surface area of fire; a higher value may be applicable to 'difficult' solvents or with alcohol-resistant foams.[69] The most economical application rate is then 3 to 5 times the critical rate.

Medium expansion foams, at a 1% to 3% concentration, may be used indoors or outdoors for fires in flammable liquids or solids. High expansion foams are used mainly to fill volumes in which flammable liquid or solid fires may exist; they are unsuitable for use outdoors due to rapid dispersal by wind.

Recommended foam solution application rates for various hydrocarbon liquids and for water-miscible solvents have been published.[71] Both medium- and high-expansion foams are difficult to use with other agents, e.g. dry powders cause breakdown by impact.

Aqueous film forming foam (AFFF) extinguishes fire by a different mechanism to standard foams. Foam solution drains from the foam blanket and is released in the form of a film that floats on the liquid fuel due to surface tension. This film spreads rapidly ahead of the foam blanket – cooling the fuel surface and achieving a quick 'knock-down'; the foam blanket serves as a reservoir which continuously releases the film. However, there is a disadvantage with AFFF; it is the water which provides foam bubbles with their heat resistance and its deliberate release, to achieve fast knock-down, results in poor heat resistance. Alcohol-Resistant AFFF is also available. On hydrocarbon fuels it forms a film similar to that described above but on a polar solvent or water-miscible fuel it releases a polymer which reacts with the fuel to form a floating membrane which separates the foam from the fuel.

A comparison of foam application methods is given in Table 4.12 (after ref. 39).

The selection and design of an effective foam system depends upon:
- The type of flammable liquid. This will determine the type of foam to be used.
- The type of hazard. This will determine the method of application and the application rate.
- The size of the hazard. This will indicate the overall requirement for foam concentrate and water supplies.

Carbon dioxide

Carbon dioxide acts by reducing the oxygen content of the air at the fire. It can be supplied from portable or fixed systems and is stored in liquid form at a pressure of approximately 750 psi. As a relatively inert, non-corrosive, non-conductive gas it is useful where the minimum of damage should be caused to the materials or equipment at risk.

Carbon dioxide may be used on fires in liquids or solids and on electrical fires but it should not be used where there is a high risk of re-ignition. It must not be used on materials containing or producing their own oxygen supply, e.g. cellulose nitrate, or able to utilise oxygen, e.g. reactive metals such as magnesium, or on metal hydrides.

Carbon dioxide is particularly effective indoors, as either a local application or total flooding system, but is likely to be ineffective outdoors due to rapid dispersal once application ceases.

Table 4.12 Foam application methods

Method	Advantages	Disadvantages
Fixed pourers	(a) Can be brought into action quickly (b) Manpower demands are small (c) Foam tests can be done without contaminating tank contents (d) Men need not enter bund area	(a) For a large tank farm the method is costly (b) Requires regular maintenance (c) Vulnerable to serious damage in the event of explosion and/or buckling of tank plates (d) Installation in existing tanks necessitates gas-freeing
Mobile monitors	(a) Extremely flexible in operation, i.e. can be used for other risks, e.g. jetties (b) Can be operated from a safe position outside the bund (c) Even for large tank farms, two or three only are sufficient (d) Maintenance is simple and not expensive (e) Not affected by tank explosions	(a) Foam wastage is high, due to winds and updraughts (b) Requires high water pressures and flow (c) Supporting mobile compound-carrying vehicle normally required
Portable foam towers and portable sectional tubes	(a) Inexpensive compared with fixed foam pourers for a large number of tanks (b) Not affected by tank explosions	(a) Portable foam pourers difficult to manoeuvre over tank bunds (b) Correct placement against tank and exact positioning of swan-neck over hand rails not easy in high winds
Fixed semi-subsurface system	(a) Foam delivered to liquid surface without loss or deterioration (b) Not likely to suffer damage through explosion (c) No need to enter tank bund for operation	(a) System can be inspected and tested only when tank is empty (b) Existing tanks to be emptied and gas-free before installing

For materials which burn at the surface the 'design concentration' of carbon dioxide is not much greater than the theoretical concentration, e.g. for gasoline 34% by volume compared with 28% by volume. However, the design concentration for deep-seated fires may be very high, e.g. 65% by volume for paper storages.[69] Thus there are two potential difficulties:

- The requisite volume of carbon dioxide needs to be introduced very carefully into a building, and it must be vented, to avoid excessive pressure rise.

- The hazardous concentration for personnel is 5% to 10% (see Ch. 5) so that a positive lock-off system is necessary to avoid discharge until the building is evacuated. Accumulation in basements may also result in an asphyxiation risk to personnel.

Systems may be 'high' or 'low' pressure, single or multi-shot. The value with portable appliances tends to be limited because of the restrictions on capacity. On the other hand carbon dioxide may be used in conjunction with water sprays, foams, dry powders or Halons without mutual interference.

Chemical agents

Liquid halogen compounds, e.g. the Halons 1211 and 1301, vaporise in air and extinguish fires by:
- Dilution of the atmosphere in the immediate vicinity of the burning material below the critical oxygen concentration.
- Interference with the chemical reaction of flame propagation.

The proportion of extinguishing medium to achieve these effects, in combination, is termed the 'inhibitory factor' – expressed as a percentage of the specific medium in dry air.[72]

Halons 1211 and 1301 are liquefiable gases with boiling points of $-4\ °C$ and $-58\ °C$ at atmospheric pressure respectively. They are colourless, odourless and electrically non-conductive and are effective on a wide range of combustibles, particularly flammable liquids and electrical fires. However, they do not provide 'permanent' sealing of the fuel as with foam; the hazard of re-ignition must be countered by the provision of adequate extinguishers, training and usage.[72] They may be used from fixed systems for local or total flooding applications, their best application being for flammable liquids or on solid fires which have not become too deeply involved. Discharges on to a deep-seated solid fire may promote significant thermal degradation into toxic acid gases and, since the normal extinguishing concentrations are approximately 5% by volume, a 'lock-off' system is recommended on major applications to avoid placing personnel at risk.

Halons may be used with water sprays, foams, carbon dioxide or dry powders. Their primary uses are in compartmented electrical plant or in unoccupied buildings, e.g. oil rig modules: they may also be used for floating roof protection.

Portable fire extinguishers and fire blankets are normally provided at strategic points in laboratories, in offices and in corridors. Types of extinguisher, and the type of fire for which they are suitable, are summarised in Table 4.13. Fire blankets are primarily used for extinguishing clothes on fire but they are also effective as shields when

Table 4.13 Portable fire extinguishers

Extinguisher / Class of fire	Water	Carbon dioxide (CO2)	Dry powder	Foam*	Vaporising liquids	Fire blankets	Sand
	Red	Black	Blue	Cream	Green	Red canister	Red bucket
CLASS A Wood, cloth, paper or similar combustible mtl. Cooling by water most effective	S	NS	NS	NS	NS	NS except for personal clothing on fire	NS
CLASS B Flammable liquids, petrol, oils, greases, fats. Blanketing or smothering most effective	NS Dangerous	S	S	S	S For small fires	S	S For small fires
CLASS C Electrical plant, non conductivity of extinguishing agents most important	NS Dangerous	S	S	NS	S	NS	NS

* Special foam required for liquids that mix with water.
NS = Not suitable
S = Suitable

approaching or passing a fire, or for smothering small oil fires, e.g. chip pans or oil baths.

Fire prevention and protection on process plants

Fire prevention and protection measures for process plants, tank farms and other facilities (e.g. loading bays, jetties) should be integrated into their design to the degree necessary to prevent a fire occurring, to detect any fire which does occur, and to restrict its spread.

A summary of the requirements during design and engineering is given in Table 4.14.

Further information on fixed fire protection is given in Chapter 12, page 723–30.

Such measures require, of course, that during operation there is also the development, and effective use, of comprehensive operating procedures and the training of all personnel in all aspects of normal and

Table 4.14 Design and engineering requirements for fire protection in process plants

---

The correct selection and use of materials and methods in the plant's construction.

Establishment of correct process conditions (e.g. maximum and minimum). Incorporation of process control and safeguarding systems, e.g. protection against over pressure (safety relief valves), instrumentation providing automatic alarms and trips, system for safe emergency disposal of excess hydrocarbon gases.

Attention to the layout and spacing of plant and facilities.

Correct choice of electrical equipment for use in hazardous areas (by defining such areas in accordance with area classification codes, e.g. in the UK BS 5345 and IP Electrical Safety Code).

Detection of leakages, e.g. gas detection on LPG pump seals, or of fire, e.g. hot oil pumps, or of overheating, e.g. pump and motor bearings.

Provision of fire protection to load-bearing elements of steel structures.

Provision of adequate drainage facilities for the removal of fire-fighting water.

Provision of fire-fighting water systems, fire detection and alarm systems, fire-fighting fixed and mobile systems, etc.

---

emergency situations. A procedure is also necessary to ensure that the installed protective systems continue to perform to the specified standards of safety.

Close liaison is advisable with the local fire authorities in order to obtain their advice on fire prevention and control, e.g. the preferred type of fire extinguisher. It is important that they are familiarised with factory layout, chemical and fire hazards, and evacuation procedures; thus regular site visits and discussions are valuable.

Plant layout and area classification

Layout is an important factor in fire prevention and plants should be isolated from each other as far as practicable. Common industrial practice is to avoid having a plant within 15 m of any road and to have a minimum separation distance of 36 to 46 m between plants.

An area classification system is used for electrical equipment; the UK system is explained in Table 12.17. An alternative is the US National Code and API Standard 520.

In addition to this, strict control is, of course, applied to potential sources of fire or explosion which can be temporarily introduced into a plant area, e.g. 'hot work' or internal combustion engines, as discussed earlier.

Plant water supplies – process plants

Application of water is frequently the only practicable measure to contain, isolate or extinguish a major fire. This water has to be instantly available in quantity, since it must be applied within minutes of a fire breaking out. It is not necessarily used for extinguishing a fire; frequently a fire is allowed to burn and the water is used to cool adjacent structures, or tanks, and prevent the fire spreading. The supply should, wherever possible, be independent of process water or cooling water supplies.

The fire ring main system may provide continuously a minimum of 14,000 to 18,000 l/min and this may be supplemented by hoses from a stored source of supply or a river. A major works may use up to 2.5×10^6 l/h, either from storage, or make-up, or combined. This amount may be for a 5- or 6-hour period, i.e. 15×10^6 litres; if the fire is not then under control the plant would probably be completely destroyed.

As to pressure, a system must be designed to give a minimal residual pressure of 6 to 7 bar. An adequate number of hydrants is essential, e.g. hydrants about 45 m apart in plant areas and 90 m apart elsewhere, e.g. tank farms. (As an approximation, at an angle of 45°, 0.07 bar (1 psig) gives 0.3 m throw.) There must be a grid system with a sufficient number of isolation valves at intersections to allow for repairs of hydrants (e.g. if to isolate an individual hydrant takes out 75% of the hydrants around a plant, the nearest one to a fire could be 150 to 180 m away).

Fixed installations using water

Fixed installations are used for those potential hazards where large quantities of water may be needed without delay. Their water supply should be from the fire water recirculation main which must be sized to cope with both duties.

> One factor in the Feyzin disaster [Ch. 15, page 879] was the failure of the spray protection system on the propane spheres because so much water was drawn off the main for other fire-fighting purposes.

Major installations protected by fixed sprays include those containing LPG, or any flammable materials of similar vapour pressure. It may be difficult to provide sufficient water, e.g. to protect three large spheres simultaneously could result in a demand for 23,000 l/min. Therefore, in the past top-half protection alone has been considered adequate, providing the contents of the sphere were rarely below the equator of the sphere.

However, problems may then arise if there is frequent operation for considerable periods of time with low liquid levels; all-over protection is then required. A second advantage of cooling the wetted portion of the tank may be to reduce the rate of vaporisation to such an extent that the vapour pressure falls below the pressure relief valve setting; the valve then reseats preventing unnecessary discharge of vapour. The rate of water application should be of the order of 10 l/min/m^2.

All LPG rail and road loading bays, which should be separated by a distance of 50 ft (15.2 m), will normally be provided with a manually operated fixed water spray installation. Coolers and transformers in the power house will be similarly protected. Drencher systems are also used to protect steelwork on process plants and the support structures for pipe racks. Controlled water injection can be used to deal with leaks from flanges or valves at the bottom of a sphere.

So far as pressure vessels are concerned there are in fact four main ways of providing protection against fire, namely:

Sloping the ground
Insulation
Water drenching
De-pressuring (and the sizing of relief valves for fire conditions).

These are illustrated in Fig. 4.10.

The pressure vessel should stand on an impervious base, sloped so that any spillage is collected. A slope of 1 in 40 is recommended so that spillage of flammable liquid cannot accumulate under the vessels. The

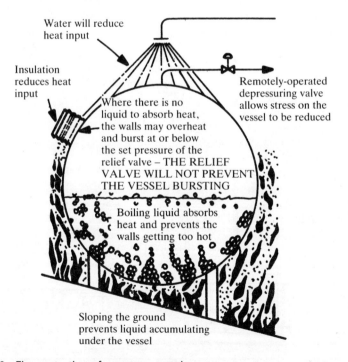

Fig. 4.10 Fire protection of pressure vessels

collecting area should be such a distance away from the tanks and of such a surface area that the flame from a fire in the collecting pit will not impinge on the tank. This procedure can be applied generally to pressure storage vessels. In the case of process vessels the principle of sloping the ground should be applied. The gradient of 1 in 40 may not be practicable, but should not be less than 1 in 60; it should lead to a safe controlled drainage, or collecting, point.

Insulation is recommended in some codes particularly of vessels containing high-risk materials, e.g. LPG. Tests, and fire experience, have shown that insulation can withstand the jets from fire hoses. Insulation is an immediate barrier to heat input before the water drench is applied. This is helpful if fire, or any associated explosion, has damaged water supply lines or injured key personnel. Nevertheless, it should only be used in addition to water drenching, not as an alternative. Insulation slows down the rate of heat input to the vessel, e.g. by a factor of about ten, but cannot ultimately prevent the vessel getting too hot. There is more time to commission water drenching and augment fixed systems with portable monitor sprays.

Water drenching should only be complementary to insulation and to sloping of the ground. Clearly any fixed installation must be periodically tested and serviced. Provision must also be made for controlled drainage.

Reducing the pressure in a vessel exposed to fire reduces the stress on the metal and removes the danger of the vessel bursting. It also prevents damage to items of equipment and reduces the size of any leak. A rule of thumb is that means for vapour de-pressuring should be provided on all vessels handling flammable liquids above 3 bar. The de-pressuring

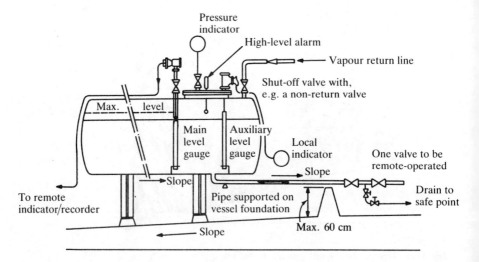

Fig. 4.11 LPG storage tank

Flammable materials handling 179

valve should allow pressure to fall to half of design pressure in 10 minutes. The API recommends that the pressure should be reduced to 8 bar or 50% of design, whichever is lower in 15 minutes.

A typical LPG storage tank installation is shown in Fig. 4.11.

Local and mobile equipment

Any fire appliances must be tailored to meet the needs of the site. Many refineries have water monitors mounted on trailers and bipods which are located at strategic points so that they can be quickly manoeuvred into position and fed through lines of standard hose.

Very large quantities of foam are needed to maintain a successful attack on a large oil storage tank fire. Very large foam monitors using mechanical protein foam produce 20,000 l/min of foam. Foam compound may be supplied in mobile 9,000 litre tankers or bowsers; at some plants, compound is pumped along piping to strategic points for connection to hose and foam-making branch pipes.

Portable dry powder extinguishers varying from 1 to 14 kg may be provided to deal with oil spillage fires. Mobile dry powder appliances are also available to deal with more serious spill fires, e.g. varying in size from 68 kg up to self-contained vehicles of 3.05 tonnes.

Steam can be used on fires involving vapour or gas leaks from seals or flanges of pipework. It is also suitable for fires involving hot vapour lines. Steam lances and/or steam spars are used, fed by armoured hose from the plant steam mains, at a pressure of about 4 bar.

Standards for equipment, installations and other measures related to fire protection

All equipment and other measures relevant to fire protection should be standardised. A list of appropriate British Standards is reproduced in Table 4.15. These can be adapted to suit a particular geographical location or environment.

With the exception of fires following explosions, it may be possible in many cases for a fire to be extinguished in a few minutes by one man; provided action is taken quickly. Therefore a quick-acting, reliable communications system is required.

The ideal method is a detection system which gives an alarm and possibly activates a suppression system e.g. a block and vent arrangement. This may not always be practicable on a large open site. Therefore, some combination of break-glass alarms, telephones connected to a multi-line emergency panel, and pocket radios is required.

Table 4.15 Some British Standards related to fire protection in process plants

BS 138	Portable fire extinguishers of the water type (soda acid)
BS 229	Flameproof enclosure of electrical apparatus
BS 336	Fire hose couplings and ancillary equipment
BS 476	Fire tests on building materials and structures Part 1. Surface spread of flame tests for materials
BS 740	Portable fire extinguishers of the foam type Part 1. Portable fire extinguishers of the foam type (chemical) Part 2. Portable fire extinguishers of the foam type (gas pressure)
BS 750	Underground fire hydrants and dimensions of surface box openings
BS 799	Oil burning equipment Part 2. Vaporising burners and associated equipment Part 3. Automatic and semi-automatic atomising burners up to 36 litres per hour and associated equipment Part 4. Atomising burners over 36 litres per hour and associated equipment for single and multi-burner installations Part 5. Oil storage tanks
BS 1250	Domestic appliances burning town gas
BS 1251	Open fireplace components
BS 1382	Portable fire extinguishers of the water type (gas pressure)
BS 1710	Identification of pipelines
BS 1721	Portable fire extinguishers of the halogenated hydrocarbon type
BS 1782	Hose couplings ($1\frac{1}{2}$ in to 8 in nominal sizes) other than fire hose couplings
BS 1821	Class I oxy-acetylene welding of steel pipelines and pipe assemblies for carrying fluids
BS 1906	Hose couplings (air and water) $\frac{1}{8}$ in to $1\frac{1}{4}$ in nominal sizes.
BS 2050	Electrical resistance of conductive and anti-static products made from flexible polymeric material
BS 2633	Class I arc welding of ferritic steel pipework for carrying fluids
BS 2654	Vertical steel welded storage tanks with butt-welded shells for the petroleum industry
BS 2773	Domestic single-room space heating appliances for use with liquefied petroleum gases
BS 2915	Domed metallic bursting discs and bursting disc assemblies
BS 2929	Safety colours for use in industry
BS 3116	Automatic fire alarms in buildings Part 1, 1970, Heat sensitive (point) detectors Part 4, 1974, Control and indicating equipment
BS 3187	Electrically conducting rubber flooring
BS 3251	Hydrant indicator plates
BS 3300	Kerosine (paraffin) unflued space heaters, cooking and boiling appliances for domestic use
BS 3326	Portable carbon dioxide fire extinguishers
BS 3351	Piping systems for petroleum refineries and petrochemical plants
BS 3398	Anti-static rubber flooring
BS 3465	Dry powder portable fire extinguishers
BS 3492	Electrically bonded road and rail tanker hose and hose assemblies for petroleum products

Table 4.15 Some British Standards related to fire protection in process plants

BS 3561	Non-domestic space heaters burning town gas
BS 3601 to BS 3604	Steel pipes and tubes for pressure purposes
BS 3709	Portable fire extinguishers of the water type (stored pressure)
BS 3811	Glossary of maintenance terms in terotechnology
BS 3980	Boxes for foam inlets and dry risers
BS 4096	Non-domestic space heaters burning liquefied petroleum gases
BS 4137	Guide to the selection of electrical equipment for use in Division 2 areas
BS 4422	Glossary of terms associated with fire Part 1. The phenomenon of fire Part 2. Building materials and structures Part 3. Means of escape Part 4. Fire protection equipment Part 5. Miscellaneous terms
BS 4533	Electric luminaires (lighting fittings) Part 2. Detail requirements: Section 2.1 Lighting fittings for Division 2 areas
BS 4547	Classification of fires (EN 2)
BS 4683	Electrical apparatus for explosive atmospheres
BS 5000	Rotating electrical machines of particular types for particular applications Part 16. Type N electric motors
BS 5146	Inspection and test of steel valves for the petroleum, petrochemical and allied industries
BS 5266	Emergency lighting of premises
BS 5274	Fire hose reels for fixed installations
BS 5306	Code of Practice for fire extinguishing installations and equipment on premises; Part 1, 1976, Hydrants systems, hose reels and foam inlets Part 2, 1979, Sprinkler systems Part 3, 1980, Portable fire extinguishers Part 4, 1979, Carbon dioxide systems Part 5, Section 5.1, 1982, Halon 1301 total flooding systems
BS 5345	Code of Practice for the selection, installation and maintenance of electrical apparatus for use in potentially explosive atmospheres (other than mining applications or explosive processing and manufacture) Part 1. Basic requirements for all parts of the code Part 2. (in preparation)
BS 5376	Code of Practice for selection and installation of gas space heating (1st and 2nd family gases)
BS 5410	Oil firing
BS 5445	Components of automatic fire detection systems Part 1. Introduction Part 5. Heat sensitive detectors, point detectors containing a static element
BS 5423	Specification for portable fire extinguishers
BS 5451	Electrically conducting and antistatic rubber footwear
BS 5458	Safety requirements for indicating and recording electrical measuring instruments and their accessories
BS 5839	Fire detection and alarm systems in buildings Part 1, 1980, Code of Practice for installation
BS 6651	Code of Practice for protection of structures against lightning

Table 4.15 Some British Standards related to fire protection in process plants

CP 3	Code of basic data for the design of buildings Chapter IV Precautions against fire; Part 3. Office buildings
CP 310	Water supply
CP 413	Design and construction of ducts for services
CP 1003	Electrical apparatus and associated equipment for use in explosive atmospheres of gas or vapour other than mining applications
CP 1008	Maintenance of electrical switchgear
CP 1013	Earthing
CP 1019	Installation and servicing of electrical fire alarm systems
CP 2001	Site investigations
CP 2008	Protection of iron and steel structures from corrosion
CP 2010	Pipelines
CP 3005	Thermal insulation of pipework and equipment (in the temperature range of $-100\,°F$ to $+1500\,°F$ ($-73\,°C$ to $+816\,°C$))
CP 3008	The use of transportable industrial space heaters
CP	The control of undesirable static electricity (in draft)

References

(*All places of publication are London unless otherwise stated*)
1. Haessler, W. M., *Chem. Engng*, 26 Feb. 1973.
2. Meidl, J. H., *Flammable Hazardous Materials*. Glencoe, Encino, California, 1978.
3. Health and Safety Executive, *Repair of Drums and Small Tanks: Explosion and Fire Risk*, Health and Safety at Work, No. 32, 1975. HMSO.
4. Lewis, B. & Von Elbe, G., *Combustion, Flames and Explosions of Gases*. Academic Press 1961.
5. Department of Energy, *Report of the Enquiry into Serious Gas Explosions*, June 1977, 38. HMSO.
6. E. Nilsson, The propane explosion in Gothenburg, 8th May 1981, *Instn Chem. Engrs, Symp. Series*, 1983 No. 80.
7. Matheson, D., Symposium on Chemical Process Hazards with Special Reference to Plant Design. Institution of Chemical Engineers 1960.
8. *Annual Report of H. M. Chief Inspector of Factories, 1971*, 3. HMSO.
9. Department of Employment and Productivity, *Dust Explosions in Factories*, Health and Safety at Work No. 22, 1970. HMSO.
10. Anon., *FPA Journal* 1969 (85), 433–40.
11. Factory Mutual System, *The Handbook of Property Conservation*. (2nd edn). Factory Mutual. Eng. Corp., Norwood USA, 1978.
12. Anon., *Loss Prevention Bulletin*. Institution of Chemical Engineers, 1984 (055), 14.
13. Health and Safety Executive, *Health and Safety: Industry & Services*, 1975, 18. HMSO.
14. *Manufacturing Chemists Assoc. Case History 1175*, 1970.
15. Dennett, M. F., *Fire Investigation*. Pergamon 1980.

16. Universities Safety Association, *Safety News*, Feb. 1975, **6**, 3.
17. G. D. Muir, *Hazards in the Chemical Laboratory*. The Chemical Society 1977, 37.
18. Imperial College, *Safety in Chemical Laboratories and in the Use of Chemicals*, (3rd edn). 1971, 13.
19. Anon., *Fire Prevention*, (Aug. 1981), 143, 29–30.
20. *Shell TRC Health and Safety Bull*, 1981 (Aug.), 1.
21. Health and Safety Executive, *Health and Safety: Manufacturing & Services*, 1976, 30. HMSO.
22. Anon., *Loss Prevention Bulletin*. Institution of Chemical Engineers, 1975 (002), 1–13.
23. Napler, D. H., *Instn Chem. Engrs Symp. Series*, 1971, No. 34, 170–4.
24. Heron, P. M., *Chem. Eng. Prog.*, Nov. 1976, 72–5.
25. Palmer, K. N., reported in *Chem. Processing*, July 1978, 21.
26. Anon., *Loss Prevention Bulletin*. Institution of Chemical Engineers, 1976 (029), 134–8.
27. *National Fire Protection Association* in C. H. Vervalin (ed.) *Fire Protection Manual for Hydrocarbon Processing Plants*, (2nd edn). Gulf, Houston, Texas, 1973.
28. Petetsky, B. & Fisher, J. A., *Loss Prevention*, 1975, **9**, 114–15.
29. Rees, W. D., *Chem. Eng*, Jan.–Feb. 1967, 205, CE7–11.
30. Searson, A. H., *Instn Chem. Engrs Symp. Series*, 1983 No. 82.
31. British Standards Institution, *BS 5958 Part 1: 1980 Code of Practice for Control of Undesirable Static Electricity*.
32. Institute of Petroleum, *Model Code of Safe Practice in the Petroleum Industry: Part 1, Electrical Safety Code*, 1965.
33. American Petroleum Institute Recommended Practice 2003, *Protection Against Ignitions Arising out of Static, Lightning and Stray Currents*, Mar. 1982.
34. *Preparation for Maintenance*, Hazards Workshop Module 004. Institution of Chemical Engineers, 1981.
35. Anon., *Loss Prevention Bulletin*. Institution of Chemical Engineers, 1982 (047), 15–19.
36. Pavlin, M. S., *Tappi*, Apr. 1983, **66** (4), 91–2.
37. Anon., *Loss Prevention Bulletin*. Institution of Chemical Engineers, 1976 (012), 1.
38. Universities Safety Association, *Safety News*, No. 8 (Feb. 1977), 16.
39. Hughes, J. R., *Storage and Handling of Petroleum Liquids: Practice and Law*. Griffin 1967.
40. Gibson, N., Lloyd, F. C. & Perry, G. R., *Instn Chem. Engrs Symp. Series*, No. 25, 1968, 26–35.
41. *Fires & Explosions*, Hazard Workshop Module 003. Institution of Chemical Engineers 1980.
42. *Original Types of Miners Flame Safety Lamps*. E. Thomas & Williams, p. 2.
43. Health and Safety Executive Report, *The Fire and Explosions at Permaflex Ltd, Trubshaw Cross, Longport, Stoke-on-Trent, 11 February 1980*. HMSO 1981.
44. Health and Safety Executive, *Guidance Note CS4; Code of Practice for the Keeping of LPG in Cylinders and Similar Containers*, 1986.

45. British Standards Institution, *BS 6651: 1985, Code of Practice for Protection of Structures Against Lightning*.
46. Imperial Chemical Industries, *Accidents Illustrated*, Case History No. 5. ICI, May 1971.
47. Imperial Chemical Industries, *Furnace Failures, Accidents Illustrated No. 7*, Case History 38. ICI, Oct. 1978.
48. Schofield, C., *Guide to Dust Explosion Prevention and Protection*. Part 1 – *Venting*, Institution of Chemical Engineers 1984.
49. Health and Safety Executive, *Health and Safety: Industry & Services*, 1975, 33
50. *Annual Report of H. M. Chief Inspector of Factories, 1969*, 1. HMSO.
51. Health and Safety Executive, *Health and Safety: Industry & Services*, 1975, 70. HMSO.
52. Mumford, C. J., *Loss Prevention Bulletin*. Institution of Chemical Engineers, 1984 (057), 1–6.
53. Berkovitch, I., *Manufacturing Chemist*, 1983, **54** (9), 71.
54. British Standards Institution, *BS 6336: 1982, Guide to Development and Presentation of Fire Tests and Their Use in Hazard Assessment*.
55. Read, R. E. H. & Morris, W. A., *Aspects of Fire Precautions in Buildings*, 1983. HMSO.
56. Fire Protection Association, *Fire Facts and Figures*, Fire Safety Data Sheet MR3, 1983.
57. Fire Protection Association, *Fire Safety Data Sheet MR3* (revised), 1986.
58. Ward, R., Fire Protection Association, *Fire Prevention*, 1984 (175), 20–7.
59. Health and Safety Executive Report, *The Fire and Explosions at B and R Hauliers, Salford, 25 September 1982*. HMSO.
60. Fire Protection Association, *Security Against Fire-Raisers*, Fire Safety Data Sheet MR5 (undated).
61. Carson, P. A., Mumford, C. J., & Ward, R., *Loss Prevention Bulletin*. Institution of Chemical Engineers, Oct. 1985 (065), 1–14.
62. Carson, P. A. & Mumford, C. J., *Loss Prevention Bulletin*. Institution of Chemical Engineers, Aug. 1986 (070), 15–29.
63. Fire Protection Association, *Security Equipment and Systems*, Fire Safety Data Sheet PE14, 1981.
64. Fire Protection Association, *Fire Safety Patrols*, Fire Safety Data Sheet MR4, 1986
65. Health and Safety Executive Report, *The Fire and Explosion at Braehead Container Depot, Renfrew, 4 January 1977*. HMSO.
66. Factories Act 1961, ss.40–9. HMSO.
67. Vervalin, C. H., *Fire Protection Manual for Hydrocarbon Processing Plants* (2nd edn). Gulf, Houston, Texas, 1973, 162–8.
68. National Fire Protection Association. *Fire Protection Handbook* (16th edn), 1986.
69. Nash P., *Instn Chem. Engrs. Symp. Series*, 1977 No. 49, 131.
70. Herzog, G. R., *Hydrocarbon Processing*, Feb. 1975, 1–4.
71. National Fire Protection Association, Pamphlet 11.
72. British Standards Institution, *BS CP 3031: 1974: Code of Practice for Fire Precautions in Chemical Plant*.

CHAPTER 5

Toxicology

Introduction

Man encountered the phenomenon of 'toxicity' at the dawn of his evolution. Certain natural 'foods' proved satisfying, nourishing and even of medicinal quality while others resulted in discomfort, illness and sometimes death. Such properties were eventually harnessed for therapeutic reasons, or for added impact in hunting or warfare. Indeed the term 'toxic' originates from the Greek 'toxon', the bow used to deliver poisoned arrows by hunters and warriors.

Early examples of occupational poisoning include the high death tolls among prison and slave labour in the lead mines of antiquity. Similarly, because of the known hazards of mining mercury, the Romans employed only slave labour to work the Spanish mines at Almaden. During the seventeenth century French hatters discovered the value of using mercuric nitrate in felting fur; exposure resulted in widespread chronic mercury poisoning throughout the trade, giving rise to the term 'mad as a hatter'. However, it was not until Ramazzini's treatise *De Morbis Artificum Diatriba* in 1700 that the adverse effects of the occupational environment on man's health were seriously considered. With the advent of the Industrial Revolution the number of occupational hazards grew significantly, including the range of toxic chemicals to which people were exposed. About 60,000 chemicals are now in common use. Some of the main toxic contaminants associated with a range of processes and industries are given in Table 5.1 (based on data from ref. 1); Tables 5.8 (after ref. 2), 5.14[2] and 5.15[3] are also relevant. Table 16.7 identifies prescribed occupational diseases in the UK caused by chemicals.

The true extent of ill health caused, or exacerbated, by occupational exposure to toxic chemicals is unknown, but some figures are given below[4-6].

- Recent figures suggest that in Britain more persons currently die as a result of occupational disease than from accidents at work. Thus in 1983 there were 998 deaths qualifying for awards under the industrial disease schemes compared with 443 industrial fatal injuries.

Table 5.1 Important selected examples of potentially toxic contaminants which may be encountered in a variety of occupational operations and industries (based on ref. 1)

Additional contaminants may arise from the presence of additives, impurities, by-products, residues, etc.; Often the contaminant is a mixture of substances. The materials to which operatives can be exposed will be influenced by the process and this list is obviously selective.

Process/Industry	Key possible contaminants	Process/Industry	Key possible contaminants
Abrasive blasting	Silica, metal dust	Cements and concrete	Silica, nuisance particulates, hexavalent chromium compounds, carbon monoxide
Abrasives	Dusts such as silicon carbide, aluminium oxide, clays, feldspars, rubber, resins, gases, e.g. carbon monoxide	Chlorine	Mercury, chlorine, caustic soda
Acid and alkali treatment	Mists of chromic, nitric, phosphoric sulphuric, nitric–sulphuric acids, hydrogen fluoride, oxides of nitrogen, cyanide and alkali mists	Construction	Solvents, asbestos acids and alkalis, cement dust, tar and oil fumes, cleaning agents, preservatives, fumigants, weed-killers, carbon monoxide, lead
		Cotton	Cotton dust
Acids		Degreasing	Aliphatic and aromatic hydrocarbons, chlorocarbons, esters, ketones, cellosolves, creosote, cresylic acid, caustic soda, soda ash, phosphates and soaps
– hydrochloric	Hydrogen chloride		
– nitric	Oxides of nitrogen, ammonia		
– sulphuric	Oxides of sulphur, ammonium vanadate, vanadium pentoxide		
Agriculture	Pesticides, fertilisers	Detergents	Oxides of sulphur, sulphuric acid, caustic soda, surfactant, bleach, nuisance dusts, proteolytic enzymes
Aluminium	Silica (in ore), fluorides, alumina, aluminium, carbon dioxide and carbon monoxide		
		Dry cleaning	Organic solvent vapours
Ammonia	Ammonia	Electroplating	Formaldehyde, ammonia, mists of tin salt, fluoroborate cyanide, alkali, hydrochloric and sulphuric acids, nickel sulphate, and zinc chloride, chromic acid mist
Asbestos	Asbestos dust		
Asphalt	Mineral dust, asphalt, hydrocarbon fumes		
Batteries	Electrolytes, lead, lead oxide, cadmium dust or fume, manganese oxide, dusts of ammonium and zinc chloride		
		Fertilisers – natural	Micro-organisms, hydrogen sulphide, ammonia and carbon dioxide
Beryllium	Dusts and fumes containing beryllium compounds or metal, acid fumes		

Table 5.1 (continued)

Process/Industry	Key possible contaminants	Process/Industry	Key possible contaminants
– mineral	Gases of silicon tetrafluoride, sulphur dioxide, hydrogen fluoride, ammonia, ammonium chloride, oxides of nitrogen, phosphorus-containing dusts	Metal machining	Metal dust, grinding and cutting oils
		Metallising	Metal powder or fume
		Milling/baking	Flour, dust, bleaches
		Mining	Particulates and gases depending upon type of mine. For example, quartz, asbestos, mercury, oxides of nitrogen, methane carbon monoxide, hydrogen sulphide, aldehydes, together with an oxygen deficiency
Forging	Carbon monoxide, oil mist, trace metals (e.g. vandium), poly-nuclear aromatics, sulphur dioxide		
Foundry	Silica and other particulates, aldehydes, ammonia, carbon dioxide, carbon monoxide, amines, acids, phenols, hydrogen cyanide, metal fumes		
		Paint	Pigment, solvent, isocyanates
		Painting	Solvent or thinner, paint mist or liquid
		Paper and pulp	Hydrogen sulphide, methyl mercaptan, dimethy sulphide, sulphur dioxide, chlorine, solvents, lime, aldehydes, e.g. acrolein
Garages	Carbon monoxide, lead and plastic fillers, degreasing solvents, petroleum products, welding fumes, epoxy resins		
Glass	Silica, lead, arsenic, barium, cadmium, manganese	Petroleum	Carbon monoxide, formaldehyde, hydrogen chloride, hydrogen sulphide, vanadium, arsenic, oxides of nitrogen, hydrocarbon gases and vapours
Grinding and polishing	Metal dust, fused sand, lead paint		
Iron and steel	Carbon dioxide, carbon monoxide, sulphur dioxide, hydrogen sulphide, fumes of iron oxide, lead, zinc, mists of sulphuric acid and dusts containing ore, coal and silica		
		Pharmaceuticals	Wide range of raw materials, organic solvents, antibiotics, steroids and potent drugs
		Plastics	Range of monomers and curing agents such as vinyl chloride, hydrogen cyanide, epichlorohydrin, phenol, formaldehyde, ethylene oxide, maleic anhydride, styrene, isocyanates, catalysts, organic solvents
Leather	Hydrogen sulphide, carbon monoxide, chromic acid, alkalis, borax, oxalic acid, formaldehyde		
Metal heat treatment	Carbon monoxide, lead fume, caustic fume or mist, sodium carbonate fume, ammonia, methane		
		Pottery	Silica dust, lead

Table 5.1 (continued)

Process/Industry	Key possible contaminants	Process/Industry	Key possible contaminants
Printing and office works	Solvents, ink and oil mist, plate-making chemicals, trichloroethylene, methanol, ammonia, carbon monoxide, ozone, sodium dichromate mist	Ship-building	Asbestos, gasoline, oxides of nitrogen, metal fumes, paint spray, oxygen deficiency
		Smelting	Ore dust, sulphuric acid mist, oxides of sulphur, metal fumes
Quarrying	Particulates depending upon quarry, explosives (e.g. TNT)	Welding	Metal fume, carbon monoxide, carbon dioxide, ozone, hydrogen fluoride, oxides of nitrogen, solvent and paint decomposition products, e.g. lead oxide soldering
Rayon	Carbon disulphide, hydrogen sulphide		
Rubber	Dusts of rubber, carbon black, sulphur, zinc oxide, phenols, amines, anti-oxidants, vulcanising agents, accelerators, pigments, and solvent vapours		
		Wood-working	Wood dust, solvent vapours, glues, wood preservatives

- In the UK occupational skin diseases, and their treatment, are responsible for around 630,000 lost working days each year and are the commonest reason for payment of industrial injury benefit.[1] In the USA dermatoses are second to occupational deafness in terms of compensation.[7]
- In 1979/80 the DHSS awarded in respect of disease attributable to occupational exposure to chemicals (excluding lead and asbestos) over 1,560 new injury benefits and 710 death benefits amounting to about 187,000 working days.[8] This underestimates the true position since the data relate only to clearly established cases of prescribed diseases.
- In 1980 in the UK over 20,000 deaths[9] and 30 million lost working days[10] were attributable to chronic bronchitis, asthma and emphysema, although the proportion contributed to by occupational factors is unknown. One estimate claims that 5% of deaths from bronchitis/emphysema each year are work related.[11]
- In 1980 in Britain there were 130,000 deaths due to cancer.[9] The DHSS statistics do not include cancers (with the exception of mesothelioma or bladder cancer) but one source[12] suggests 2–8% are occupationally linked; others put the figure much higher.

Also, estimated rates of fatality or incidence of disease attributed to types of chemicals or physical exposure (Table 1.3) in general are high compared with accident rates. Incidents involving accidental poisoning, dust disease and dermatitis in the agriculture industry are exemplified by Table 6.18.

A range of legislative measures aim to protect employees against occupational accidents or ill health due to occupational exposure to toxic chemicals, for example, in the UK, the Carcinogenic Substances Regulations, Chemical Works Regulations, The Chromium Plating Regulations, Ionising Radiation Regulations, the Control of Lead at Work Regulations, Agriculture (Poisonous Substances) Act, The Factories (Notifications of Diseases) Regulations, The Asbestos Regulations, The Packaging and Labelling of Dangerous Substances Regulations, Hydrogen Cyanide (Fumigation) Act, etc. Some materials are covered by legislation in depth, as exemplified by the Control of Lead at Work Regulations which is supported by a detailed Code of Practice to provide advice on how to meet the legal requirements. Where no relevant legislation exists to cover an individual material, or situation, certain general sections of the Factories Act may be applicable as may sections of the Health and Safety at Work Act. These are discussed further in Chapter 16.

Moreover legislation such as the Solvents in Food Regulations 1967 regulate the use of chemicals in consumer products. Chapter 13 refers to legal obligations in marketing chemicals.

In general, a separate range of legislation covers the control of toxic emissions or discharges to the environment, e.g. in the UK, The Control of Pollution Act, The Alkali Act, The Clean Air Act, etc.

In the US both State and Federal legislation has been passed to protect the safety of employees, customers and the environment. The Occupational Safety and Health Act (1970), like the UK HASAWA, is enabling legislation which places responsibility on the employer. This is supported by additional laws including the Toxic Substances Control Act (1976). Pollution and Transport legislation relevant to toxic substances is also well established. Recent or proposed legislation by the EEC on notification of new substances, asbestos, lead, and general regulations on control of substances hazardous to health at work, though differing in detail follow UK legislation in principle.

As with any chemical the appropriate precautions for toxic substances should, in addition to satisfying legal requirements, reflect the risk. This entails knowing the identity of the material, the nature and potency of its toxic effects, and the nature and level of exposure to the substance. The purpose of this chapter, therefore, is to summarise the elements of toxicology as a basis for the precautionary measures discussed in Chapter 6 and the monitoring techniques outlined in Chapter 7. A description is given of the body's defence mechanisms against invasion by chemicals and how these can be by-passed, or over-burdened, by exposure to substances with subsequent damage to selected organs. Examples are given of the effects of specific chemicals. Also included is a summary of test methods for determining the toxicity of chemicals. It is not intended to provide a comprehensive list of the toxic effects of all the chemicals likely to be encountered occupationally or in the home. For this, reference is recommended to standard texts (see refs 10, 13–16), or to the information sources discussed in Chapter 18.

A toxic substance is one which possesses the inherent ability to cause adverse biological effects at susceptible sites in, or on, the body. This inherent ability to cause adverse biological effects represents the **hazard** associated with the chemical. **Risk** on the other hand reflects the probability that injury will result as a consequence of the circumstances of use.

Hazard recognition

General

To ensure that toxic materials are handled safely the risk associated with a particular process or operation involving them must be assessed. The first stage relies on recognition of the hazards involved, i.e. the possible implication of exposure, taking cognisance of idiosyncrasies of individual reactivity. Since toxic substances are classified in a variety of ways, e.g. according to the nature of the toxic effect (carcinogen, irritant, sensitiser, etc.) or the principal target organ affected, an awareness of physiology of certain organs and the fundamentals of industrial toxicology and diseases is important.

Adverse biological effects result from exposure to a toxic substance only if the chemical (or its metabolites) reach appropriate receptors in sufficient quantities, and for sufficient length of time, to promote their influence. Therefore, factors which influence the nature and magnitude of any adverse health effect include the inherent properties (chemical and biological activity), the mode of entry, duration and level of exposure (or more correctly the body burden), individual susceptibility, etc.

Toxic effects occurring within a short time of administration (see page 252) of a single dose, or multiple doses given within say 24 hours are termed 'acute'. Those adverse effects developing after repeated exposure to a substance at concentrations significantly lower than those required to cause acute toxicity are termed 'chronic'. Symptoms may develop gradually (e.g. degeneration of peripheral nerves due to exposure to organophosphorus pesticides) or after a long dormant period (e.g. carcinogens). The site of attack (target organ) may be either the point of contact of the substance, or a remote site (e.g. as a result of absorption into the blood stream with subsequent transport to a susceptible organ). The former effects are referred to as 'local' and the latter as 'systemic'.

Routes of entry

For chemicals to exert their toxic influence they must enter the body and reach susceptible sites. Figure 5.1 summarises routes of entry of chemicals into the body and their distribution and excretion.[2] The principal routes of entry are, in order of significance inhalation, skin absorption (including the mucous membranes), and gastro-intestinal absorption following ingestion. Injection is another route which is more commonly encountered in hospital and laboratory work rather than in the chemical industry in general.

> A man died from trichloroethylene poisoning after the chemical splashed over him while he was helping to drain a degreasing tank. The evidence was that he was drenched in the chemical and some went into his mouth and eyes.[17]

Inhalation

The respiratory system is equipped with a variety of defence mechanisms to deal with invasion by chemicals (see page 200). However, they do not represent a perfect barrier and inhalation is the dominant route by which substances gain entry into the body in the occupational context.

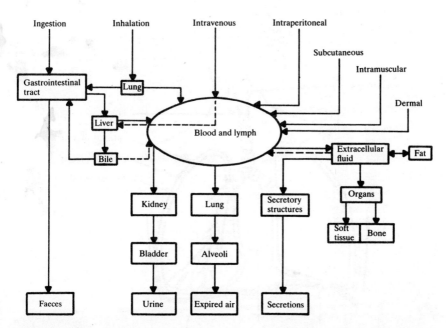

Fig. 5.1 Routes of absorption, distribution and excretion of toxicants in the body (Courtesy L. J. Casarett)

192   The Safe Handling of Chemicals in Industry

The respiratory system encompasses the conducting airways and respiratory units identified in Fig. 5.2. The gas–tissue interface of the adult human is vast (70 m² alveoli surface and 90 m² total surface area) and the normal breathing capacity of an adult is 18–20 l/min. With each inspiration about 500 ml of air passes deep into the lungs.

For liquid or solid particulates, size and shape of the particle are among the key factors which influence the site of deposition, retention, distribution and ultimate health effect (see Ch. 3). Generally particles larger than 50 μm aerodynamic diameter (i.e. the diameter of a uniform sphere which has the same terminal velocity as the aggregate or other irregular particle) are prevented from entering the system as a result of inadequate suction power.

Particles between 10 and 50 μm are effectively filtered in the nose. Particles of 7–10 μm on impact with the mucous surface are carried outwards by the ciliary escalator (between terminal bronchioles and throat) up into the pharynx within a few hours where they are either expectorated or swallowed. Hence inhaled matter may eventually also involve exposure of the intestinal tract.

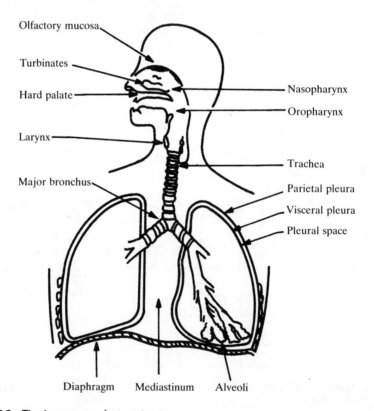

Fig. 5.2 The human respiratory tract

Particles of 0.5–7 μm aerodynamic diameter (particularly between 2.7 μm) are deposited in the respiratory bronchioles and alveoli. The prime clearance mechanism for soluble material is solubilisation in the respiratory tract fluids. Very soluble particles (e.g. potassium chromate) pass through the lungs in minutes. Less soluble matter trapped in the alveolar region is scavenged by large phagocytic cells, termed macrophages, which either cross the alveolar membrane or exit via the ciliary escalator to be ultimately swallowed or expectorated. Sometimes particles are cleared by direct intercellular penetration. Particles smaller than 0.5 μm and gases remain airborne and are exhaled, or diffuse and come into contact with membranes of the airway or lung and are dealt with as described above. Gases of low water solubility but high fat solubility pass through the lung to be distributed to other organs for which they may show a special affinity. Typical systemic poisons, which may be toxic by inhalation, include carbon disulphide, volatile organic chemicals such as hydrocarbons, chlorocarbons, ketones, alcohols, etc.

The rate of absorption of an 'inert' gas from alveoli into the blood stream is dictated by solubility in the blood plasma and the volume of blood flowing through the lungs in unit time. Blood flow is increased with physical exertion. If gas is taken up by organs or tissues, absorption will continue until these and the blood become saturated. The complete removal of gas from the body may take many hours.

Skin absorption

The main structural features of the skin are shown in Fig. 5.3. It comprises two main strata namely the epidermis (outer) layer and dermis. The former consists of two layers the stratum corneum or keratin (composed of dead cells) and living epidermal cells. The dermis contains nerves, blood and lymph vessels, hair follicles, oil and sweat glands and their ducts, plus a matrix of fibres and ground substances.

One of the prime physiological functions of the skin is to provide a protective barrier for the body against invasion by foreign substances. The stratum corneum provides the greatest barrier to absorption. Nevertheless, the skin is not a perfect barrier and its large surface area (about 1.7 m^2 for the average adult) and its direct contact with the external milieu render it vulnerable to hostile environments, either as a target organ or as an entry route for certain systemic poisons as illustrated above.

For toxicants to be absorbed through the skin into the systemic circulation they must penetrate the stratum corneum and the dermis. Absorption may occur via epidermal cells, cells of sweat glands or sebaceous glands, or through hair follicles. In general, the epidermal route is of greatest importance, since the sweat glands and follicles amount to only a small fraction of the total surface area of the skin.

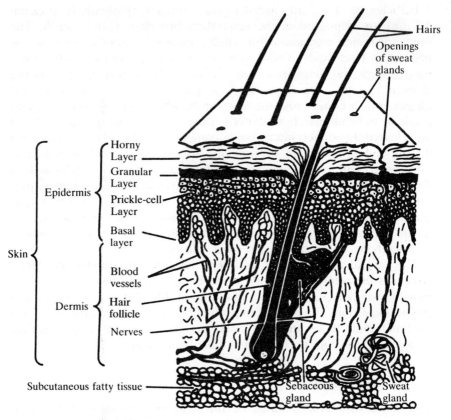

Fig. 5.3 Skin structure

The skin surface is covered by a protective oily acid film which can easily be wiped off or removed by solvents and emulsifying agents.

The stratum corneum offers reasonable protection against many chemicals with the notable exception of alkalis at high pH. It is a hydrated multicellular layer composed of dead cells. Diffusion of chemicals through this section of the skin is the rate-determining step in percutaneous absorption. Polar substances diffuse through the hydrated outer protective surface; non-polar chemicals can dissolve in the lipid zone at a rate which is related to the lipid solubility and which is inversely related to molecular weight.

The dermis is composed of porous, non-selective watery diffusion material and provides a poorer barrier to diffusion than does the stratum corneum.

The skin differs in structure according to the part of the body. For example the stratum corneum is thicker on the soles of the feet and palms of the hand than on the scrotum. The latter also possesses greater diffusivity. Thus the ease of absorption of chemicals through the skin varies

Table 5.2 Penetration of organophosphates and carbamate insecticides through human skin*

Site	Penetration potential
Forearm	1
Palm of hand	1
Dorsum of hand	2
Abdomen	2
Follicle-rich sites (e.g. scalp, forehead, angle of jaw)	4
Intergenious axilla	4–7
Scrotum	almost complete

* Using the forearm as a frame of reference, i.e. penetration potential of 1.

widely according to the anatomical region as indicated by Table 5.2 (based on ref. 18).

Penetration is clearly facilitated by the presence of skin abrasions. Skin absorption is often an underestimated problem when handling chemicals. For example, the corrosive properties of phenol and its homologues tend to be appreciated.

> A chemical operator was preparing a mixture of phenol and liquid caustic soda. Because of layering in the reactor, when the agitator was switched on a violent reaction caused half of the contents to be projected on to the operator. He suffered extensive phenol burns to head, eyes, trunk and legs, severe disfigurement and blindness in both eyes.[19]

However, the rapidity with which phenolic compounds penetrate the skin and affect the central nervous system, the kidneys and the liver, and the speed with which death can result is less well appreciated.

> An operator was sprayed with cresol. Despite having his contaminated clothing removed and having a shower within 30–60 seconds (and medical aid within 15 minutes), 25–30% of his body surface had been burned. Sufficient of the chemical had also been absorbed through his skin to cause mental confusion and signs of kidney damage.[20]

> A technical assistant had the trousers on both thighs soaked with phenol which splashed from a loosened cover on a filter (due to inadequate isolation – see Ch. 17). He was not wearing special protective clothing other than rubber gloves. This incident proved fatal, despite emergency washing of his legs with tap water and, after walking/being carried 165 m to the first-aid centre, for the administration of oxygen and medical treatment.[21]

Indeed toxicity by skin penetration can be higher than that by ingestion as illustrated by Table 5.3, although problems of interspecies comparisons are acknowledged.

Table 5.3 Skin penetration toxicity v. oral toxicity

Chemical	Ratio of skin to oral toxicity*
Propylamine	10
5-Indanol	7
Tridecyl acrylate	7
4-Methyl pyridine	5
Di-hexyl ether	4
2-Ethyl butyric acid	4
Butyl vinyl ether	3
Formaldehyde	3
Methyl heptanol	2

* Based on ratio of $\frac{\text{oral-rat LD}_{50}}{\text{skin-rbt LD}_{50}}$ mg/kg from reference 16.

Table 5.4 identifies those substances assigned an 'Sk' notation in the 1986 list of UK occupational exposure limits to indicate those materials which can be absorbed through the skin as a result of local contamination (e.g. splash on the skin), or in certain cases from exposure to high atmospheric concentrations of vapour.[22]

Ingestion

As far as the working environment is concerned ingestion does not constitute a significant route of entry of industrial substances because:
- Fewer materials can enter via this route. For example it is almost impossible for gases and vapours to be ingested unless significant mouth breathing occurs which can be exacerbated by chewing gum, etc. Nevertheless, for particulates this mode of entry is second only to inhalation.
- The duration of exposure via ingestion is usually shorter than by other routes.
- For many substances oral toxicity is lower than inhalation toxicity or skin penetration as illustrated by Table 5.3.
- The hazard can be significantly reduced by the prohibition of eating or drinking in the workplace and good personal hygiene.

In the UK 'poisons' are specified in the Poisons List made under the Poisons Act 1972. Other materials, no matter how toxic, are not poisons in law. The sale of poisons for non-medicinal purposes is governed by the Poisons Act. Special restrictions relating to storage, conditions of sale, and record-keeping apply under the UK Poison Rule (1972) for Schedule S1 poisons. Selected examples of S1 poisons of industrial importance are given in Table 5.5 and the reader is advised to consult the original Rules for a comprehensive listing of S1 poisons.

Table 5.4 Materials with an 'Sk' notation in list of occupational exposure limits

Compound	Compound
Acrylamide	Hexachloroethane
Acrylonitrile	Hexahydro-1,3,5-trinitro
Aldrin	1,3,5-triazine
Allyl alcohol	Hexan-2-one
Allyl 2,3-epoxypropyl ether	Hydrazine
Aniline	Hydrogen cyanide
Anisidines, o- and p-isomers	2-Hydroxypropylacrylate
Azinphos-methyl	2,2'-Iminodi(ethylamine)
Aziridine	Iodomethane
Butan-1-ol	Malathion
2-Butoxyethanol	Mercury alkyls
Butylamine	Methanol
γ-BHC	Methomyl
Bromoform	2-Methoxyethanol
Bromomethane	2-Methoxyethyl acetate
Captafol	Methyl acrylate
Carbon disulphide	2-Methylcyclohexanone
Carbon tetrachloride	2-Methyl-4,6-dinitrophenol
Chlordane	4-Methylpentan-2-ol
Chlorinated biphenyls	4-Methylpentan-2-one
2-Chlorobuta-1,3-diene	N-Methyl-N,2,4,6-tetranitroaniline
1-Chloro-2,3-epoxy propane	Mevinphos
2-Chloroethanol	Morpholine
1-Chloro-4-nitrobenzene	Nicotine
Chlorpyrifos	4-Nitroaniline
Cresols, all isomers	Nitrobenzene
Cumene	Nitrotoluene
Cyanides	Octachloronaphthalene
Diazinon	Parathion
2,2'-Dichloro-4,4'-	Parathion-methyl
methylene dianiline	Pentachlorophenol
1,3-Dichloropropene	Phenol
Dichlorvos	p-Phenylenediamine
Dieldrin	Phenylhydrazine
2-Diethylaminoethanol	Phorate
Di-isopropylamine	Picric acid
N,N-Dimethylacetamide	Propan-1-ol
N,N-Dimethylaniline	Propan-2-ol
Dimethyl formamide	Propylene dinitrate
Dimethyl sulphate	Prop-2-yn-1-ol
Dinitrobenzene	Sodium fluoroacetate
2,4-Dinitrotoluene	Sulfotep
1,4-Dioxane	Tetraethylpyrophosphate
Dioxathion	Tetramethyl succinonitrile
Endosulfan	Thallium, soluble compounds
Endrin	Tin compounds, organic
2-Ethoxyethanol	Toluene
2-Ethoxyethyl acetate	Tricarbonyl (eta-cyclopentadienyl)
Ethyl acrylate	manganese
Ethylene dinitrate	Tricarbonyl (methylcyclopentadienyl)
4-Ethylmorpholine	manganese
2-Furaldehyde	1,1,2-Trichloroethane
Furfuryl alcohol	Trichloroethylene
Glycerol trinitrate	Xylene
Heptachlor	Xylidine

198 The Safe Handling of Chemicals in Industry

Table 5.5 Selected examples of S1 poisons of industrial importance.

(The Rules should be consulted in individual cases since the definition is often dependent on concentration and the Rules contain a more comprehensive listing.)

- Alkaloids – a wide range of alkaloids and their derivatives
- Amino alcohols esterified with benzoic acid, phenylacetic acid, phenylpropionic acid, cinnamic acid or the derivatives of these acids, their salts
- Antimonial poisons
- Arsenical poisons
- Barbituric acid; its salts; derivatives of barituric acid; their salts; compounds of barbituric acid, its salts, its derivatives, their salts, with any other substance
- Barium, salts of
- Bromomethane
  Chloroform
  Chloropicrin
- Dinitroceresols (DNOC); their compounds with a metal or a base
- Dinitronaphthols; dinitrophenols; dinitrothymols
- Ethylmorphine; its salts; its esters and ethers; their salts
- Hydrocyanic acid; cyanides, other than ferrocyanides and ferricyanides
- Hydroxyurea
- Lead, compounds of, with acids from fixed oils
- Mercuric chloride
  mercuric iodide
  nitrates of mercury
  organic compounds of mercury
- Monofluoroacetic acid; its salts
- m-Nitrophenol; o-nitrophenol; p-nitrophenol
- Organo-tin compounds, the following:
  Compounds of fentin
- Paraldehyde
- Paraquat, salts of
- Phosphorus compounds, the following:
  Amiton
  Azinphos-ethyl
  Azinphos-methyl
  Chlorfenvinphos except sheep dips containing not more than 10% weight in weight, of chlorfenvinphos
  Demephion
  Demeton-methyl
  Demeton-O
  Demeton-S
  Demeton-O-methyl
  Demeton-S-methyl
  Demeton-S-methyl sulphone

Table 5.5 (continued)

Dichlorvos
Diethyl 4-methyl-7-courmarinyl phosphorothionate
Diethyl p-nitrophenyl phosphate
Dimefox
Dioxathion
Disulfoton
Ethion
Ethyl-*p*-nitrophenyl phenylphosphonothionate
Fonofos
Mazidox
Macarbam
Methidathion
Mevinphos
Mipafox
Omethoate
Oxydemeton-methyl
Parathion
Phenkapton
Phorate
Phosphamidon
Pirimiphos-ethyl
Schradan
Sulfotep
TEPP (HETP)
Thiometon
Thionazin
Triphosphoric pentadimethylamide
Vamidothion
• Thallium, salts of
• Tetamine; its salts
• Zinc phosphide

For certain materials ingestion can become problematic when personal hygiene is poor. Indeed for slow-acting cumulative poisons such as arsenic, lead, etc., regulations exist to minimise the body burden by prohibiting smoking and the consumption of food and drink in the work area. Even so, frequency of subconscious hand-to-mouth contact throughout a normal day can be high and an awareness of the hazards is essential to minimise accidental contact by contaminated skin or 'protective' gloves.

Accidental, careless or irresponsible contamination of the food chain can also lead to ingestion hazards.

> After the Second World War inhabitants along the lower stream of the Jinzu River, Japan, were found to be suffering from an unrecorded osseous disease, termed the itai-itai disease. The disease, which mostly affected farm women above middle age who had lived for more than 30 years in the area, was identified as chronic cadmium poisoning. It resulted from contamination of river water, soil and daily diet by cadmium released through mining activities.[23]

Minamata disease – in which methyl mercury contaminates the environment, is transmitted through the food chain, and causes mass human poisoning of regional scope – was officially 'discovered' in 1956. By 1959 it was demonstrated to have been caused by ingestion of fish contaminated by mercury discharged from a chemical plant. However, about 40 earlier cases of organic mercury poisoning had previously been reported including workers involved with mercury-treated seeds and timber, or people who had eaten the disinfected seeds.[24]

Natural defence systems

The body is equipped with a variety of systems for protection against invasion by chemicals. An awareness of these mechanisms is important in understanding the hazards associated with toxic substances.

Primary barriers tend to prevent foreign substances from entering the body. Examples are the skin, respiratory filtration (including the ciliary escalator – see page 192), and reflexes such as sneeze, cough, and apnoea. Secondary defences deal with substances once penetration of the primary system has occurred. Examples are:

1. *Phagocytosis* – a process whereby the substance is engulfed by wandering cells such as macrophages (see page 193); in certain cases (e.g. asbestos) the scavenging cell is destroyed by the chemical.
2. *Immune response* – a mechanism whereby foreign chemicals and redundant cells are recognised by the body. The immune system consists of three main components: antigens, lymphocytes and immunoglobulins/antibodies. The degree of sensitivity to this response can be altered by chemicals termed sensitisers or allergens. A useful review of occupational chemicals capable of causing allergies has recently been published.[25]
3. *Inflammation* – a reaction of tissue to toxic substances typified by changes in the blood vessels and the accumulation of defensive cells such as lymphocytes and macrophages. Pus may form and discharge, and repair occurs by scar tissue.
4. *Repair* – a process involving removal of dead cells or foreign matter and the regeneration of supporting tissue and in some cases regeneration of specialised cells. Where damage is excessive fibrous protein called collagen is formed as scar. Extensive overgrowth is termed fibrosis which can hamper the functioning of internal organs. Extensive scarring typifies certain pneumoconioses. The disorder of the repair process can also lead to conditions such as emphysema, where over-distension and destruction of the alveoli (air sacs) of the lung leads to widespread loss of function.
5. *Excretion* – many materials may leave the body quite rapidly unchanged in the urine, faeces or exhaled breath; others may be excreted after metabolism.

6. *Metabolism* – in the present context describes biochemical processes whereby the body detoxifies foreign substances (though in some instances more toxic metabolites may result). These reactions, which usually occur in the liver but also in the gut-wall and kidneys, either degrade the chemical (transformation) or involve complex formation (conjugation). In both cases the metabolic products are usually rendered more water soluble and are either excreted directly via the kidneys or act as intermediates in further biochemical reactions.

Chemicals entering the body in sufficient quantity may harm these defence mechanisms or natural body functions such as growth.

Disorders of growth

Classes of material causing disorders of growth are teratogens, mutagens and carcinogens:
- *Mutagens* cause interference with genetic cellular material so as to produce inheritable abnormalities. There is evidence to suggest that mutations are involved in the action of many carcinogens.
- *Teratogens* produce congenital abnormalities or other adverse effects in the offspring following exposure of a mother during the period of formation of foetal organs. A precautionary measure is to prevent women of child-bearing age from working with such materials. Some chemicals are known to be toxic in the human foetus (see Table 5.6) though examples of teratogenic effects tend to be associated in the main with drugs.[2] A detailed review[26] of the hazardous effects of chemicals on the reproductive system was undertaken recently at the request of the UK HSE. Table 5.7 summarises the effects reported in female workers, male workers and in animals, though the causal

Table 5.6 Drug and chemical toxicity in the human foetus

Substances/Drugs	Toxic effect
Alcohol	Muscular hypotonia, withdrawal
Antibacterials	Various, depending upon the drug
Anticoagulants (e.g. Coumarin and Sodium warfarin)	Haemorrhage, death
Antidiabetics	Various, depending upon the drug
Ammonium chloride	Acidosis
Antihistamines	Infertility
Magnesium sulphate	Central depression and neuromuscular block
Lithium	Cyanosis and flaccidity
Smoking	Premature birth, small babies, perinatal loss
Vaccinations	Foetal vaccinia
Salicylates, large amounts	Bleeding
Anaesthetics (general anaesthetics)	Newborn depression
Solvents	Newborn depression

Table 5.7 Effects of chemicals on the reproductive system

Effects reported in male workers

Decreased libido and impotence
Chloroprene, manganese, organic lead, inorganic mercury, toluene diisocyanate, vinyl chloride.

Testicular damage or infertility
Chlordecone (Kepone), chloroprene, dibromochloropropane, organic lead.

Effects reported in female workers

Menstrual and other gynaecological disorders
Aniline, benzene, chloroprene, formaldehyde, inorganic mercury, polychlorinated biphenyls (PCB), styrene, toluene.

Abortions or infertility
Anaesthetic gases, aniline, arsenic, benzene, ethylene oxide, formaldehyde, lead.

Decreased foetal growth, low birth weight or poor survival
Carbon monoxide, formaldehyde, PCB, toluene, vinyl chloride

Teratogenic effects
Organic mercury

Maternal death related to pregnancy
Beryllium, benzene.

Effects reported in animal studies but with no human evidence

Testicular damage or reduced male fertility
Benzene, benzo(a)pyrene, boron, cadmium, epichlorohydrin, ethylene dibromide, polybrominated biphenyl (PBB).

Fetotoxic or embryolethal effects
Chloroform, dichloromethane, ethylene dichloride, ethylene oxide, inorganic mercury, nitrogen dioxide, PBB, selenium, tetrachloroethylene, thallium, trichloroethylene, vinylidine chloride, xylene.

Teratogenic effects
Arsenic, benzo(a)pyrene, chlorodifluoromethane, chloroprene, monomethyl formamide, acrylonitrile, methyl ethyl ketone, tellurium, vinyl chloride.

Transplacental carcinogenesis
Arsenic, benzo(a)pyrene, vinyl chloride.

---

relationship is firmly established in only a few cases. The significance of findings from experimental animals to human risk is difficult to assess. Genetic mutations can result from exposure to ionising radiation, with effects on the offspring of exposed workers.
- *Carcinogens* cause cancer which is essentially a disorder of the growth of cells which may involve a variety of tissues in the body. The growing mass fulfils no beneficial purpose to the host. New cells may be benign or malignant. The former are simple and localised while the latter are capable of spreading and destroying neighbouring tissues and of being transported to distant organs via the blood stream and by lymphatics. This production of multiple small tumours is termed metastasis.

Cancer is associated with a host of factors including smoking, diet, lifestyle, geographical location, drugs, radiation and exposure to chemicals. These factors can interact, e.g. the risk of lung cancer in workers exposed to asbestos is greatly increased in smokers. Exposure to chemicals merits special attention because of the, often long, latent period between exposure and the realisation of disease. However, the subject is too vast to cover in this text and the following is restricted to identifying some chemicals with carcinogenic potential. Other sources are given in references 27–31.

Industries that have been linked with cancer include asbestos manufacture and use, certain dyestuffs, rubber, engineering, cotton and other industries using certain types of oils, industries using radioactive chemicals, certain metal industries, the furniture and the boot and shoe industry. Table 5.8 attempts to classify carcinogens but this general approach should be used with care as the mechanisms differ between classes.

Asbestos is the most hazardous of the fibrous silicates. It exists in several forms shown in Table 5.9 and in Chapter 7. In addition to causing asbestosis, it can give rise to cancer of the lung and also cancer of the lining of the lung and abdominal organs (mesothelioma). Both the asbestos industry and user industries are at risk. The main route of entry is inhalation.

Table 5.8 Classes of carcinogens

Certain aromatic amines	– see Fig. 5.4(a) and 5.4(b)
Certain electrophilic species	– see Table 5.10
Certain polynuclear aromatics	– see Table 5.10
Certain nitro and other heterocyclics	
Certain azo dyes	
Certain nitrosamines	
Certain nitrosamides and ureas	
Certain dialkylhydrazines	
Certain thioamides	
Certain vinyl derivatives (e.g. vinyl chloride, acrylonitrile, vinyl carbamate)	
Certain chlorinated alpthatic hydrocarbons (e.g. carbon tetrachloride, tetrachloroethane, ethylene bromide, bis(chloromethyl)ether)	
Dioxane	
Certain inorganics (e.g. asbestos, uranium, radium, some nickel compounds, chromates)	

Table 5.9 Forms of asbestos

Chrysotile (white asbestos)	$(OH)_6Mg_6Si_4O_{11}.H_2O$
Actinolite	$Ca(Mg\ Fe)_3(SiO_3)_4.xH_2O$
Amosite (brown asbestos)	$(FeMg)\ SiO_3.1.5\%\ H_2O$
Anthophyllite	$(MgFe)_7Si_8O_{22}(OH)_2$
Crocidolite (blue asbestos)	$NaFe\ (SiO_3)_2FeSiO_3.xH_2O$
Tremolite	$Ca_2Mg_5Si_8O_{22}(OH)_2$

Aromatic amines, which have been used in the dyestuffs and rubber industries, are as a class hazardous but of varying degrees of potency. In the UK The Carcinogenic Substances Regulations (1967) prohibit the manufacture and use of the aromatic amines shown in Fig. 5.4(a) and severely restrict those listed in Fig. 5.4(b).

Other general classes of carcinogens are electrophilic compounds (which are capable of reacting directly with nucleophilic agents) and polycyclic aromatics – examples are given in Table 5.10 (after ref. 2).

Many substances giving rise to cancer are themselves inactive but produce carcinogenic metabolites. Examples of these include benz (α) pyrene, 2-acetylaminofluorene, 4-dimethylaminoazobenzene, pyrrolizidine alkaloids, aflatoxin B_1, safrole, 3-hydroxyxanthine, dimethyl nitrosamine, dimethyl hydrazine, ethionine, thioacetamide, acetamide and cyasin. These are termed procarcinogens.

Carcinogens can be divided into:
- Proven human carcinogens, i.e. those for which clear evidence exists that man developed cancer as a consequence of exposure to the material, e.g. β-naphthylamine.
- Putative human carcinogens, i.e. where animal and/or other data are such that the material is highly likely to be a human carcinogen if sufficient exposure occurs, e.g. carbon tetrachloride.
- Questionable carcinogens, i.e. where some evidence exists that the material may possibly have carcinogenic properties, though the information is not substantial, e.g. trichloroethylene.

Using data from the International Agency for Research on Cancer Table 5.11 lists chemicals/processes according to the above three-tier categorisation scheme.[32]

Exposure to ionising radiation may also induce cancer as discussed in Chapter 8.

Potency

Whether carcinogenic effects are produced directly or via reactive intermediates is immaterial with regard to handling precautions. More important considerations are the carcinogenic potency (see page 251) associated with the substance and the process conditions under which it is to be used. Guidelines have been given by the American Conference of Governmental Industrial Hygienists (ACGIH).[33] Thus a substance is not considered to be an occupational carcinogen of any practical significance which reacts by the respiratory route at or above 1,000 mg/m^3 for the mouse, 2,000 mg/m^3 for the rat; by the dermal route, at or above 1,500 mg/kg for the mouse, 3,000 mg/kg for the rat; by the gastrointestinal route at or above 500 mg/kg/d for a lifetime, equivalent to about 100 g total dose (TD) for the rat, 10 g TD for the mouse. These dosage

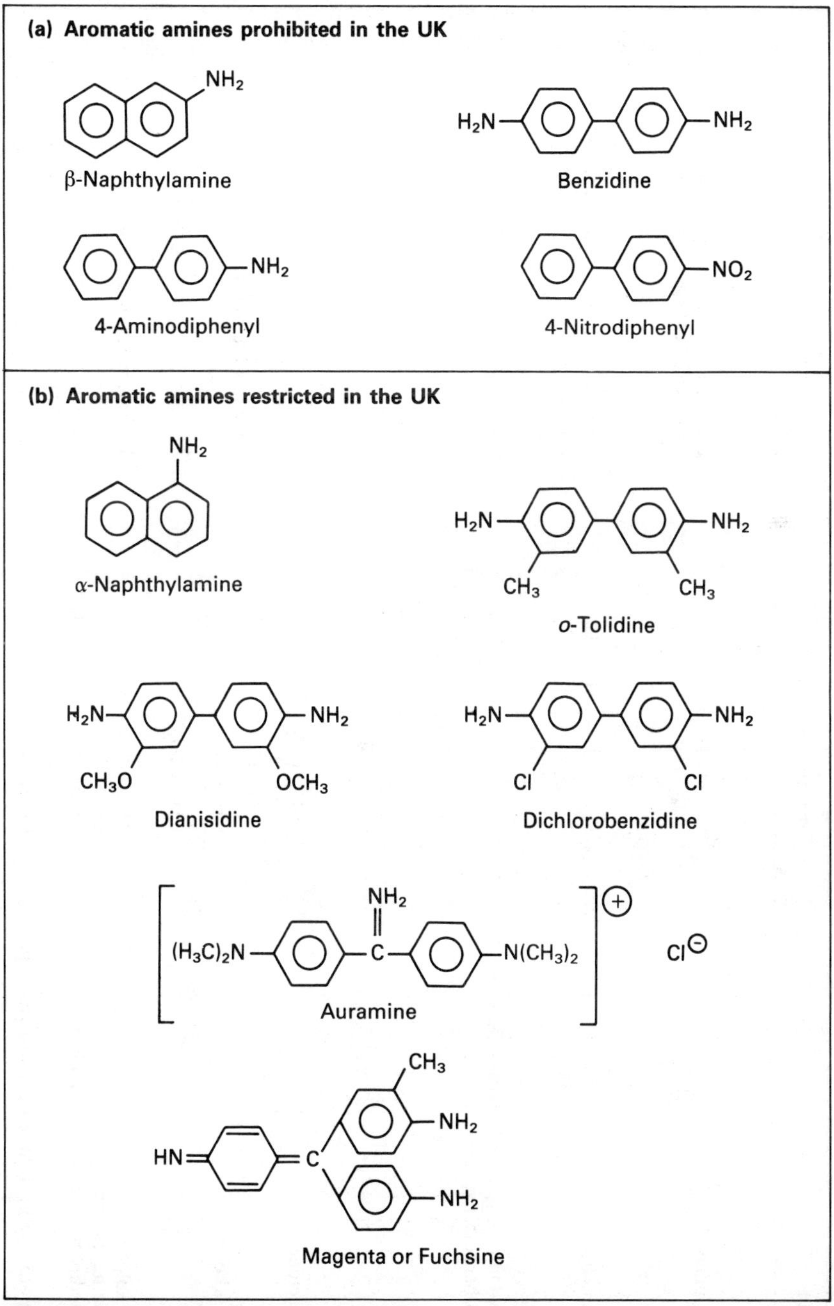

Fig. 5.4 Aromatic amines
(a) Aromatic amines prohibited in the UK
(b) Aromatic amines restricted in the UK

Table 5.10 Some electrophilic and polycyclic aromatic carcinogens (after ref. 2)

Compound	Structure	Use/source
Electrophilic carcinogens		
Ethylene imine	$H_2C\!-\!CH\!-\!R$ with $N\!-\!H$	Polymers, textiles, paper, adhesives, water treatment, coatings
1,2,3,4-Butadiene epoxide	$H_2C\!-\!CH\!-\!CH\!-\!CH_2$ with epoxides	Polymers, solvents, fumigants, and sterilising agents
(+)-7β-8α-Dihydroxy-9⁴,10a-epoxy-7,8,9,10-tetrahydrobenzo(a)pyrene	(polycyclic epoxide structure with HO, OH)	
β-Propiolactone	$H_2C\!-\!CH_2$ / $O\!-\!C\!=\!O$	
Propane sultone	$CH_2\!-\!CH_2$ / $CH_2\!-\!SO_2$ / O	Chemical intermediates
Dimethyl sulphate Methyl methanesulphonate 1,4-Butanediol dimethanesulphonate (Myleran)	$CH_3OSO_2OCH_3$ $CH_3SO_2OCH_3$ $CH_3SO_2O(CH_2)_4OSO_2CH_3$	Intermediates in drug and agrochemical industry; Speciality solvents; pharmaceuticals, perfumes and flavours
Bis (2-Chloroethyl) sulphide (mustard gas, Yperite)	$ClCH_2CH_2\!-\!S\!-\!ClCH_2CH_2$	

Compound	Structure	Use
Bis (2-Chloroethyl) amine nitrogen mustards, R=H or R=CH$_3$	ClCH$_2$CH$_2$\N—R ClCH$_2$CH$_2$/	
Cytoxan (Endoxan)	ClCH$_2$CH$_2$\ \ \ \ \ \ \ \ \ \ O\ \ N—CH$_2$\ N—P\ \ \ \ \ \ \ \ \ \ \ \ \ \ CH$_2$ ClCH$_2$CH$_2$/\ \ \ \ \ \ \ \ \ \ \ \ \ O—CH$_2$/	
2-Naphthylamine mustard (Chlornaphazine)	ClCH$_2$CH$_2$\N—(naphthalene) ClCH$_2$CH$_2$/	Chemotherapy
Triethylenemelamine	H$_2$C\ \ \ \ \ \ \ \ \ \ \ \ \ \ \ \ \ \ CH$_2$ \ \ \ \ N—(triazine)—N H$_2$C/\ \ \ \ \ \ \ \ \ \ \ \ \ \ \ \ CH$_2$ \ \ \ \ \ \ \ \ \ \ N \ \ \ \ H$_2$C—CH$_2$	
Bis (chloromethyl) ether	ClCH$_2$OCH$_2$Cl	Intermediate, ion-exchange resin
Ethylene bromide (or dibromide)	BrCH$_2$CH$_2$Br	Soil fumigant
Benzyl chloride	C$_6$H$_5$CH$_2$Cl	Intermediate in manufacture of plasticisers, pharmaceuticals perfumes and flavours
Methyl iodide	CH$_3$I	Intermediate
Dimethylcarbamyl chloride	(CH$_3$)$_2$NCOCl	
Dimethyl nitrosamine	(CH$_3$)$_2$N.NO	
N-Methylnitrosourea	CH$_3$N—CONH$_2$ $\longrightarrow$ CH$_3^+$ \ \ \ \ | \ \ \ \ NO	

Table 5.10 (continued)

Compound	Structure	Use/source
N-Methylnitrosourethan	$CH_3-N-COOC_2H_5 \longrightarrow CH_3^+$ $\quad\quad\;\;\mid$ $\quad\quad\;\;NO$	
N-Methyl-N'-nitro-N-nitrosoguanidine	$\quad\quad\;\;\;NH$ $\quad\quad\;\;\;\parallel$ $CH_3-N-C-NHNO_2 \longrightarrow CH_3^+$ $\quad\quad\;\;\mid$ $\quad\quad\;\;NO$	
N-Methyl-N'-acetyl-N-nitrosourea	$CH_3-N-CONHCOCH_3 \longrightarrow CH_3^+$ $\quad\quad\;\;\mid$ $\quad\quad\;\;NO$	
	Polynuclear aromatic carcinogens	
Benz (α) anthracene		Pitch/coal tar
Diben (a,h) anthracene		

Benzo (α) pyrene

3-Methylcholanthrene 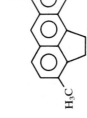 Coal tar and shale oil

5-Methylchrysene

Dibenz [a,] acridine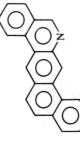

Table 5.11 Chemicals associated with cancer in humans
(Though not all are still of industrial importance)

A. Chemicals and industrial processes which are carcinogenic for humans

Substance or process	Site affected and type of neoplasm	Confirming animal test
4-Aminobiphenyl	Bladder – carcinoma	+
Arsenic and certain compounds	Skin, lung, liver – carcinoma	–
Asbestos	Respiratory tract – carcinoma	+
	Pleura and peritoneum – mesothelioma	
	Gastrointestinal tract – carcinoma	
Auramine manufacture	Bladder – carcinoma	Not applicable
Benzene	Blood – leukaemia	–
Benzidine	Bladder – carcinoma	+
Bis (chloromethyl) ether and technical grade chloromethyl ether	Lung – carcinoma	+
Chlornaphazine	Bladder – carcinoma	±
Chromium and certain compounds	Lung – carcinoma	+
Diethylstilbestrol	Female genital tract – carcinoma (transplacental)	+
Haematite mining (underground)	Lung – carcinoma	Not applicable
Isopropanol manufacture (strong acid process)	Respiratory tract – carcinoma	Not applicable
Melphalan	Blood – leukaemia	+
Mustard gas	Respiratory tract – carcinoma	+
2-Naphthylamine	Bladder – carcinoma	+
Nickel refining	Respiratory tract – carcinoma	Not applicable
Soots, tars, and mineral oils	Skin, lung, bladder – carcinoma	+
Vinyl chloride	Liver – angiosarcoma	+
	Brain	
	Lung – carcinoma	
	Lymphatic system – lymphoma	

B. Chemicals which are probably carcinogenic in humans

Substance	Site affected (human)	Confirming animal tests
Acrylonitrile	Colon, lung	+
Aflatoxins	Liver	+
Amitrole	Various sites	+
Auramine	Bladder	+
Beryllium and certain compounds	Bone, lung	+
Cadmium and certain compounds	Kidney, prostate, lung	+
Carbon tetrachloride	Liver	+
Chlorambucil	Blood	+
Cyclophosphamide	Bladder, blood	+
Dimethylcarbamoyl chloride	?	–
Dimethyl sulphate	Lung	+
Ethylene oxide	Gastrointestinal tract, blood	±
Iron dextran	Connective tissue	+
Nickel and certain compounds	Respiratory tract	+

Table 5.11 (continued)

Substance or process	Site affected and type of neoplasm	Confirming animal test
Oxymetholone	Liver	–
Phenacetin	Kidney, bladder	±
Polychlorinated biphenyls	Skin, various site	+
Thiotepa	Blood	+

C. Substances which may be linked to cancer in humans

	Animal tests
Chloramphenicol	No data
Chlordane/heptachlor	Limited
Chloroprene	Inadequate
Dichlorodiphenyltrichloroethane	Limited
Dieldrin	Limited
Epichlorohydrin	Limited
Haematite	Negative
Hexachlorocyclohexane (lindane)	Limited
Isoniazid	Limited
Isopropyl oils	Inadequate
Lead and lead compounds	Adequate
Phenobarbital	Limited
N-Phenyl-2 naphthylamine	Inadequate
Phenytoin (diphenylhydantoin)	Limited
Reserpine	Inadequate
Styrene	Limited
Trichloroethylene	Limited
Tris(aziridinyl)-p-benzoquinone	Limited

limitations exclude such substances as dioxane and trichloroethylene from consideration as carcinogens. However, the status of dioxane as a carcinogen is currently under review in the UK.

Criteria by which ACGIH judge whether an industrial substance classifies as having high, medium or low carcinogenic potency in experimental animals is given in ref. 33.

The ACGIH in its publication of Threshold Limit Values[33] provides a list of carcinogens as Appendix A to the document and classifies them as 'Human carcinogens' and 'Industrial substances suspect of carcinogenic potential for MAN' (see Table 5.12).

Target organs

Once chemicals enter the body (i.e. through the lung, skin or gut barriers) they are transported around the body via the blood stream to their target organs as illustrated by Fig. 5.1.

Water-soluble materials may be excreted in urine or faeces while volatile material may be discharged in expired air. The site of accumulation depends upon the properties of the chemical and the properties of

Table 5.12 Chemicals listed as carcinogens by the ACGIH[33]

Carcinogens
The Committee lists below those substances in industrial use that have proven carcinogenic in man, or have induced cancer in animals under appropriate experimental conditions. Present listing of those substances carcinogenic for man takes two forms: those for which a TLV has been assigned (1a) and those for which environmental conditions have not been sufficiently defined to assign a TLV (1b).

A1a. Human carcinogens. Substances, or substances associated with industrial processes, recognized to have carcinogenic or cocarcinogenic potential, with an assigned TLV:

Substance	TLV
Asbestos	
Amosite	0.5 fiber[k]
Chrysotile	2 fibers[k]
Crocidolite	0.2 fiber[k]
Other forms	2 fibers[k]
bis-(Chloromethyl) ether	0.001 ppm
Chromite ore processing (chromate)	0.05 mg/m^3, as Cr
Chromium (VI), certain water insoluble compounds	0.05 mg/m^3, as Cr
Coal-tar pitch volatiles	0.2 mg/m^3, as benzene solubles
Nickel sulfide roasting, fume and dust	1 mg/m^3, as Ni
Vinyl chloride	5 ppm
†Zinc chromates	0.01 mg/m^3, as Cr

A1b. Human carcinogens. Substances, or substances associated with industrial processes, recognized to have carcinogenic potential without an assigned TLV:

    4-Aminodiphenyl – Skin
    Benzidine – Skin
    β-Naphthylamine
    4-Nitrodiphenyl – Skin

For the substances in 1b, no exposure or contact by any route–respiratory, skin or oral, as detected by the most sensitive methods–shall be permitted. The worker should be properly equipped to insure virtually no contact with the carcinogen.

A2. Industrial substances suspect of carcinogenic potential for MAN.
Chemical substances or substances associated with industrial processes, which are suspect of inducing cancer, based on either:
(1) limited epidemiologic evidence, exclusive of clinical reports of single cases; or
(2) demonstration of carcinogenesis in one or more animal species by appropriate methods.

Acrylamide – Skin	0.03 mg/m^3
Acrylonitrile – Skin	2 ppm
Antimony trioxide production	—
Arsenic trioxide production	—
Benzene	10 ppm
Benzo(a)pyrene	—
Beryllium	2 µg/m^3
‡1,3-Butadiene	10 ppm
Carbon tetrachloride – Skin	5 ppm
Chloroform	10 ppm
Chlormethyl methyl ether	—
**Chromates of lead and zinc, as Cr	0.05 mg/m^3
Chrysene	—
3,3'-Dichlorobenzidine – Skin	—

Table 5.12 (*continued*)

Dimethyl carbamoyl chloride	—
1,1-Dimethylhydrazine – Skin	0.5 ppm
Dimethyl sulfate – Skin	0.1 ppm
Ethylene dibromide – Skin	—
Ethylene oxide	1 ppm
Formaldehyde	1 ppm
Hexachlorobutadiene	0.02 ppm
Hexamethyl phosphoramide – Skin	—
Hydrazine – Skin	0.1 ppm
4,4'-Methylene bis(2-chloroaniline) – Skin	0.02 ppm
†Methylene chloride	50 ppm
‡4,4'-Methylene dianiline	0.1 ppm
Methyl hydrazine – Skin	C 0.2 ppm
Methyl iodide – Skin	2 ppm
2-Nitropropane	10 ppm
N-Nitrosodimethylamine – Skin	—
N-Phenyl-beta-naphthylamine	—
Phenylhydrazine – Skin	5 ppm
Propane sultone	—
β-Propiolactone	0.5 ppm
Propylene imine – Skin	2 ppm
o-Tolidine – Skin	—
o-Toluidine – Skin	2 ppm
‡p-Toluidine – Skin	2 ppm
Vinyl bromide	5 ppm
Vinyl cyclohexene dioxide – Skin	10 ppm

For the above, worker exposure by all routes should be carefully controlled to levels consistent with the animal and human experience data (*see* Documentation), including those substances with a listed TLV.

** See Notice of Intended Changes
‡ 1986–1987 Adoption.
† 1986–1987 Revision or Addition.
(k) Fibers longer than 5 μm and with an aspect ratio equal to or greater than 3:1.

the organs. For example halogenated hydrocarbons have an affinity for lipid-rich organs such as the liver and brain. Once the body's defence mechanisms are overburdened critical organ concentrations are reached and disease results. Target effects are influenced by route of entry and level of exposure; many substances exhibit multiple target specificity, illustrated[34] by Table 5.13.

A physiological taxonomy of airborne contaminants is reproduced in Table 5.14 (after ref. 35). It should be noted that 'low' and 'high' concentrations depend on the toxicity of the material. A lethal concentration of a gas such as phosgene would be a very low concentration in comparison with a lethal concentration of, say, carbon dioxide.

There is hardly an organ that is not susceptible to disease on exposure to certain chemicals in sufficient dosage; the following small selection serves as an illustration.

Table 5.13 Selected high-volume substances and effects of exposure

Substance	Kidney	Liver	Central nervous system	Reproductive system	Pulmonary system	Skin
Acetone	x	x			x	x
Acrylonitrile					x	x
Ammonia	x	x			x	x
Asbestos					x	
Cresol	x	x	x		x	x
Dichloromethane	x	x	x		x	x
Diethylene glycol	x	x	x	x	x	x
2-Ethoxyethanol	x	x	x	x	x	x
Ethylene glycol	x	x			x	
Lead	x	x	x	x		
Mcthyl cthyl kctonc	x	x	x		x	x
2-Methyl-2, 4-pentanediol	x	x			x	x
Oxalic acid	x		x	x	x	x
Phenol	x	x	x		x	x
Sodium hydroxide					x	x
Sulphuric acid	x	x			x	x
Talc					x	x
1,1,1-Trichloroethane	x	x	x		x	x
Trichloroethylene	x	x	x		x	

Table 5.14 Physiological classification of airborne contaminants

Type of contaminant	Effect	Examples
Irritants At low concentrations produce discomfort and reddening of the body surface with which they are in contact. At high concentrations may produce corrosion and blistering of the surface with liberation of fluid from it.	***Respiratory tract*** *Short-term* Low concentrations: irritation of nose and throat, coughing exacerbations of naturally occurring lung disease. Higher concentrations: severe irritation, coughing, difficulty in breathing, pulmonary oedema (fluid in the lungs), death. *Long term* Permanent lung damage. ***Eyes*** Low concentration: slight stinging in eyes, tear production. High concentration: marked production of tears and or pus, damage to membranes and cornea.	Acrolein; sulphur dioxide; hydrogen chloride; chromic acid; formaldehyde; fluorine; chlorine; bromine; ozone; cyanogen chloride; phosgene; nitrogen dioxide; arsenic trichloride.

Table 5.14 (continued)

Type of contaminant	Effect	Examples
	Skin	
	Short-term Low concentration: itching, redness, soreness. High concentration: severe reddening, thickening (oedema), blistering.	Inorganic acids (chromic, nitric); organic acids (acetic, butyric); inorganic alkalis (sodium hydroxide, ammonium hydroxide, sodium carbonate); organic bases (amines); organic solvents; detergents; salts (nickel sulphate, zinc chloride); acids, alkalis, chromates.
	Long-term Skin thickening, cracking and dermatitis.	
Allergic sensitisers Single and repeated exposure is without hazard in many workers but a few react to the materials in an unusual way or to an unusual degree.	*Respiratory tract* Asthma-like symptoms	Epoxy-resins, picryl chloride, chlor-2, 4-dinitrobenzene; isocyanates; p-phenyl diamine; proteolytic enzymes.
	Skin Allergic dermatitis	Isocyanates; p-phenylene diamine; complex salts of platinum; cyanuric chloride.
Asphyxiants Interfere with the use of oxygen by tissue (inert)	*Simple anoxia* Substances produce a deficiency of oxygen in inhaled air. Causes rapid unconsciousness and death.	Argon; methane; hydrogen; nitrogen, helium.
	Toxic anoxia Substances interfere with the body's ability to transport or utilise oxygen in the tissues. Causes rapid unconsciousness and death.	Carbon monoxide; cyanogen, hydrogen cyanide; arsine; nitrites; aniline; dimethyl aniline; toluidine; nitrobenzene.
Anaesthetics and narcotics Depress the activity of the central nervous system	Headache, dizziness, loss of consciousness, respiratory or cardiac depression, death.	Acetylene, olefins ethers, paraffins, ketones, alcohols, esters.
Systemic poisons Substances which cause injury at sites other than, or as well as, the site of contact	Effect depends on the site of principal injury.	Halogenated hydrocarbons (affect many organs)

Table 5.14 (continued)

Type of contaminant	Effect	Examples
		Benzene, phenols (brain and bone marrow) ionising radiations (skin, gut, bone marrow). Carbon disulphide; (nervous system, heart) methanol; n-hexane; methyl-n-butyl ketone (nerves, brain); organophosphorus compounds, tetra-alkyl lead compounds (brain).
		Lead (bone marrow, brain); managanese (lungs); cadmium (lungs, testes); beryllium (lungs); mercury (kidneys).
		Arsenic (blood and many other tissues); phosphorus (bone); selenium (liver); fluorides (many organs).
Respiratory fibrogens On prolonged exposure produce fibrosis (stiffness) of the lungs.	Difficulty in breathing, shortage of breath, chronic cough, death	Free crystalline silica (quartz, tridymite, cristobalite); asbestos, talc, mica.
Carcinogens On frequent and prolonged exposure cancers may be produced	If untreated will cause death	Coal-tar pitch dust; crude anthracene dust; crude mineral oil; arsenic (skin cancers); asbestos; polycylic aromatic hydrocarbons; nickel ore; bis (chloro-methyl) ether; mustard gas (lung cancer); β-naphthylamine; benzidine; 4-amino-diphenylamine (bladder cancer); vinyl chloride monomer (liver cancer); mustard gas; wood dust (nasal cancer); benzene (blood cell cancer).
Inert	At low concentrations are without effect. At high concentrations inert gases produce a simple asphyxia while dusts increase airway resistance and cause difficulty in breathing	Methane; hydrogen; nitrogen; helium. Nuisance dusts – Portland cement, calcium silicate; corundum; sucrose; limestone; tin oxide.

Respiratory system

Since inhalation is the prime entry route of chemicals into the body exposure can readily lead to pulmonary disease. Persons with chronic lung disorders such as asthma, bronchitis, emphysema and pulmonary fibrosis may be especially at risk. Table 5.15 lists principal industrial toxicants causing lung disease as a result of inhalation.[2] Many gases or vapours are irritant to the respiratory tract, the actual site depending upon solubility and aggressiveness of the substance. Initially coughing may result but more prolonged exposure may lead to chronic bronchitis. The more soluble gases such as chlorine, sulphur dioxide, and ammonia and other fast-acting substances attack the upper airway passages. The toxicological effects of some of these gases are summarised in Chapter 15, Tables 15.7 and 15.8.

> A 19-year-old worker at a fertiliser plant was drenched in ammonia liquid as he mixed the alkali with phosphoric acid. The agitator on the mixer had failed causing a crust to develop on the liquid surface. As this was disturbed a violent reaction took place. The accident proved fatal and the cause of death was broncho-pneumonia and inhalation of toxic fumes.[36]

Less soluble gases such as other halogens (chlorine, bromine and iodine), ozone and phosphorus chlorides attack both the upper respiratory tract and the pulmonary tissue. Relatively insoluble compounds such as arsenic trichloride, nitrogen dioxide and phosgene penetrate deeper in the tract affecting the bronchioles and alveoli spaces; it is this phenomenon which results in their effects often being delayed. High concentrations of certain irritants cause severe damage to tissue resulting in inflammation, e.g. hydrogen fluoride. Accumulation of lung fluid (pulmonary oedema) can result within a few hours. High exposures can produce emphysema and fibrosis.

Polymer fume fever This is characterised by a sharp attack of chest tightness, choking and a dry cough and sometimes rigours may follow several hours after exposure to the thermal degradation products of PTFE (polytetrafluoroethylene) plastic. This is due to the strongly irritant mixture of aliphatic and cyclic fluorocarbon compounds produced between 250 and 300 °C.

> A mechanic suddenly developed mystery fevers. Blood tests, X-rays and scans proved negative. Further studies revealed that his temperature rose each afternoon but only on workdays and coincident with his starting maintenance work on a conveyor. Investigation showed that while he used grease to lubricate the conveyor his colleagues used PTFE aerosols. The man also smoked and rolled his own cigarettes. It transpired that PTFE contaminated his tobacco when making cigarettes and he subsequently inhaled polymer thermal degradation products when smoking. The mechanic

Table 5.15 Principal industrial toxicants producing lung disease through inhalation[2]

Toxicant	Chemical composition	Occupational source	Pulmonary damage
Asbestos	Fibrous silicates (Mg, Ca, and others)	Mining, construction, ship-building, manufacture of asbestos-containing materials	Asbestosis, lung cancer
Aluminium dust	Aluminium metal and small amount of Al_2O_3	Manufacture of aluminium products, fireworks, ceramics, paints, electrical goods, abrasives	Fibrosis
Aluminium	Al_2O_3	Manufacture of abrasives, smelting	Fibrosis initiated from short exposures
Ammonia	NH_3	Ammonia production, manufacture of fertilisers, chemical production, explosives	Irritation
Arsenic $Pb_3(AsO_4)_2$	As_2O_3, AsH_3 $Pb_3(AsO_4)_2$	Manufacture of pesticides, pigments, glass, alloys	Lung cancer, bronchitis, laryngitis
Beryllium	Be, $Be_2Al_2(SiO_3)_6$ (beryl, ore), Be(II) salts	Ore extraction, manufacture of alloys, ceramics	Dyspnoea, interstitial granuloma, fibrosis, cor pulmonale, chronic disease
Boron	B_2H_6, B_4H_{10}, B_5H_9	Chemical process	Acute CNS
Cadmium oxide (fume dust)	CdO	Welding, manufacture of electrical equipment, alloys, pigments, smelting	Emphysema
Carbides of tungsten titanium tantalum	WC TiC TaC	Manufacture of cutting edges on tools	Pulmonary fibrosis
Chlorine	Cl_2	Manufacture of pulp and paper, plastics, chlorinated chemicals	Irritation
Chromium (IV)	Na_2CrO_4 and other chromate salts	Production of Cr compounds, paint pigments, reduction of chromite ore	Lung cancer
Coal dust	Coal plus SiO_2 and other minerals	Coal mining	Pulmonary fibrosis
Coke oven emissions	Polycyclic hydrocarbons, SO_x, NO_x, and particulate mixtures of heavy metals	Coke production	Lung cancer

Table 5.15 (continued)

Toxicant	Chemical composition	Occupational source	Pulmonary damage
Hydrogen fluoride	HF	Manufacture of chemicals, photographic film, solvents, plastics	Irritation, oedema
Iron oxides	Fe_2O_3	Welding, foundry work, steel manufacture, haematite mining jewellery making	Diffuse fibrosis
Kaolin	$Al_4Si_4O_{10}(OH)_8$ plus crystalline SiO_2	Pottery making	Fibrosis
Manganese	MnO, Mn(II) salts	Chemical and metal industries	
Nickel	NiCO (nickel carbonyl), Ni, Ni_2S_3 (nickel subsulphide), NiO	Nickel ore extraction, nickel smelting, electronic electroplating, fossil fuel	Nasal cancer, lung cancer, acute pulmonary oedema (NiCO)
Osmium tetraoxide	OsO_4	Chemical and metal industry	
Oxides of nitrogen	NO, NO_2, HNO_3	Welding, silo filling, explosive manufacture	Emphysema
Ozone	O_3	Welding, bleaching flour, deodorising	Emphysema
Phosgene	$COCl_2$	Production of plastics, pesticides, chemicals	Oedema
Perchloro-ethylene	C_2Cl_4	Dry cleaning, metal degreasing, grain fumigating	Oedema
Silica	SiO_2	Mining, stone cutting construction, farming, quarrying	Silicosis (fibrosis)
Sulphur dioxide	SO_2	Manufacture of chemicals, refrigeration, bleaching, fumigation	
Talc	$Mg_6(SiO_2)OH_4$	Rubber industry, cosmetics	Fibrosis, pleural sclerosis
Tin	SnO_2	Mining, processing of tin	
Toluene 2,4-diiso-cyanate	2,4-diisocyanato-1-methylbenzene (CH_3-C_6H_3(NCO)$_2$)	Manufacture of plastics	Decrement of pulmonary function (FEV_1)
Vanadium	VO_5	Steel manufacture	Irritation
Xylene	dimethylbenzene ($C_6H_4(CH_3)_2$)	Manufacture of resins, paints, varnishes, other chemicals, general solvent for adhesives	Oedema

had suffered from polymer fume fever which disappeared once the cause had been traced and precautions taken.[37]

A man used an oxyacetylene torch to burn off a seized bearing; soon afterwards he and a colleague were rushed to hospital with high fever, headaches, breathing difficulties and other symptoms of distress which proved to be polymer fume fever. The bearings were plastic-coated with PTFE which decomposed when subjected to the cutting flame. A manufacturer's warning label on the bearings had become obliterated with time.[38]

Pneumoconiosis A group of lung diseases, termed the pneumoconioses, is caused by the inhalation over a period, and retention in the lungs, of various industrial dusts. The main diseases are[39]:

- Coalworkers' pneumoconiosis — Inhalation of coal dust. May be simple or complicated (i.e. with fibrosis).
- Asbestosis — Inhalation of fibres from chrysotile, amosite, crocidolite or anthophyllite asbestos.
- Silicosis — Inhalation of free silica dust, e.g. in foundries, pottery industry, mining.
- Mixed-dust pneumoconiosis — Inhalation of mixed dusts (e.g. iron and silica) or talc.

Siderosis due to the inhalation of iron oxide has in the past been considered relatively harmless but such dust/fume rarely arises in isolation, i.e. it may be accompanied by other dusts, such as silicates, or gases as in welding.

Inhalation of the dust from cobalt, used as a binder for tungsten carbide, may eventually result in a progressive diffuse interstitial pulmonary fibrosis termed 'hard-metal disease'.

A granulomatous disease, in which multiple accumulations of cells are distributed in nodules throughout the lungs (and other organs), has been caused by long-term exposure to beryllium, e.g. in the past among workers in the manufacture of fluorescent lights.

Granulomatous lesions, fibrosis and functional impairment of the lungs are also found in:

- Farmer's lung — e.g. among agricultural workers exposed to spores from mouldy hay.
- Byssinosis[39] — e.g. among workers exposed to dust from cotton, flax or hemp.
- Bagassosis (sugarcane lung) — e.g. due to exposure to dust from bagasse fibre.

Asthma Isocyanates, e.g. toluene di-isocyanate (TDI) used in polyurethane foam manufacture and less volatile isocyanates used in polyurethane paints,[40] can act as respiratory sensitisers.

A lady operator was employed in a factory manufacturing foam cushions using TDI. She worked for two months at the point where moulds were waxed after the foam cushions had been removed; in theory, therefore, there should have been no exposure to TDI. She suffered acute bronchospasm and was off sick at various times over a period of three weeks. The major attack occurred while assisting with cleaning plant near the mixing head and TDI storage drums.[41]

Diphenylmethane di-isocyanate has also been identified as a cause of asthma among workers in an iron and steel foundry where it was a constituent of a binder for mould making.[42] Less volatile isocyanates, or pre-polymers, have in general replaced TDI in paints, but the hazard remains if spray application is used, i.e. due to inhalation of mists.

Fumes from solder fluxes containing colophony (pine resin) may cause asthma[43] as may formaldehyde from the use of formalin solutions or other sources. Amine fumes from fluxes used in aluminium soldering have also caused asthma.

In the UK occupational asthma is now a prescribed disease for workers exposed to platinum salts; isocyanates in inks, paints, foam, etc.; epoxy resin hardeners; animals in research laboratories; colophony in soldering flux; flour in agriculture and food processing; and proteolytic enzymes in detergent manufacture.

Cancer, mesothelioma Cancer of the lung or pleura may result from repeated inhalation of certain substances such as fumes from nickel refining, coal tar, pitch, bischloromethyl ether, or forms of asbestos.

As far as the rare mesothelioma of the pleura is concerned, e.g. resulting from asbestos, due to exposures very much less than those associated with asbestosis,

> '... crocidolite has been more dangerous than chrysotile and anthophyllite. The position of amosite may be intermediate between crocidolite and chrysotile.' [44]

The current control limits for asbestos in the UK are chrysotile (white) 0.5 fibres/ml of air, amosite (brown) 0.2 fibres/ml and crocidolite (blue) 0.2 fibres/ml.

(In the UK, under the Asbestos (Prohibitions) Regulations 1985 the following are now prohibited.
- The import of crocidolite and amosite fibre.
- The supply of those minerals and products containing them as an article or substance for use at work.
- The use of those minerals and products containing them in the manufacture and repair of any other product.
- Asbestos spraying.
- Installation of new asbestos insulation.)

Skin

The scale of the problem associated with occupational skin diseases and their treatment was discussed on page 188. Some causes of occupational dermatoses are listed in Table 5.16 (after ref. 3) and the major disorders are prescribed diseases in the UK.

Table 5.16 Dermatoses related to occupational exposures to chemicals

Abrasive wheel makers
Carborundum
Emery
Resin glues

Aircraft workers
Adhesives (resins)
Alkalis
Bichromates, chromates
Chromic acid
Cutting fluids
Cyanides
Epoxy resins
Flame retardants
Glass fibre
Hydraulic fluids
Hydrofluoric acid
Nitric acid
Lubricants
Oils
Paints
Plastics
Rubbers
Solvents, thinners

Animal handlers
Antibiotics
Cleaners, detergents
Deodorants
Feeds
Germicides, biocides
Insecticides
Medicaments
Pesticides

Artists/painters
Acrylic resins
Epoxy resins
Paint removers
Pigments
Plasticisers
Thinners, solvents

Artists/sculptors
Dusts
Plaster of Paris
Polishes

Athletes
Adhesives
Antibiotics

Lime
Medications
Soaps

Bakers
Benzoyl peroxide
Cinnamon
Dough
Dusts
Flavours (oils)
Flour
Spices
Sugar

Barbers/hairdressers
Ammonium thioglycolate
Antiseptics
Bleach, e.g. ammonium persulphate
Cosmetics
Depilatories
Detergents
Dyes, e.g.
    p-phenylene diamine
    p-tolylene diamine
    o-nitro-p-phenylene diamine
Hair conditioners
Hair sprays
Hair straighteners
Hair tonics
Hydrogen peroxide solns.
Perfumes
Shampoos
Shaving creams
Wave solutions

Barmen
Citrus fruit
Detergents
Disinfectants
Flavours
Moisture
Soaps

Bath attendants
Deodorants
Hypochlorite solution
Linaments
Lotion
Oils
Soaps

Table 5.16 (continued)

Battery makers
Alkali
Cobalt
Epoxy sealer
Fibre glass plates
Mercury
Nickel
Pitch
Plastics
Solvents
Sulphuric acid
Zinc chloride

Bleachers
Borax
Chlorine compounds
Hydrochloric acid
Hydrogen peroxide
Oxalic acid
Per-salts
Potassium hydroxide
Sodium hydroxide
Solvents

Bookbinders
Formalin
Glues (natural and resin)
Inks
Shellac
Solvents

Brick masons
Cement
Chromates
Epoxy resins
Lime

Bronzers
Acetone
Ammonia
Ammonium sulphides
Amyl acetate
Antimony sulphide
Arsenic
Arsine
Benzine, benzol
Cyanides
Hydrochloric acid
Lacquers
Mercury
Methyl alcohol
Petroleum hydrocarbons
Phosphorus
Resins
Sodium hydroxide
Sulphur dioxide
Turpentine (or substitutes)
Varnishes

Broom/brush makers
Bleaches
Dust, vegetable
Dyes
Glues (natural and resin)
Pitch
Plastics
Rubber
Shellac
Solvents
Tar
Varnish
Woods

Butchers
Antibiotics
Brine
Detergents
Enzymes

Button makers
Dusts – animal, vegetable, mineral
Dyes
Hydrogen peroxide
Plastics

Cable workers
Chlorinated diphenyls
Chlorinated naphthalenes
Dyes
Epoxy resins
Solvents

Candle makers
Ammonium salts
Borax
Boric acid
Chlorine
Chromates
Hydrochloric acid
Potassium nitrate
Sodium hydroxide
Stearic acid
Waxes

Canners
Citrus oil
Dyes
Enzymes
Fruit acids and sugars
Resins
Sodium chloride
Vegetable juices

Cardbox box makers
Anti-flame agents
Dyes
Glues (natural and resin)
Mildew proofers
Waxes

Table 5.16 (continued)

Caretakers
Detergents, synthetic
Disinfectants
Polishes (essential oils)
Soaps
Solvents
Waxes

Carpenters
Bleaches
Glues (resin and natural)
Insulation agents
Oils
Polishes
Rosin
Shellac
Solvents
Stains
Woods

Carpet makers
Alizarine dye
Aniline dyes
Bleaches
Chlorine
Fungicides
Glues
Insecticides
Jute
Loom oils
Solvents

Carroters – felt hat
Acids
Mercury
Quinones

Case hardeners/metal treatment workers
Oils (quench)
Sodium carbonate
Sodium cyanide
Sodium dichromate
Sodium nitrate

Cellulose workers
Acids
Alkalis
Bleaches
Carbon disulphide
Finishing oils

Cement workers
Cement
Chromates
Cobalt
Epoxy resins
Lime
Pitch
Resins

Chemical workers
Acids
Alkalis
Amines
Chlorinated naphthalenes
Chlorinated solvents
Chromates
Cresols, phenols
Detergents
Formaldehyde
Nickel salts
Organic chemicals
Petroleum solvents

Chrome platers
Chromium compounds
Degreasers (solvents)
Metal cleaners (alkali)
Nickel compounds
Sulphuric acid

Clerks/typists
Adhesives
Carbon paper
Copy paper (carbonless)
Duplicating fluids
Duplicating materials
Indelible pencils
Ink removers
Inks
Rubber
Solvents
Type cleaner
Typewriter ribbons

Cloth preparers
Acids
Alkalis
Amino resins
Detergents, synthetic
Dyes
Flame retardants
Formaldehyde
Potassium salts
Soaps
Sodium metasilicate
Sodium salts
Sodium silicate

Coal-tar workers
Anthracene oil
Benzol
Coal tar
Creosote
Cresol
Naphtha
Pitch
Solvents

Table 5.16 (continued)

Confectioners
Chocolate
Citric acid
Dyes – food
Essential oils (flavours)
Flour
Fruits
Pineapple juice
Spices
Sugar
Tartaric acid

Construction workers
Adhesives, resin
Cement
Concrete
Creosote
Gasoline
Glass fibre
Oils
Paints
Pitch
Sealers
Solvents
Wood preservatives

Cooks
Fruit acids
Spices
Sugars
Vegetable juices

Dairy workers
Antibiotics
Deodorants
Detergents

Degreasers
Alkalis
Chlorinated hydrocarbon solvents
Petroleum solvents

Dentists
Anaesthetics, local
Antibiotics
Disinfectants
Eugenol
Mercury and metallic amalgams
Oil of clove
Resins
Soaps
Waxes

Disinfectant makers
Carbolic acid
Chloride of lime
Chlorinated phenols
Chlorine
Cresol
Formaldehyde
Iodine
Mercurials
Quaternary ammonium compounds
Surfactants
Zinc chloride

Dispensing chemists
Acids
Alkalis
Antibiotics
Bleaching powder
Detergents (synthetic)
Drugs
Iodoform
Soaps
Sugar

Dock workers
Castor bean pomace
Chemicals
Fumigants
Insecticides
Petroleum
Tar

Dry cleaners
Acetic acid
Ammonia
Amyl acetate
Dusts
Methanol
Nitrobenzene
Perchloroethylene, trichloroethylene
Sizing chemicals
Stoddard solvent
Turpentine (or substitutes)
Waterproofing chemicals

Dye makers
Acids
Alkalis
Antimony compounds
Benzine
Calcium salts
Coal-tar products
Cresol
Dextrins
Dye intermediates
Ferrocyanides
Formaldehyde
Gums
Hydroquinone
Lead salts
Potassium chlorate

Table 5.16 (continued)

Dyers
Acids
Alkalis
Bleaches
Detergents (synthetic)
Dyes
Mercurial salts
Solvents
Zinc chloride

Electric apparatus makers
Acids
Asbestos
Chlorinated diphenyls
Chlorinated naphthalenes
Enamels
Epoxy resins
Phenolic resins
Pitch
Rubber
Solder fluxes/colophony rosin
Solvents
Synthetic waxes
Varnishes

Electricians
Chlorinated diphenyls
Chlorinated naphthalenes
Epoxy resins
Solder fluxes, colophony rosin
Solvents
Waxes (synthetic)

Electroplaters
Acids
Alkalis
Benzine
Chromic acid
Lime
Nickel
Potassium cyanide
Soaps
Waxes (chlorinated)
Zinc chloride
Zinc cyanide

Enamellers
Acids
Alkalis
Antimony
Arsenic
Chromium
Cobalt
Nickel

Engravers
Acids
Alkalis
Chromic acid
Ferric chloride
Potassium cyanide
Solvents

Etchers
Alkalis

Explosive workers
Ammonium salts
Mercury compounds
Nitroglycerine
PETN
Picric acid
Tetryl
TNT

Farm workers
Antibiotics
Detergents (synthetic)
Disinfectants
Feeds
Fertilisers
Fungicides
Lubricants
Oils
Paints
Pesticides
Solvents
Tar
Weed killers
Wood preservatives

Felt hat makers
Acids
Bacteria
Dyes
Glauber's salt
Hydrogen peroxide
Mercuric nitrate
Sodium carbonate

Fertiliser makers
Acids
Ammonium compounds
Calcium cyanamide
Castor bean pomace
Fluorides
Lime
Nitrates
Pesticides

Table 5.16 (continued)

Phosphates
Potassium salts

Fish dressers
Brine
Detergents
Sodium chloride

Florists
Fertilisers
Herbicides

Flour mill workers
Chemical bleaches
Dusts
Flour
Pesticides

Food preservers
Bleaches
Brine
Resins
Spices
Sugar
Vinegar
Waxes

Foundry workers
Acids
Lime
Resin binder systems
Solvents

Fur processors
Acids
Alkalis
Alum
Bleaches
Chromates
Dyes
Formaldehyde
Lime
Oils
Salt

Furniture polishers
Acids
Alkalis
Benzine
Essential oils in polish
Methyl alcohol
Naphtha
Pyridine
Rosin
Soaps
Solvents

Stains
Turpentine
Waxes

Galvanisers
Acids
Alkalis
Zinc chloride

Garage workers
Antifreeze
Detergents (synthetic)
Epoxy resins
Gasoline
Gasoline additives
Glass fibre
Greases
Oils
Paint removers
Paints
Solvents

Gardeners
Fertilisers
Fungicides
Herbicides
Insecticides
Sodium chlorate

Glass workers
Arsenic
Borax, boric acid
Glass fibre
Glass wool
Hydrofluoric acid
Lead components
Lime
Metallic oxides
Petroleum oils
Resins
Soda ash

Heat treatment workers
Oils
Sodium carbonate
Sodium cyanide
Sodium dichromate
Sodium nitrate

Histology technicians
Alcohol
Aniline
Benzol, toluene
Epoxy resins
Formaldehyde
Glyoxal

Table 5.16 (continued)

Glutaraldehyde
Mercury bichloride
Osmium tetroxide
Potassium dichromate
Stains
Waxes
Xylene

Horticultural workers
Seed dressing/dressed seed, e.g. mercury compounds

Hospital domestic staff
Allantoin
Acids
Alkalis
Benzalkonium chloride
Chlorfamine
Chlorhexidine chloride
Chromium, chromic sulphate
Detergents
EDTA
Formaldehyde
Glutaraldehyde
Hydrochloric acid
Hypobromic acid
Hypochlorite compounds
Hypobromite compounds
Nickel
Phosphoric acid
Potassium hydroxide
Rubber
Soap
Sodium carbonate
Sodium metasilicate
Sodium perborate
Sulphonic acid
Triethanolamine
Trisodium phosphate

Ink makers
Anti-skinning agents (anti-oxidants)
Chrome pigments
Cobalt compounds (driers)
Detergents, synthetic
Dyes
Mercurial pigments
Resins
Soaps
Solvents
Turpentine
Varnishes

Insecticide makers
Aldrin
Allethrin

Arsenic trioxide
Calcium arsenate
Chlordane
DDT
Dieldrin
Lindane
Malathion
Methoxychlor
Parathion
Piperonyl compounds
Pyrethrin
Stobane

Jewellers
Acids
Adhesives, resin
Chromium
Cyanides
Mercury
Mercury solvents
Nickel
Rouge
Solder flux

Kitchen staff
Detergents (synthetic)
Grease
Soaps
Water softeners

Laboratory workers, chemical
Acids
Alkalis
Chromates
Detergents (synthetic)
Dusts
Gases
Organic chemicals
Poisons
Soaps
Solvents

Laundry workers
Alkalis
Bactericides
Bleaches
Chemical dusts
Detergents (synthetic)
Enzymes
Fibre glass
Fungicides
Optical brighteners
Soaps

Linoleum makers
Asphalt
Dyes

Table 5.16 (continued)

Oils
Pigments
Resins
Solvents

Machinists
Greases
Anti-oxidants
Aqueous cutting fluids, synthetic cutting fluids
Chromates
Germicides, biocides
Chlorinated cutting oils
Cutting oils
Lubricants
Rust inhibitors
Soluble cutting fluids
Solvents
Hydraulic oils

Match factory workers
Ammonium phosphate
Chromates
Dextrins
Dyes
Formaldehyde
Glues
Gums
Phosphorus sesquisulphide
Potassium chlorate
Red phosphorus
Waxes

Mercerisers
Acids
Alkalis

Metal polishers
Abrasives
Acids
Alkalis
Ammonia
Naphtha
Pine oil
Potassium cyanide
Soaps
Soluble oils
Solvents
Triethanolamine
Waxes

Mirror makers
Acids
Ammonia
Cyanides
Formaldehyde
Lacquers

Silver nitrate
Solvents
Tartaric acid
Varnishes

Mordanters
Acids
Alkalis
Aluminium salts
Antimony compounds
Arsenates
Chromates
Copper salts
Iron Salts
Lead salts
Phosphates
Silicates
Tin salts
Zinc chloride

Motor vehicle workers (assembly)
Adhesives
Antifreeze
Brake fluids
Brake linings
Flame retardants
Gasoline
Hydraulic fluids
Oils
Rubber
Solvent

Motor vehicle workers (body)
Abrasives
Adhesives
Alkalis
Paints
Rubber compounds
Solder
Solvents

Motor vehicle workers (mechanic)
Acids
Adhesives
Alkalis
Antifreeze
Brake fluids
Brake linings
Cleansers
Detergents
Epoxy resins
Gasoline
Hydraulic fluids
Lubricants
Rubbers
Solvents, thinners

Table 5.16 (continued)

Nickel platers
Acids
Alkalis
Degreasers
Detergents, synthetic
Nickel sulphate
Zinc chloride

Nitroglycerine makers
Ethylene glycol dinitrate
Nitric acid
Nitroglycerine
Sodium carbonate
Sulphuric acid

Nurses
Anaesthetics, local
Antibiotics
Antiseptics
Detergents, synthetic
Disinfectants
Drugs
Ethylene oxide
Formalin
Rubber gloves
Soaps
Tranquillisers

Optical workers
Alkalis
Grinding fluids
Oils
Turpentine

Packinghouse workers
Antibiotics
Brine
Detergents
Synthetic enzymes

Paint makers
Anti-mildew agents
Chromates
Coal-tar distillates
Driers
Fish oils
Latex
Oil, vegetable
Petroleum solvents
Pigments
Plasticisers
Resins
Thinners
Turpentine
Zinc chloride

Painters
Acetone
Acids
Alkalis
Benzine
Chlorinated hydrocarbons
Chromates
Driers
Paint strippers
Paints, oil base
Paints, resin
Pigments
Solvents, thinners
Turpentine

Paraffin workers
Paraffin
Paraffin distillates
Solvents

Pencil makers
Aniline dyes
Chromium pigments
Glues
Gums
Lacquer
Lacquer thinners
Methyl violet
Pyridine
Resins
Solvents
Waxes

Petroleum recovery workers
Acids
Alkalis
Brine
Crude petroleum
Explosives
Ionising radiation
Lubricating oils

Petroleum refinery workers
Acids
Alkalis
Aluminium chloride
Gas oil
Gasoline
Hydrofluoric acid
Kerosene
Paraffin
Paraffin distillates
Petroleum
Petroleum solvents
Tar
Waxes

Photoengravers
Ammonium bichromate
Etching acids
Inks

Table 5.16 (continued)

Photodevelopers
Solvents

Photographers
Acids
Alkalis
Benzine
Chlorinated hydrocarbons
Chromates
Hydroquinone
Methyl *para*-aminophenol sulphate
para-Aminophenol
Para formaldehyde
para Phenylenediamines
Pyrogallic acid
Sodium hypochlorite
Sodium sulphide
Turpentine

Physicians
Adhesives
Anaesthetics (local)
Antibiotics
Antiseptics
Detergents (synthetic)
Drugs
Formalin
Soaps
Tranquillisers

Pipeline layers
Burns
Fluxes, welding
Solvents
Tar

Pitch workers
Pitch
Solvents
Tar

Plasterers
Lime

Plastics/resin makers
Catalysts
Glass fibre
Epoxy resins
Polyester resins
Solvents
Phenol-formaldehyde

Plumbers
Adhesives
Caulking compound
Fluxes, solder
Hydrochloric acid
Lead vapours

Paint fumes
Parasites
Solvents
Tar
Zinc chloride

Railroad workers (shop)
Alkalis
Antiseptics
Chlorinated hydrocarbons
Chromate (anti-oxidants)
Cutting fluids
Detergents (synthetic)
Diesel fuel oil
Greases
Insecticides
Lacquers
Lubes
Magnaflux (fluorescein)
Paint
Paint strippers
Paint thinners
Solvents

Railroad workers (track)
Creosote
Fungicides
Herbicides
Pitch
Tar

Rayon workers
Acetic anhydride
Acids
Alkalis
Ammonium sulphide
Bleaches
Calcium bisulphite
Carbon disulphide
Coning oils
Sodium cyanide
Sodium sulphide
Sodium sulphite
Solvents

Refrigeration workers
Ammonia
Brine
Chromates
Ethyl bromide
Ethyl chloride
Glass fibre
Methyl chloride
Sulphur dioxide

Road workers
Asphalt
Cement

Table 5.16 (continued)

Concrete
Epoxy resins
Herbicides
Paint
Pitch
Tar

Rocket fuel handlers
Aniline
Boron hydrides
Chlorine trifluoride
Dimethylhydrazine
Ethyl oxide
Fuming nitric acid
Gasoline
Hydrazine
Hydrogen fluoride
Hydrogen peroxide
Kerosene
Liquid oxygen

Rubber workers
Accelerators
Acids
Activators
Adhesive removers
Alkalis
Antimony
Anti-oxidants
Benzol
Chloroprene dimers
Chromium pigments
Formaldehyde
Oils
Plasticisers
Resins
Retarders
Soaps
Solvents
Tar
Turpentine
Zinc chloride

Shipyard workers
Chlorinated diphenyls
Chlorinated naphthalenes
Chromates
Fungicides
Glass fibre
Paint removers
Paint thinners
Paints
Resins
Solvents
Tar
Welding fumes
Wood preservatives

Shoe manufacturers
Adhesives
Ammonia
Amyl acetate
Amyl alcohol
Aniline dyes
Benzene
Benzol
Fungicides
Hexane
Naphtha
Resins
Rubbers
Shoe polishes
Tanning agents
Waxes

Soap makers
Alkalis
Bacteriostats
Bleaches
Detergents (synthetic)
Oils, vegetable
Perfumes

Solderers
Acids
Cyanides
Fluxes/colophony
Hydraxine salts
Rosin
Zinc chloride

Stockyard workers
Insecticides

Stone workers
Cement
Dusts
Lime

Sugar refiners
Acids
Burlap
Heat
Jute
Lime
Sugar

Tannery workers
Acetic acid
Alum
Ammonium chloride
Arsenic salts
Benzol
Brine
Calcium hydrosulphide
Chromium compounds

Table 5.16 (continued)

Dimethylamine
Dyes (mineral)
Dyes (vegetable)
Formaldehyde
Lime
Oils
Pancreatic extract
Sodium hydroxide
Sodium sulphide
Solvents
Sulphuric acid
Tannin

Tar workers
Heat
Pitch
Solvents
Tar

Taxidermists
Arsenic salts
Calcined alum
Mercuric chloride
Solvents
Tannin
Zinc chloride

Tinners
Paint
Pitch
Zinc chloride

Undertakers
Formaldehyde
Mercury
Oil of cinnamon
Oil of cloves
Phenol
Thymol
Zinc chloride

Upholsterers
Flame retardants
Glues
Lacquer
Lacquer solvents
Methyl alcohol

Veterinary workers
Anaesthetics, local
Antibiotics
Deodorants
Drugs
Mercuric chloride
Pesticides
Soaps and deodorants

Watchmakers
Acids
Chromates
Metal polishes
Nickel
Potassium cyanide
Rouge
Solvents

Waterproofers
Aluminium sulphate
Melamine formaldehyde resins
Oils
Paraffin
Pitch
Resin paints
Rubber
Solvents
Tar
Waxes

Welders
Fluxes
Metal fumes

Wire drawers
Alkalis
Drawing oils
Lime
Soaps
Sulphuric acid

Wood preservers
Chlorophenols
Chromates
Copper compounds
Creosote
Cresols
Mercuric chloride
Phenylmercuric compounds
Resins
Tar
Zinc chloride, zinc sulphate

Wood workers
Acid bleaches
Amino resin glues
Epoxy glues
Fillers
Formaldehyde
Lacquers
Oil stains
Paints
Phenolic resin glues
Rosin
Solvents
Varnishes

The commonest form of industrial skin complaint is dermatitis, characterised by an inflammatory response, and classed as irritant dermatitis, allergic contact dermatitis or photo-allergic dermatitis. The majority of industrial forms of dermatitis are caused by primary irritants.

Chemicals which attack tissue extensively are referred to as corrosive, and in some cases will also damage the materials of protective clothing or even constructional materials. An extended list of these materials is reproduced as Table 5.17 though severity will depend upon a range of factors such as chemical type, concentration, duration of exposure, temperature, etc.

Aggressive substances such as strong acids and alkalis produce effects within moments. Strong acids, such as sulphuric or nitric acid, become hydrated by the water content of the skin and can combine with skin protein to form albuminates. Mineral acids are corrosive and may cause burns and charring quickly. Other compounds may have more delayed effects.

> Phenol deposited on the skin does not give rise to an immediate burning sensation; the skin will often go white and become anaesthetised and insensitive to touch. The severity of the burn depends upon the period of contact; the true extent of it is only evident many hours afterwards.[45]

Phenolic compounds, including catechol and hydroquinone, may depigment the skin by damaging or destroying the melanocytes. Phenolics in adhesives and disinfectants have caused this effect.

Alkalis such as hydroxides of sodium and potassium, and to a lesser extent their carbonates, produce deep burns and ulceration in concentrated form. In dilute solution dermatitis may result. Sodium hydroxide is a particularly effective solvent for keratin.

Some substances react violently with the skin to produce blistering as well as inflammation. Severe exposure can kill cells and possibly lead to secondary bacterial invasion. An example of cell death is chronic ulceration of hands and nasal septum associated with chrome-plating.

Milder irritants include organic solvents and detergents. Detergents, emulsifying agents, wetting agents, anti-foaming agents and solubilisers lower the surface tension of aqueous systems and improve fat-solubility properties of the solution. Surfactants contain a hydrophobic chain linked to a hydrophilic head group as illustrated in Table 5.18. They can produce dryness, fission and dermatitis by removing the protective oily layer of the skin and aiding penetration but may require frequent or prolonged contact before clinically recognisable effects develop.

As discussed in Chapter 6 removal of irritants from the skin, i.e. reduction of contact time, is an important precaution.

Table 5.17 Examples of corrosive chemicals

Acids and anhydrides
Acetic acid, acetic anhydride, acid mixtures, battery fluids, chlorosulphonic acid, chromic acid, chloroacetic acid, dichloroacetic acid, fluoroboric acid, fluorsilicic acid, hydrobromic, hydrochloric, hydrofluoric and hydriodic acids, methacrylic acid, nitric acid, nitrohydrochloric acid, perchloric acid, phenolsulphonic acid, phosphorus pentoxide, propionic acid, selenic acid, spent acids, sulphamic acid, sulphuric acid and oleum (fuming sulphuric acid), sulphurous acid, thioglycolic acid, trichloroacetic acid.

Alkalis
Ammonium hydroxide, potassium hydroxide (caustic potash), quaternary ammonium hydroxides, sodium hydroxide (caustic soda).

Halogens and halogen salts
Aluminium bromide and chloride, ammonium bifluoride and other bifluorides, antimony trichloride, pentachloride and pentafluoride, beryllium chloride, boron trichloride, bromine, chlorine, calcium fluoride, chromic fluoride, chromous fluoride, iron chlorides (ferric chloride, ferrous chloride), fluorine, iodine, lithium chloride, phosphorus oxybromide and oxychloride (phosphoryl bromide and chloride), phosphorus trichloride and pentachloride, phosphorus sulphochloride (thiophosphoryl chloride), potassium fluoride and bifluoride, potassium hypochlorite, pyrosulphuryl chloride, sodium chlorite, sodium fluoride, sodium hypochlorite, stannic chloride, sulphur chloride, sulphuryl chloride, thionyl chloride, titanium tetrachloride, vanadium dichloride, zinc chloride.

Interhalogen compounds
Bromine trifluoride and pentafluoride, chlorine trifluoride, iodine monochloride.

Organic halides, organic acid halides, esters and salts
Acetyl bromide, allyl chloride and allyl iodide, acrylonitrile monomer, allyl chloroformate, allyl iodide, ammonium thiocyanate, anisoyl chloride, benzyl chloride, benzhydryl bromide (diphenyl methyl bromide), benzoyl chloride, benzyl bromide, butyl acid phosphate, benzyl chloroformate (benzyl chlorocarbonate), chloroacetyl chloride, ethyl chloroformate (ethyl chlorocarbonate), dibromoethane (ethylene bromide), 1,2-dichloroethane (ethylene chloride), ethylene oxide, fumaryl chloride, ethyl chloroformate (methyl chlorocarbonate), propionyl chloride, iso-propylchloroformate, diisooctyl acid phosphate, *p*-chlororbenzyl chloride, chloropropionyl chloride, sodium fluorosilicate.

Chlorosilanes
Allyl trichlorosilane, amyl trichlorosilane, butyl trichlorophenyltrichlorosilane, cyclohexyl trichlorosilane, dichlorophenyl trichlorosilane, diethyl trichlorosilane, diphenyl dichlorosilane, dodecyl trichlorosilane, hexadecyl trichlorosilane, hexyl trichlorosilane, methyl trichlorosilane, nonyl trichlorosilane, octadecyl trichlorosilane, octyl trichlorosilane, phenyl trichlorosilane, trimethyl trichlorosilane, vinyl trichlorosilane.

Miscellaneous corrosive substances
The following corrosive substances are widely used but do not fall into any of the above classes: ammonium sulphide, benzene sulphonyl chloride, benzyl dimethylamine, beryllium nitrate, catechol, chlorinated benzenes and toluenes, chlorobenzaldehyde, chlorocresols, cresols, cyclohexylamine, dibenzylamine, dichlorophenol, diethyl sulphate, diketene, dimethyl sulphate, hexamethylenediamine, hydrazine, hydrogen peroxide, organic peroxides, phenols, soda lime, sodium aluminate, sodium amide, sodium bisulphate, sodium bisulphite, sodium chromate and dichromate, sodium pyrosulphate, sodium hydride, triethyltetramine, tritolyl borate, silver nitrate.

Proprietary mixtures, e.g. cleaning, disinfecting, bleaching, degreasing solids or solutions based on these chemicals are corrosive to a degree dependent upon dilution.

Table 5.18 Examples of surfactants

Hydrophobic chain	Hydrophilic head group	Class
$CH_3-(CH_2)_n-$	$-CO_2^-Na^+$	anionic
$CH_3-(CH_2)_n-$	$-OSO_3^-Na^+$	
$CH_3(CH_2)_n\underset{CH_3}{CH}-\langle O \rangle-$	$-SO_2^-Na^+$	
$CH_3(CH_2)_n-$	$-N^+(CH_3)_3Cl^-$	cationic
$CH_3(CH_2)_n-$	$-(OCH_2CH_2)_xOH$	non-ionic

A workman was engaged on dirty work demolishing buildings and boilers. This brought him into contact with dust, metal particles and oil. Adequate washing facilities were not provided. The man eventually contracted dermatitis affecting both hands. Whilst he could not prove that the dermatitis was a direct result of his working conditions his employers were held liable.[46]

The pH of the skin can become alkaline. Sebaceous glands and hair follicles are particularly vulnerable to fat solvents. Substances such as TNT dust may become trapped by hairs and produce follicular irritation.

'Oil acne' is a common form of skin irritation which may occur on any part of the body where there is contact with oil or oily clothing. Other skin rashes, generally on the hands and arms, may occur depending on the composition of the oil. Fuel oils, e.g. diesel oil cause dermatitis while the typical local effect of kerosene and the lighter fractions is to cause damage by repeated degreasing of the skin, which also renders it prone to attack by other agents.

Allergic cutaneous sensitisation accounts for about 20% of occupational dermatitis cases. The key differences from primary irritation are ones of time and mechanism. Generally, skin sensitisers produce no effect on first exposure but induce cellular changes. On subsequent contact with the same, or a closely related material, an acute dermatitic reaction may be provoked. The reaction may develop in some workers even when exposure is slight while others may show no effect even at high exposures.

Sensitisation has been reported to gold and potassium cyanide solutions among wokers involved in plating electrical circuits.[47] Other important industrial skin sensitisers are given in Table 5.19.[48] When such reactions are stimulated by light the effect is termed photo-sensitisation. Examples include coal-tar constituents and certain antibacterials.

Once an injury resulting from exposure to an irritant heals there is eventually no increased risk from re-exposure; in contrast a person

Table 5.19 Principal skin sensitisers in industry

Coal-tar and its direct derivatives
Acridine
Anthracene
Carbazole
*Cresol
Fluorene
Naphthalene
Phenanthrene
*Phenol
Pyridine

Dyes
Amido-azo-benzene
Amido-azo-toluene
Aniline black
Auramine
Bismarck brown
Brilliant indigo, 4 G.
Chrysoidine
Crystal and methyl violet
Erio black
Hydron blue
Indanthrene violet, R.R.
Ionamine, A. S.
Malachite green
Metanil yellow
Nigrosine
Orange Y
Paramido phenol
Paraphenylendiamine
Pyrogene violet brown
Rosaniline
Safranine
Sulphanthrene pink

Dye intermediates
Acridine and compounds
Aniline and compounds
Benzanthrone and compounds
Benzidine and compounds
Chloro compounds
Naphthalene and compounds
Naphthylamines
Nitro compounds

Explosives
Ammonium nitrate
Dinitrophenol
Dinitrotoluol
Fulminate of mercury
Hexanitrodiphenylamine
Lead styphnate
Picric acid and picrates
Potassium nitrate
Sensol
Sodium nitrate

Trinitromethylnitramine (Tetryl)
Trinitrotoluene

Insecticides
*Arsenic compounds
Creosote
*Fluorides
*Lime
*Mercury compounds
Nicotine
Organic phosphates
*Petroleum distillates
*Phenol compounds
Pyrethrum
Rotenone
Tar

Natural resins
Burgundy pitch
Copal
Dammar
Japanese lacquer
Pine rosin
Wood rosin

Oils
*Cashew nut oil
Coconut oil
Coning oils (cellosolves, eugenols)
Cutting oils (the inhibitor or antiseptic they contain)
Essential oils of plants and flowers
Linseed oil
*Mustard oil
Sulphonated
Tung oil

Photographic developers
Bichromates
Hydroquinone
Metol
Para-amido-phenol
Paraformaldehyde
Paraphenylendiamine
Pyrogallol

Plasticisers
Butyl cellosolve stearate
Diamyl naphthalene
Dibutyl tin laurate
Dioctylphthalate
Methyl cellosolve oleate
Methyl phthalylethylglycola
Phenylsalicylate
Propylene stearate
Stearic acid
Triblycol di (2, ethyl butyrate)

* Compounds which also act as primary irritants

Table 5.19 (continued)

Rubber accelerators and anti- oxidants	
Guanidines	Chloro-naphthalenes
Hexamethylene tetramine	Chlorophenols
Mercapto benzo thiazole	Cumaron
Ortho-toluidine	Epoxies
Para-toluidine	Melamine formaldehyde
Teramethyl thiuram monosulphide and disulphide	Phenol formaldehyde
Triethyl tri-methyl triamine	Polyesters
	Sulphonamide formaldehyde
Synthetic resins	Urea formaldehyde
Acrylic	Urethane
Alkyd	Vinyl
Chlorobenzols	
Chlorodiphenyls	**Others**
	Enzymes derived from *B. subtilis*

exhibiting allergic cutaneous sensitisation must be protected from all further exposures to even minute quantities to prevent adverse response.

> A worker handling a rubber anti-oxidant became sensitised and was apparently removed from all possible contact with the material. His dermatitis did not disappear. However, he was still wearing garters, the elastic of which was contaminated. When these were discarded he recovered.[49]

Other skin problems include skin cancer but although more cancers occur on the skin than any other site, the number of occupational origin is uncertain. Benign lesions include asbestos, and tar, warts. Known skin carcinogens include anthracene, inorganic arsenic, asphalt, coal-tar, creosote oil, mineral oils, pitch, shale oil and tar.

> Long-term exposure to mineral oils can result in ulcers which may undergo malignant change. Such epitheliomatous ulceration is rarely found in petroleum refinery workers but cases have occurred in the engineering industries notably among tool-setters and operators exposed over a number of years to lubricating oils.[50]

The skin can be affected systemically by substances entering the body by other routes, for example, exposure to thallium can result in loss of hair and arsenic exposure can cause a proliferation of keratin.

In the routine handling and use of chemicals it is always best to err on the side of caution and avoid skin contact so far as possible. For example exposure to a degreasing agent, e.g. a common solvent, may render the skin liable to attack by another irritant. Commonly, irritant dermatitis starts at a site where skin has been damaged, i.e. a cut or lesion, so that personal protection and first aid are important preventive measures as discussed in Chapter 6.

Eyes (and ears)

The basic structure of the eye is shown in Fig. 2.7. Since chemicals come into contact with the eyes by accidental splashing, by rubbing the eye with dirty hands or contaminated protective gloves, or as a result of atmospheric exposure to gas, vapours, dusts and mists, ocular routes are also important to consider. Indeed the cornea and conjunctiva are directly exposed to the external milieu.

In order to function the cornea must remain transparent. A scar – the body's normal repair process – small enough to be insignificant on other organs, could destroy the functioning of the cornea. Thus a tiny amount of corrosive chemical reaching the cornea and conjunctiva can cause blindness. Generally the cornea and conjunctiva are more sensitive to mild irritants and more severely damaged by stronger irritants than is the skin.

Damage by acids is a function of pH and the ability of the anion to combine with protein molecules. For some concentrated acids, e.g. sulphuric acid, the severity of injury is also influenced by their power to dehydrate and their heat of dehydration. The most effective first-aid treatment is immediate irrigation of the eye with copious volumes of water followed by medical attention. Often, the delay caused by looking for special eye-wash solution can be serious.

> The normal method of transferring sulphuric acid to acid tanks on a plant was by pumping it up 4-m pipes. When the automatic pump failed a hand-operated bellows pump was substituted. However, the joint between the pipe and pump failed under the pressure of acid and the operator's spectacles were knocked off by the stream of escaping acid. He sustained burns to his head, neck and shoulders and lost the sight of his left eye: the sight of his right eye was also threatened.[51]

Strong alkalis are widely encountered in industry and in the home and these can cause severe damage if splashed in the eye. Effects can be more delayed than with acids with infiltration, ulceration and perforation occurring up to 1 week from the time of exposure. Severity is influenced by pH and duration of exposure and is generally insensitive to the nature of the cation.

Many neutral organic solvents such as ethyl alcohol, ether, ethyl acetate, hexane, etc., cause pain on contact with the eye as a result of their ability to de-fat epithelial cells. Damage is rarely extensive or long lasting except in extreme cases or if hot solvents are involved.

Substances such as dinitrophenol cause opacity in the lens while methanol and thalium can damage the optic nerve. Methanol can cause permanent blindness by destruction of the retina; this may be caused by formaldehyde liberated by the alcohol dehydrogenase in the retina. Some substances are classed as inert yet produce systemic effects when absorbed in sufficient quantities (e.g. fluoroacetates).

Detergents can cause eye irritation though the effect is apparently not related to their ability to reduce surface tension. In general, cationic surfactants are more damaging to the eye than are anionic actives and both are more aggressive than non-ionics. Some surfactants, especially cationics at high concentrations, can cause severe burns with permanent opacity and vascularisation. In accidents rapid washing with water prevents permanent damage.

Vesicant war gases such as the thio mustards and nitrogen mustards can produce severe eye damage.

Heavy-metal ions can combine with protein groups and in high enough concentrations metal salts can cause tissue destruction. Splashes of such materials into the eye could cause corneal opacity and ulceration.

Some materials are claimed with varying degrees of confidence[52] to have systemic effects on the eyes, e.g. inorganic arsenic and certain drugs.

Table 5.20 lists reported minimum concentrations of certain airborne chemicals causing eye effects.[53]

An authoritative text[52] lists the toxic effects involving the eyes for hundreds of individual chemicals.

Table 5.20 Reported minimum concentrations at which 'eye effects' of certain substances have been observed[53]

Compound	Concentration (ppm)
Acetaldehyde	30
Acetone	500
Ammonia	140
Bis (2-chloroethyl) ether	260
Carbon dioxide	> 20%
Carbon disulphide	30
Chlorine	3
Ethyl alcohol	0.75%
Fluorides	5 mg/m^3
Formaldehyde	20
Hydrochloric acid	0.05 mg/l
Isopropyl alcohol	800
Methyl ethyl ketone	350
Ozone	1
Phosgene	4
Phosphorous trichloride	2
Styrene	200
Sulphur dioxide	20
Sulphur	6
Toluene	300
Trichloropropane	100
Turpentine	175
Xylene	200

Note that exposure to butyl alcohol is reported to aggravate the degree of occupational deafness to workers simultaneously exposed to noise, compared with a control group exposed to the same noise levels in the absence of alcohol exposure.[54,55]

Blood

Blood is composed of three main types of cells, viz. red cells, white cells and platelets.

- Red cells (erythrocytes) transport oxygen from alveoli of lungs to peripheral tissue by loose union with the blood pigment haem (shown in Fig. 5.5) which links with the protein, globin. These blood cells also transport carbon dioxide in the reverse direction for excretion via the lungs.
- White cells (leucocytes) aid recognition of foreign matter and provide a defence mechanism by phagocytosis and antibody production.
- Platelets (thrombocytes) assist blood coagulation and thereby provide the first line of defence against blood loss.

In healthy human blood there are about 4.9×10^6 red cells/mm^3, 5,000–10,000 white cells/mm^3 and 250,000 platelets/mm^3. However, blood and its cellular elements are susceptible to disease and they exhibit changes as a consequence of disease in most other organs. Table 5.21 (pp. 245–6) lists some occupational poisons detectable by blood studies. Cell deficiencies are termed anaemia, leucopenia or thrombocytopenia for red, white or platelet cells respectively. In some cases, chemical exposure affects one type of cell only. For example, arsine gas destroys red cells (haemolysis) and lead causes anaemia by affecting the blood-forming process in the bone marrow. Other materials such as benzene, antimetabolites, mustards, arsenic, chloramphenicol, trinitrotoluene, gold, hydantoin derivatives, phenyl butazone, and ionising radiations are toxic to bone marrow and can deplete the level of all three cells, a condition known as pancytopenia. Some chemicals such as benzene may induce cancer of blood cells (leukaemia).

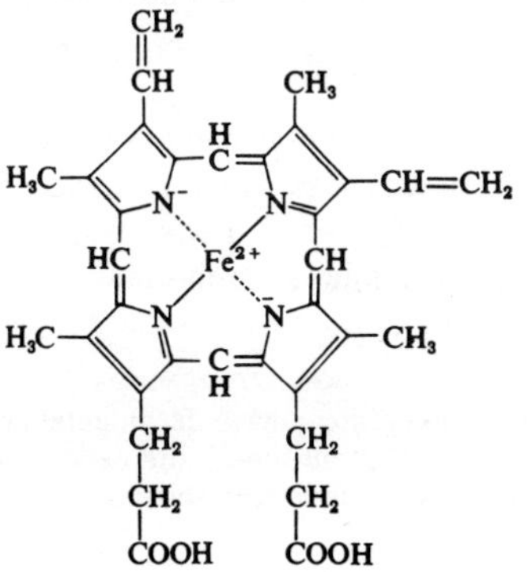

Fig. 5.5 Haem

Asphyxia occurs when as a result of interference with the respiratory function there is a diminished supply of oxygen to the blood and tissues, and usually a diminished elimination of carbon dioxide from the lungs. This may be caused by a simple asphyxiant gas, a category including most of the 'inert' gases (e.g. nitrogen, argon, helium) and some flammable gases (e.g. hydrogen, methane, ethylene, butane, propane, acetylene) and is termed 'simple anoxia'.

The inert gases have no colour, smell or taste and deficiency of oxygen may arise from, for example[56]:

- Use of nitrogen or argon to exclude air from vessels.
- Use of carbon dioxide fire extinguishers in a confined space.
- Excessive generation of, e.g. nitrogen or helium gas from cryogenic liquids.
- Leakage of argon from an argon arc welding set in an unventilated enclosure.
- Formation of rust inside a closed steel tank (oxygen is removed from the atmosphere by the oxidation of iron).
- Neutralising vessel contents with carbonate or bicarbonate, displacing the air with carbon dioxide.

Responses at given concentrations of oxygen in the atmosphere at sea level have been summarised as[57]:

16–21%   No noticeable effect.*
12–16%   Increased respiration, slight diminution of co-ordination.
10–12%   Loss of ability to think clearly.
6–10%   Loss of consciousness, death.

Thus to reduce the oxygen content to a fatal level requires a simple added asphyxiant gas concentration of about 50%.

Data have been published on the physiological effects of reduced oxygen concentrations.[58] The insidious nature of asphyxiation from oxygen starvation has been emphasised.[59]

> A welder was using electric welding gear inside a compartment of an 18,000-litre road tanker. The compartment measured 85 cm wide and was ventilated by an extractor fan. During the lunch break he switched off the ventilator fan and left the welding gear in the compartment. The inerting gas, argon, leaked during this time and displaced the air. The welder returned to work but was soon seen to have collapsed and was lying face upwards displaying signs of sickness. He was hauled from the tank by a rescue team and eventually recovered. It was estimated that with the fan off it would have taken only ten minutes for the air to become unbreathable.[60]

---

* Air containing less than 19.5% oxygen can have detrimental effects, if the body is already under stress (e.g. low pressure at high altitudes) and exposure to air containing less than 18% should not be permitted under any circumstances.

A horizontal cylindrical vessel of 1.8 m diameter and 6,800 litre capacity had been pressure tested with carbon dioxide. A charge-hole cover joint was leaking and was therefore removed to remake the joint. In doing so a fitter's mate slipped and fell into the vessel which contained residual gas from the pressure test. Although he was rescued soon afterwards he was found to be dead.[61]

An experienced supervisor climbed about 24 metres to the top manhole platform of a column. He went to investigate whether there was an adequate flow of water from a hosepipe into the column through the open manhole. Sodium nitrate solution in the column was being roused with nitrogen so the column had a nitrogen atmosphere. Shortly afterwards the man was found with his head and shoulders inside the manhole. He had died of asphyxiation.[62]

Certain amines, nitrites and carbon monoxide also produce asphyxia by reacting with haemoglobin, thereby reducing the blood's capacity to transport oxygen. This is called 'toxic anoxia'. Thus, carbon monoxide, a colourless odourless gas produced in the incomplete combustion of fossil fuel (e.g. inadequately maintained or badly flued boilers), like oxygen, also unites with haemoglobin in erythrocytes to form carboxy-haemoglobin. As a consequence this depletes the blood's oxygen content (anoxia) and leads to fainting and death within minutes of exposure to a high concentration of carbon monoxide. The clinical symptoms associated with various percentage degrees of saturation of haemoglobin with carbon monoxide are[63]:

0–10%	No symptoms.
10–20%	Tightness across the forehead, possibly headache, flushed skin, yawning.
20–30%	Headache, dizziness, palpitation on exertion.
30–40%	Severe headache, weakness, dizziness, nausea, possibly collapse.
40–50%	Ditto, with increased respiration and pulse, with more possibility of collapse and syncope.
50–60%	Syncope, coma, Cheyne–Stokes respiration.
60–70%	Coma, weakened heart and respiration, possible death.
70–80%	Respiratory failure and death.
90%	Prompt cardiac arrest.

The limit of safety is 18–20%. The percentage carboxyhaemoglobin can be assessed roughly in terms of concentration in the atmosphere, time of exposure, initial blood concentration and degree of activity.[64]

The body of a charwoman was discovered in a fume-filled kitchen by office staff arriving for work. Normal emergency procedures failed to resuscitate her. Death was due to carbon monoxide poisoning due to inhalation of fumes from a coal-fired boiler in a kitchen at the rear of the office block. The boiler was found to be in good working order but the exhaust duct was obstructed by detached cement and soot causing slow accumulation of fumes.[65]

A company commenced manufacture of a product for stomach-upsets. The instructions from the previous makers of the substance were followed but when the ethyl nitrate ingredient was added to the solution the mixture bubbled and gave off fumes. The vapour affected the man directly involved in the process and four girls in an adjoining room all turned 'a blue-green colour' because the chemical affected the ability of the blood to take up oxygen. Those affected were given oxygen and transferred to hospital.[66]

Methylene chloride, which is used in paint removers and degreasers, has also resulted in carbon monoxide poisoning – due to conversion within the body.[67]

Cyanosis can be produced by other materials such as aniline and other aromatic compounds due to the oxidation of haemoglobin to methaemoglobin.

First-aid measures include removal from the contaminated atmosphere and the application of artificial respiration to increase the blood's oxygen level by conversion of carboxyhaemoglobin into oxyhaemoglobin. Mouth to mouth resuscitation is without danger to the first-aider so long as the victim's skin is contamination-free.[65]

Cyanides and cyanogen also interfere with oxyhaemoglobin by inhibiting cytochrome oxidases and again toxic anoxia results. Immediate first aid includes moving the patient to fresh air and providing amyl nitrate to inhale.

It has been suggested that blood and plasma are the target organs in connection with metal fume-fever, since this may well be the site of the immune response. This disease also termed 'brass-founders ague' is characterised by chills, fever, nausea, vomiting, muscular pains, dryness in the mouth and throat, fatigue, headache and weakness. It results from the inhalation of minute particles of various metal oxides. The most common causes arise from copper, zinc and magnesium; manganese, tin, iron nickel, selenium, and antimony have also been implicated. The operations involved include welding operations, smelting of zinc and copper and their alloys, and galvanising.

A man had been engaged in melting gold-plated springs using a fluoride flux to recover gold. Unbeknown to him the springs were made of beryllium/copper. Shortly after completing the task tightness of the chest developed and he felt cold and perspired. He became cyanosed, dyspnoeic, developed bronchospasm and was admitted to hospital for treatment. He was discharged 7 days later and the diagnosis rested between pneumonitis due to metal fumes or to the fluoride flux.

Table 5.21 Occupational poisons which may produce abnormalities detectable by studies of the blood

Substance	Findings	Comments
Acrylonitrile	Anaemia, leucocytosis	Reported but not definitely established
Allylisopropyl acetyl carbamide	Thrombopenia	
Aniline	Anaemia, stippled red cells, leucocytosis	Findings differ in acute and chronic exposure
Antimony	Leucopenia	Effects may resemble those of arsine
Arsenic	Anaemia, leucopenia	Industrial poisoning may be caused by arsine
Benzene	Decrease in all formed elements	Death may result from depression of the bone marrow
Carbon disulphide	Anaemia, leucocytosis, immature WBC*	Conflicting reports in medical literature
Carbon tetrachloride	Anaemia, leucopenia	Questionable
Cobalt	Increased RBC*	Animal experiments only
DDI	Anaemia, leucopenia	Rarely occurs
Ethyl silicate	Anaemia, leucocytosis	Animal experiments only
Ethylene glycol monomethyl ether	Decrease in all formed elements, increased percentage of immature WBC*	Based on human observations
Ethylene oxide		Questionable
Fluorides	Anaemia	
Lead	Mild anaemia, stippling of RBC*	Well established
Manganese	Leucopenia	Questionable
Mercury	High haemoglobin	Reported but unconfirmed
Methylchloride	Decrease in formed elements	Animal experiments
Nitrobenzenes (nitrophenols)	Reduced RBC* with signs of regeneration	Due to blood destruction
Nitrous fumes	Decreased WBC*	Reported but unconfirmed
Phenylhydrazine	Anaemia with signs of regeneration	Due to blood destruction
Radium	Decrease in all formed elements	Well established in humans
Selenium	Anaemia	Animal experiments
Tetrachloroethane	Signs of blood destruction	Not fully established in humans
Thallium	Increased WBC*	Reported but unconfirmed
Thorium	Decrease in all formed elements	Due to radioactivity

* WBC = white blood cells;
  RBC = red blood cells.

Table 5.21 (*continued*)

Substance	Findings	Comments
Toluene	Anaemia, leucopenia	Mild when compared with benzene
Toluene diamine	Anaemia	
Toluidine	Anaemia	Questionable
Trichloroethylene	Anaemia	Reported but unconfirmed
Trinitrotoluene	Anaemia, leucopenia	Well established
Uranium	Anaemia, leucopenia	Due to radioactivity
Vanadium	Anaemia	Questionable
Xylene	Anaemia, leucopenia	Questionable

Gastrointestinal tract and liver

As mentioned earlier the gastrointestinal tract can be the route of entry for chemicals into the body, e.g. by accidental swallowing. It can also be a target organ. Any substance which can cause skin burns, e.g. corrosive materials, can damage the oesophagus and stomach if ingested.

Gastroenteritis and increased motility of the intestines result from ingestion of arsenic, barium, chloromethane or fluorides.

Liver functions are mainly associated with metabolism. It is the first organ to be exposed to a chemical absorbed from the gastrointestinal tract but its vulnerability depends upon the nature of its metabolic products. Thus, though the liver normally detoxifies chemicals, some parent materials or their metabolic products may damage this organ. The acute effects of chemicals on the liver are either to produce an accumulation of lipids (fatty liver) or cell death (necrosis). Chronic exposure to hepatotoxic substances can cause alterations of the whole liver structure with degenerative and proliferative changes associated with the different forms of cirrhosis. Thus some chemicals induce morphological changes and also depress metabolic activity of the liver, bile flow and excretion of bilirubin. The latter is essential for control of haemoglobin. Compounds such as carbon tetrachloride, selenium, arsenic or tetrachloroethane cause severe liver damage and jaundice. Vinyl chloride can cause liver cancer (angiosarcoma).

Table 5.22 lists materials known, or suspected of, possessing hepatotoxic properties.

Kidneys and urinary tract

Though the kidney has several roles the main function of this organ is excretion of non-nutrients including detoxified chemicals produced in the

Table 5.22 Some known or suspected hepatotoxins

Acetaminophen	Cobalt	Methyl formate
Acrylonitrile	Cycloheximide	Methylene chloride
Aflatoxin	Cycloparaffins	Nitrobenzol
Allylalcohol	DDT	Phenacetin
Allylformate	Dioxane	Phenol
Antimony	Dimethyl aminoazobenzene	Phenylhydrazine
Arsenic	Dimethyl formate	Phosphorus
Beryllium	Dimethyl nitrosamine	Pyrrolizidine alkaloids
Bromobenzene	Dinitrophenol	Tannic acid
Cadmium	Diphenyl	Tetracycline
Carbon disulphide	Ethanol	Tetrachloroethane
Carbon tetrachloride	Ethylene chlorohydrin	Thioacetamide
Cerium	Ethylene dichloride	Trichloroethylene
Chlordane	Galactosamine	Trifluorochloroethylene
Chlorinated diphenyls	Hydrazine	Trinitrotoluene
Chlorinated naphthalenes	Methyl bromide	Uranium
Chloroform	Methyl chloride	Urethane

liver. However, the kidney is particularly vulnerable because it is a dynamic organ often insulted by high concentrations of chemicals. Thus a series of compounds can cause injury, notably a range of organics (such as carbon tetrachloride, chloroform, bromobenzene, trichloroethylene and trichlororethane) and most heavy metals. Indeed cadmium with a half-life of 10–40 years in the kidney means this organ can be affected for a lifetime as a result of short-term exposure. Uranium has caused kidney damage. Aromatic mercury compounds are used in some fungicides and can release inorganic mercury in tissue. The toxicity of ethylene glycol (used in certain antifreeze products) may be attributable to its metabolism into oxalic acid. Aromatic amines such as β-naphthylamine and benzidene can cause bladder cancer, and in the UK strict legislation prohibits or controls the use of these compounds.[68]

Bone and muscle

Chronic and acute inflammation of bone can be induced by some chemicals. An important example of necrosis of jawbone is that caused by phosphorus. Excessive accumulation of fluoride and phosphorus increases bone fragility. In contrast cadmium uptake can cause depletion of the calcium content resulting in bone-softening and deformation. Compression of nerves by deformed bones can be painful. Lead may be deposited in bone, though it is usually the peripheral and central nervous system which is affected and in some cases kidney damage may result. Examples of substances which can attack the bone marrow are chloramphenicol, sulphonamides, phenylbutazone, perchlorates, benzene and certain alkylating agents.

As a result of carbon monoxide poisoning, loss of muscle power may lead to the victim becoming trapped at the location of exposure.

Nervous system

The nervous system may be considered in two parts:
1. Central nervous part incorporating the brain and spinal cord.
2. Peripheral part consisting of the motor and sensory nerves.

The central nervous system, especially the brain, is protected by the blood–brain barrier. However, this barrier is penetrated easily by primary anaesthetics and certain drugs. The effect depends upon concentration in the brain together with specific pharmacologic action. The potency of simple alcohols increases with molecular weight, peaking with amyl alcohol, which reflects solubility in the brain. The presence of polyhydroxy groups reduces solubility and anaesthetic effect. Substitution of hydrogen by halogen increases anaesthetic properties and also toxicity in other organs.

> A worker was found slumped over a degreasing tank and did not respond to attempts to revive him. The tank contained 1,1,1-trichloroethane solvent and although the concentration of the vapour in the workplace atmosphere was below the Threshold Limit Value (350 ppm) the concentration immediately below the rim of the degreasing tank was 6,000 ppm when not operational and over 70,000 ppm when the liquid solvent was disturbed. Inhalation of trichloroethane at levels above 70,000 ppm is sufficient to be rapidly fatal. The operator had decided to wash his hands in the solvent before going for lunch and had been overcome by the vapour.[69]

For compounds causing anoxia (see page 243) such as carbon monoxide and cyanides the brain, the tissue most dependent on a continuous supply of oxygen for survival, is the prime target organ. If the exposure is not fatal, recovery is usually complete and any residual disability is dictated by the degree of cerebral structural damage following the hypoxia. Azides and nitrogen trifluoride also affect the central nervous system (CNS).

Hydrogen sulphide is a gas whose toxicity tends to be under-rated: This gas may create a hazard, from sour crudes, in petroleum recovery and refining and when anaerobic decomposition of sulphur-containing materials occurs, e.g. in sewers and wells. Very low concentrations can be recognised by the characteristic smell of rotten eggs but only when first encountered; even at low concentrations prolonged or repeated exposure tires the sense of smell so that odour is no longer a reliable warning. Higher concentrations which have a sweetish odour immediately paralyse the olfactory nerves. The asphyxiant action of the gas is related to its ability to paralyse the respiratory centres of the brain resulting in cessation of respiration. The physiologic response to various concentrations varies with individuals but typical effects are[10,33]:

Parts per million H_2S	Response
0.2	Detectable odour
20–150	Conjunctivitis
150	Olfactory nerve paralysis
250	Prolonged exposure may cause pulmonary oedema
500	Systemic symptoms may occur in 0.5 to 1 hour
1,000	Rapid collapse, respiratory paralysis imminent
5,000	Immediately fatal

Three tannery workers became unconscious, one for over 48 hours, after collapsing beside a drum of sodium sulphide solution in which hides were being immersed to remove surface hair. Overnight build-up of fumes were such that the first worker to lift the drum lid was overcome by concentrated fumes. The initial concentration was such that these undispersed fumes caused two rescuers to be overcome. A second rescue party arrived within minutes but were not affected, presumably because by that time the vapours had dispersed. The hides were apparently acidic which resulted in an interaction with the sulphide solution to liberate hydrogen sulphide gas.[65]

The central nervous system is particularly susceptible to attack by organometallic compounds such as triethyl tin, organophosphorus and organomercury compounds, and certain carbamates. Some can provoke neurological damage.

Certain organophosphorus compounds (e.g. triorthocresyl phosphate) inhibit cholinesterase and disturb nerve function. They can also produce delayed structural damage in cells of the central nervous system. When triorthocresyl phosphate was inadvertently used as cooking oil in Morocco a mass epidemic of paralysis occurred. Tetraethyl lead, an anti-knock agent added to petrol, is soluble in lipids and can produce striking changes to the brain function and behaviour.

Carbon disulphide acts mainly on the nervous system; it is more selective on the central nervous system than most other solvents. It interferes with metabolism of catecholamines – substances aiding the transmission of an impulse from one nerve cell to the next. Most occupational exposures have been in the rayon and rubber industries. The part of the brain affected is the same as that damaged by manganese. The latter affects the basal nuclei and produces a syndrome like Parkinson's disease. More selective interaction between substances and brain biochemical processes occur from exposure to a range of substances. Effects include cerebral oedema from triethyl tin, encephalopathy from lead, ataxia, blindness, deafness, etc., from methyl mercury, and behavioural disorders including tremor from mercury vapour. Exposure to mercury vapour may result in reversible damage to the CNS while organomercury compounds (e.g. methyl and ethyl derivatives used as fungicides) have a more lasting effect on the cerebral and cerebellar cortex.

Other substances capable of affecting the CNS include thallium, tellurium, hexanedione, bromophenylacetylurea, DDT.

In addition to affecting the CNS some chemicals may affect the extremities to produce peripheral neuropathy. Examples include carbon disulphide and hexane.

Risk evaluation

General

The first steps in Risk Evaluation involve determining the materials in a process, the nature and potency of their toxic effects, and the levels and duration of human exposure. Exposures are deduced from a knowledge of the process and by monitoring. Thus the first stage is to assess what specific chemical, or combinations of chemicals, are involved or likely to be generated in the process or product use under consideration. This may require consideration of the basic chemistry and physics (as in Ch. 3) and confirmation by monitoring (Ch. 7).

> Most cases of industrial poisoning involving nitrogen dioxide arise in the manufacture of nitric and sulphuric acids and of explosives and other nitro-compounds and various nitrates, from nitric acid spillages, and from dipping of copper or brass articles in nitric acid. Nitrous fumes are also liberated by the slow burning or incomplete combustion of nitro-explosives and in the combustion of nitrated compounds. Nitrogen dioxide is generated in maize silos (silo-fillers' disease) and by the combination of atmospheric oxygen and nitrogen in electric arc welding.[70] Nitrous fumes are also produced when metals are welded or cut by oxy-acetylene, propane or butane flames; they are produced, in fact, whenever such a flame is burning whether it is in contact with steel or just burning in air.[71]

> Although trichloroethylene and perchlororethylene are classed as non-flammable solvents, another danger can arise when the vapours of these solvents are exposed to sources of high temperature in the presence of air. Heat from a flame or red-hot surfaces decomposes the vapour with the formation of corrosive and toxic gases.[72] These gases, including phosgene and chlorine dioxide, may also be produced by decomposition associated with welding operations.

A guide to incompatible chemicals which give off toxic gases is given in Table 10.3 and a guide to the toxic thermal degradation products of polymers in Table 15.8.

The next stage is an evaluation of the potential for harm, or toxic properties, of the range of substances. Having recognised the nature of potential adverse effects, an assessment of the risk associated with a specific operation or process is required to determine the appropriate control measures. The level of risk is a function of the 'probability' of exposure and the likely 'consequences' of such an event. Sources and levels of emission of chemicals from a range of unit operations and processes are identified in Chapter 6. Formal techniques for evaluating the 'likelihood' of exposure included Hazard and Operability Studies (HAZOPS) and Hazard Analyses (HAZANS) (see Ch. 12). These are normally reserved for complex arrangements such as chemical plants. With regard to assessing the 'implications' of exposure there are critical concentrations of the substance in question which cause the body functions to fail and the onset of disease: individual susceptibility must be borne in mind.[73]

Such critical concentrations can include a deficiency of essential chemicals, such as oxygen. The composition of air at sea level is approximately 78% vol. nitrogen, 21% vol. oxygen and 1% vol. argon plus other inert gases, carbon dioxide, water vapour and impurities such as dust, fluorocarbons, etc. The consequence of exposure to oxygen-deficient atmospheres is considered dangerous below 16% (see page 242).

Conversely, excess of essential and non-essential substances provokes adverse responses. This concept of dose was recognised around AD 1500 by Paracelsus who wrote,

> 'All things are poisons, for there is nothing without poisonous qualities. It is only the dose which makes a thing poison.'

Thus all materials entering the body in sufficient quantity can exert a toxic effect, even 'essential' materials such as water.

> A hospital patient who believed his food and medicines to be poisoned virtually starved himself and started compulsively drinking up to 35 pints of water a day in an attempt to clean his system. He died of water on the brain and water intoxication.[74]

Toxicity tests

Thus, having identified the type of toxic effect associated with a specific chemical the next stages in assessing the risk are an appreciation of the 'potency' of the substance and an awareness of the actual levels of material to which people are exposed (see Ch. 7). Toxicity data are derived from epidemiology studies, occasionally from experiments involving humans, but most commonly from studies involving experimental animals. Sources of data include the literature, suppliers, or in-house toxicity studies. It is inappropriate here to discuss the ethics of *in vivo* tests. Nevertheless, an awareness of test procedures is essential as an aid to the correct interpretation of the results.

The route of entry used to administer chemicals in such tests must be carefully considered since it can influence the concentration reaching the target organ. Thus, intravenous (IV) injection introduces the material directly into the circulation and the organs receive the initial impact almost immediately. If administered by another route the body may have time to develop some resistance, physiological adjustment then resulting in a modified response. For example, intraperitoneal (IP) injection of the substance into abdominal fluids is subject either to metabolic transformation in the liver or is excreted via the bile before reaching the general circulation. Higher LD_{50} values (lower toxicity) by IP compared to IV injection suggest these two routes play a significant role by the body in handling the chemical.

A large battery of tests is available and the following serves only as an illustration; the reader is referred elsewhere for more detailed considerations (e.g. ref. 75 which also provides guidelines on test procedures for evaluating the risk to the environment as a result of accumulation or degradation).

Acute toxicity

These tests are used to determine the short-term toxic effects of a chemical and study the relationship between dose and adverse effects, usually with the aim of enabling a median lethal dose to be calculated as illustrated by Fig. 5.6.

Whether this relationship holds for all substances and all effects is questioned. It has been suggested that for carcinogens a linear response passing through the origin is more appropriate than the sigmoid curve illustrated in Fig. 5.6 thus denying the existence of a 'no-effect' level.

The ED_{50}, the medium effective dosage, is that dose required to produce a specified effect in 50% of test animals. The most commonly encountered notation is the LD_{50}. This is the calculated dose which would cause the death of 50% of the entire population of an experimental

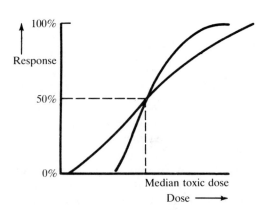

Fig. 5.6 Dose/response relationship for toxic substances

species, as determined from the exposure of a significant number of that population to the substance. (Extremes of the curve indicate at one end the dose at which all animals would die and at the other a dose which all animals would survive.) The LD_{50} value is usually expressed in terms of the weight of 'poison' per unit of body weight, usually mg of chemical per kg of animal. The reported value should always be accompanied by an indication of the species of animal used, route of administration and vehicle used to dissolve or suspend the material if applicable, and the time period over which the exposure was observed.

For example, Acetic acid: oral; rat; rodents; LD_{50} 3,310 mg/kg

For gases the dose is normally expressed as ppm or mg/m^3 which represents the concentration of material in the exposure chamber. Therefore, the term adopted for gases is usually LC_{50}.

For example, Ammonia: inhalation; mouse; rodents; LC_{50} 4,230 ppm/one hour.

The nature of the toxic effect can also vary with dose as illustrated by Fig. 5.7 for different exposure levels of chlorine.[76]

The tests identify highly toxic substances and provide data on the possible hazards which could arise as a result of human exposure.

The relative hazards of different exposure pathways are assessed using different routes of administration of the test substance and attempts have been made to link data from acute toxicity studies using different routes of exposure to assign chemicals a hazard rating (see Table 5.23).

In the UK the following scheme is used in the Classification, Packaging and Labelling of Dangerous Substances Regulations, 1984 and based on EEC guidelines.

Category	LD_{50} absorbed orally in rat mg/kg	LD_{50} absorbed percutaneously in rat or rabbit mg/kg	LC_{50} absorbed by inhalation in rat mg/litre (4 hours)
Very toxic	≤25	≤50	≤0.5
Toxic	>25 to 200	>50 to 400	>0.5 to 2
Harmful	>200 to 2000	>400 to 2000	>2 to 20

Reference 16 quotes toxic doses (including LD_{50}, LC_{50}, eye and skin irritation, etc.) for thousands of substances. However, the LD_{50} test is increasingly under criticism as a meaningful indicator of human toxicity (after ref. 2) because:

- Compound metabolism in test animals may differ between species and indeed from that in man.
- Test conditions may vary between compounds.
- Unlike human communities, the homogeneity of the experimental animal colony is assured by strict control.
- At best the test indicates lethality and is useless as an index for health protection.

- Substances of equal LD_{50} value may differ significantly in their toxicity spectrum either side of the median lethal dose, as shown in Fig. 5.6.
Acute oral toxicity – usually determined within a short period after the

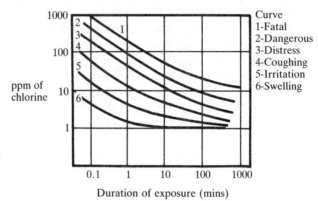

Fig. 5.7 Effects of exposure to chlorine

Table 5.23 Toxicity rating system

Toxicity rating	Commonly used term	LD_{50} Single oral dose for rats (g/kg)	4 hr Vapour exposure causing 2 to 4 deaths in 6-rat group (ppm)	LD_{50} Skin for rabbits (g/kg)	Probable lethal dose for man
1	Extremely toxic	0.001 or less	Less than 10	0.005 or less	Taste (1 grain)
2	Highly toxic	0.001 to 0.05	10 to 100	0.005 to 0.043	1 teaspoon (4 cc)
3	Moderately toxic	0.05 to 0.5	100 to 1,000	0.044 to 0.340	1 oz (30 g)
4	Slightly toxic	0.5 to 5.0	1,000 to 10,000	0.35 to 2.81	1 pint (250 g)
5	Practically non-toxic	5.0 to 15.0	10,000 to 100,000	2.82 to 22.6	1 quart
6	Relatively harmless	above 15.0	above 100,000	above 22.6	above 1 quart

test substance has been administered as a single dose or multiple doses given over 24 hours. The test compound is usually administered by gavage using a stomach tube. The observation period lasts at least 14 days.

Acute dermal toxicity – determined by applying the test substance in graduated doses to the skin of several groups of experimental animals. The substance is held in contact with the skin for 24 hours and again the animals are observed for a minimum of 14 days.

Acute inhalation toxicity – for gases, vapours and particulate matter of respirable dimensions is evaluated by single uninterrupted exposures to the chemical for up to 24 hours in special inhalation chambers.

Acute dermal irritation and corrosion – determined by application of the test substance in a single dose to the skin of experimental animals. The degree of irritation is scored at specified intervals using a 0 to 4 scale where 0 represents no effect and 4 the most severe effect. The observation period is not normally in excess of 14 days. Irritation is considered as reversible inflammation whereas corrosion is the production of irreversible tissue damage.

Acute eye irritation or corrosion – determined by a single dose of the test substance to the anterior surface of one of the eyes of the experimental animal and the magnitude of the effects described by a score.

Allergic sensitisation – these tests are designed to determine the potential for compounds to illicit sensitisation reactions following entry of the material in to the body via skin absorption or inhalation. Though the allergic response may arise from a variety of mechanisms, all involve at least one exposure to initiate the sensitisation process. Various methods are available for assessing skin sensitisation potential. In general these involve subjecting the animal to the test substance and then after a period of not less than 1 week to a challenge exposure to the test compound to establish whether a hypersensitive state has been induced. Sensitisation is determined by examining the reaction to the challenge insult after 24 hours and 48 hours.

Short-term repeated dose and subchronic toxicity – a more common form of human exposure to many chemicals involves repeated doses which do not produce immediate toxic effects. However, delayed effects may arise from, e.g. accumulation of the substance in body tissues. These effects are evaluated by subchronic testing usually over periods of 14, 28 and 90 days. They are designed to provide data on toxic effects, target organs, the reversibility or otherwise of adverse reactions, and an indication of the 'no effect concentration'. The most common tests are for oral toxicity (feeding trials) though tests are available for assessing subchronic toxicity via other routes of entry such as inhalation.

Reproductive toxicity – More and more attention is being directed at assessing the potential for harm to the reproductive system including fertility and teratogenicity as a result of exposure to chemicals. Accordingly, tests are continuously being devised and refined for assessing the hazards in this area.

Carcinogenicity – The potential for chemicals to induce cancer in experimental animals is determined by observing the test animals for a major part of their lifespan for the development of neoplastic lesions during or following exposure to the test substance by an appropriate route. Some aspects of the potency were discussed on page 204.

The use of experimental animals is the most reliable test method available for determining carcinogenic potential of chemicals. However, because of the vast number of chemical developments coupled with the fact that these tests can take two years or so, this approach is impracticable for evaluating every new substance prepared. Moreover, for weak carcinogens the resolving ability of these tests is low and large numbers of animals are required to demonstrate statistical significance. As a consequence mutagenicity tests are used to screen for potential carcinogens. Thus the 'Ames test' is based on the reverse mutation of several strains of the bacterium *Salmonella typhimurium*. These mutants require histidine for growth but after exposure to a mutagenic chemical, reverse mutations occur. These enable the bacteria to exist on a histidine-free medium. Chemicals requiring metabolism for activation can be detected by adding a mammalian liver fraction which contains the enzymes which may be necessary for metabolic activation of the chemical. A positive result is one where a significant excess number of colonies is produced compared with the number given by controls. Positive results especially if substantiated by other short-term tests such as the appearance of chromosomal abnormalities in cultured mammalian cells suggest that the chemical should undergo further long-term testing in experimental animals. Where positive tests are obtained in animal experiments critical evaluation of the data is crucial with respect to species and strain of animal used, dosage rate, route of administration, etc.

Chronic toxicity

The aim of these tests is to characterise the toxicity profile of a chemical to mammalian species following prolonged or repeated exposure. The duration of these studies for effects other than neoplasia is debatable. The general toxic effects explored include neurological, physiological, biochemical and haematological effects plus any pathological changes resulting from exposure to the test substance. Since these studies usually require dose–response relationships and no-effect levels, at least three dose concentrations are used. Frequency of exposure is normally daily though this will be influenced by the route of administration.

Hygiene standards

The next stage in assessing the risk is to determine actual levels of exposure to the chemical. The only way of guaranteeing the total safety and health of a worker is complete removal of harmful substances. Such an approach is rarely practicable. With regard to airborne chemicals some materials possess low odour thresholds and give warning of pending

danger. Reliance on the nose as an indicator, however, can be hazardous since:
- Some materials of low odour threshold may paralyse the olfactory nerves and cause the sense of smell to be lost within minutes (e.g. hydrogen sulphide).
- Some materials are odourless (e.g. nitrogen).
- Some materials such as arsine, hydrogen cyanide, nitric oxide, toluene di-isocyanate, phosphine and stibine may be present at concentrations in excess of their hygiene standard yet not detectable by smell. Table 5.24 lists a selection of substances with Threshold Limit Values (TLVs) below odour concentrations.[77]

Hygiene standards have been devised to provide guidance on the upper limit of acceptable levels of airborne substance to which it is believed most workers can be repeatedly exposed without risk of adverse effect. Because of the wide variation in individual susceptibility a small fraction of workers may experience discomfort at exposures equal to, or even less than, the standard. Indeed a small percentage may be affected more seriously by aggravation of a pre-existing condition or by development of an illness. Such values, though varying from country to country, (see Table 5.25) (after ref. 79), rely heavily on the dose–response principle and the concept of 'no-effect' thresholds. Supporting information for standards varies from chemical to chemical, both in terms of quantity and of reliability. Some are based on human experiences, whilst others are arrived at by extrapolation from animal experiments or by analogy with related substances (see Table 5.26).[80]

Also, the effect against which protection is aimed varies between substances, as illustrated in Table 5.27.[80] The magnitude of built-in safety margins varies between standards. Ranking standards clearly does not

Table 5.24 Comparison of TLV and odour thresholds for a selection of chemicals

Chemical	TLV (ppm)	Odour threshold (ppm)
Acrolein	0.1	0.61
Acrylic acid	10	110
Arsine	0.05	0.1
Diborane	0.1	2.5
Ethylene dichloride	10	88
Formic acid	5	49
Methyl acrylonitrile	1	7
Methyl formate	100	600
Methyl isocyanate	0.02	2.1
Phosgene	0.1	0.9

The 100% odour recognition concentrations based on reference 78 are much higher than those quoted above, e.g. methyl formate 100% ORC = 2,000.

Table 5.25 Comparison of selected chemical substances and their Occupational Exposure Limits for 18 countries. Values quoted are mg/m^3 and are taken from the 1977 ILO list[79]

Compound / Country	Acetaldehyde	Benzene	Chlorine	Methyl chloride	Pyridine	Toluene
Austria	180	80	3	210	15	375
Belgium	180	30	3	210	15	375
Bulgaria	—	—	1	5	5	50
Czechoslovakia	200	50	—	100	5	200
Finland	180	32	3	105	15	750
German DR	100	50	1	100	10	200
Fed. Rep. Ger.	360	—	1.5	105	15	750
Hungary	—	20	1	20	5	50
Italy	100	20	3	200	6	300
Japan	—	80	3	210	—	375
Netherlands	180	30	3	210	15	375
Poland	100	30	1.5	20	—	100
Romania	100	—	—	100	10	300
Sweden	90	30	3	—	15	375
Switzerland	180	32	1.5	105	15	380
USSR	5	5	1	5	5	50
USA, OSHA	360	—	3	—	15	—
ACGIH	180	30	3	210	15	375
Yugoslavia	360	50	2	50	15	200

Table 5.26 Distribution of procedures used to develop ACGIH TLVs[†] for 414 substances through 1968*

Procedure	% Total
Industrial (human) experience	38
Human volunteer experiments	11
Animal, inhalation – chronic	20
Animal, inhalation – acute	2
Animal, oral – chronic	4.5
Animal, oral – acute	0.5
Analogy	24

* Exclusive of inert particulates and vapours
[†] Threshold Limit Value is the trademark registered by the ACGIH

necessarily provide an index of toxicity and none represents a fine dividing line between safety and danger. All standards contain an element of compromise and an awareness of the basis on which each standard has been established is essential to apply the value properly.

Of all the hygiene standards, the American Threshold Limit Value (TLV) is possibly the most familiar. The list of values is published annually and contains three types of TLV, viz.:

1. *Time-weighted averages (TLV-TWA)* In practice employee exposure to toxic airborne substances varies both with respect to time and

Table 5.27 Distribution of criteria used to develop ACGIH TLVs† for 414 substances through 1968*

Criteria	%	Criteria	%
Organ or organ system affected	49	Cancer	1.5
Irritation	40	Fever	0.5
Narcosis	5	Visual changes halo	0.5
Biochemical changes	2	Visibility	0.5
Odour	2	Taste	0.25
Organ function changes	2	Roentgenographic changes	0.25
Allergic sensitivity	1.5	Cosmetic effect	0.25

* Exclusive of inert particulates and vapours.
† Threshold Limit Value is the trademark registered by the ACGIH.

space. These standards were devised in an attempt to provide guidance in such fluctuating situations and they refer to airborne concentrations of contaminant in the workplace averaged over a normal 8-hour working day and a 40-hour working week.

Excursions are permitted above the limit provided that they are compensated by equivalent excursions below the limit during the work-day (see Fig. 5.8). The relationship between excursion and TLV-TWA is a rule of thumb which limits exposures in excess of three times the TLV-TWA for no more than a total of 30 minutes during the work-day and under no circumstances should they exceed five times the TLV and ensuring the TLV-TWA is not exceeded.

2. *Short-term exposure limit (TLV-STEL)* Situations such as emergencies, maintenance, etc., require that for operational convenience limits

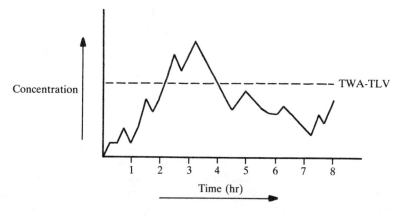

Fig. 5.8 Fluctuation of exposure during an 8 hr day

must be set for short-term exposures where normal time-weighted criteria are inappropriate and unnecessarily restrictive. TLV-STELs refer to concentrations to which workers can be exposed continuously for a short period of time without suffering from:
(a) irritation;
(b) chronic or irreversible tissue damage;
(c) narcosis of sufficient extent as to increase the chance of accidental injury, impair self-rescue or materially reduce work efficiency.

TLV-STELs are appropriate for those substances where there are recognised acute effects from a substance whose toxic effects are primarily of a chronic nature. They refer to 15-minute time-weighted average exposures which should not be exceeded during the workday and also which should not result in the 8-hour TLV-TWA being exceeded. Exposures at STEL concentrations should not be repeated more than 4 times per day and intervals of at least 1 hour should separate successive events.

3. *Ceiling values (TLV-C)* For substances which are predominantly fast-acting, and whose threshold limit is more appropriately based on this particular response, a ceiling limit that should not be exceeded is more relevant than a TWA approach.

Skin notation In the list of TLVs substances are assigned a 'skin' notation to signify potential contribution to the overall exposure by the cutaneous route (including eyes and mucous membranes) either as a result of airborne material or more particularly by direct contact with the substance.

Exposure limits for carcinogens Appendix A to the ACGIH TLV list identifies industrial substances which have proven to be carcinogenic in man, or have induced cancer in experimental animals (see page 212). Substances/processes recognised to have carcinogenic or co-carcinogenic potential with an assigned TLV are differentiated from those with no set TLV. For the latter no exposure or contact is permitted. Substances suspected of carcinogenic potential for man are listed and exposure by all routes should be carefully controlled to levels consistent with animal or human experience data.

Amendments to Threshold Limit Values

Standards are reviewed annually by the ACGIH TLV Airborne Contaminants Committee. As new dose-response data are generated or previously unknown adverse health effects are discovered, existing TLVs are revised, deleted or added to. Proposed amendments are published in the notice of intended changes at the rear of the TLV document and are normally transferred to the adopted list after two years.

Toxicology

Documentation

The ACGIH also publishes full documentation on TLVs which summarises the relevant toxic properties of the individual chemicals and highlights on what base the TLV was set. TLVs are not intended to be used directly in evaluation or control of community air pollution since they are aimed at a healthy (working) adult population whereas the general community also contains the aged, the infirmed, children and pregnant women. Also TLVs are aimed to provide protection from a duration of exposure of about 8 hours per day, 40 hours per week, as opposed to 24 hours per day, i.e. 168 hours per week.

Thus erroneous standards may arise from attempts to use TLVs to predict 'acceptable' ground-level concentrations for environmental pollutants outside the workplace.

Hygiene standards in the United Kingdom

Until 1980 the UK Health and Safety Executive (HSE) published a list of standards in Guidance Note EH/15 'Threshold Limit Values'. This was a reprint of the ACGIH TLV list with an explanatory introduction by HSE identifying those few substances for which alternative standards were applicable in the UK. No list was issued between 1980 and 1984. In 1984 a new list of Occupational Exposure Limits (OELs) was published in Guidance Note EH/40 representing advice on levels to which exposure to airborne toxic substances should be controlled in workplaces. The values are used by HSE as part of the criteria in judging whether a situation is in compliance with relevant statutory provisions.

There are two main classes of Occupational Exposure Limits in EH/40, viz. Control and Recommended Limits. The former which originate from Regulations, Approved Codes, European Directives or agreement by the Health and Safety Commission (through the Advisory Committee on Toxic Substances) refer to levels which should not be exceeded in order to ensure compliance. For certain materials, such as those without an apparent 'no-effect' threshold, additional requirements may be necessary even when the Control Limit is observed. Thus extra cost/effort may be required in judgement of what is 'reasonably practicable'. Recommended Limits represent recommended realistic performance criteria for employers in plant design and exposure control. Again, even where exposures are below the Recommended Limit further effort may be required by employers for certain substances.

As with TLVs, both short- and long-term values are quoted for OELs. The former usually refers to 8 hr TWA values though different periods exist for some materials, e.g. 12 months TWA for vinyl chloride. Short-term limits are normally expressed as 10-minute TWA concentrations.

Materials for which percutaneous absorption represents a significant mode of entry, either as a result of direct contact with the bulk chemical or in high concentrations of vapour, an 'Sk' notation is assigned to the OEL (see Table 5.4). Whenever complications result (e.g. in hot or humid conditions) adjustment may be required to account for the situation.

Additions and alterations

HSE announces proposed amendments in the *Toxic Substances Bulletin* and lists changes for the coming year in the current edition of Guidance Note EH 40.

Use of hygiene standards

Approaches for monitoring exposure to airborne toxic substances, including hardware and sampling strategies, and the application of hygiene standards is discussed fully in Chapter 7 for both single substances and mixtures.

Other hygiene limits

The legal position in other countries regarding employee exposure to toxic chemicals is summarised in Table 5.28.

Biologic standards

The exposure to certain chemicals can also be monitored by determining concentrations of the substance, or its metabolites in urine, blood, hair, nails, body tissues or fluids, or in exhaled breath as exemplified by the breathalyser for assessing alcohol consumption. Figure 5.9 illustrates a means of collecting exhaled breath for subsequent laboratory analysis. Blood lead levels have long been used for determining operator exposure to this substance and continue to complement environmental analysis.

> Two workers at a minerals firm came down with lead poisoning when lead dust exposure levels exceeded 200 times the permitted value. Twenty of the firm's twenty-three employees had high levels of lead in their blood. The company was prosecuted under the Control of Lead at Work Regulations and was fined £1,000.[81]

Alternatively, exposure to some chemicals can be gauged by measurement of responses such as changes in the amount of a critical biochemical constituent, changes in activity of a critical enzyme or changes in a physiological function, e.g. vitalograph measurements for assessing exposure to respiratory irritants or sensitisers. Measurements taken to

Table 5.28 Legal status of hygiene standards

UK
As mentioned on page 216, in the UK there are both Control Limits and Recommended Limits. The HSE will use Control Limits to determine whether, in its opinion, the requirements of the relevant legislation are being observed. Failure to comply with Control Limits, or to reduce exposure still further, where reasonably practicable, may result in enforcement action. Recommended Limits on the other hand are more of a guidance to what may be considered as representing good practice. HSE inspectors will use these Recommended Limits as part of their criteria for assessing compliance with the HASAWA and other relevant statutory provisions.

USA
The Occupational Safety and Health Act 1970, has full Federal legal status. Thus contravention of any of the OSHA permanent or emergency temporary standards promulgated under this Act is, prima facie, an infringement of the statute. The only employers exempted from the requirements of the Act are state and local governments.

Federal Republic of Germany
Under West German legislation, there is an insurance fund for each industry into which employers are required to subscribe and from which medical services and compensation to employees are paid. Because premiums relate to the amounts paid out by the funds, employers have an economic incentive to minimise potential payments by adhering to MACs* and TRCs.† The German Research Association annually issues MACs; these are not statutory but are promulgated by various industrial associations and by the Federal Ministry of Labour.

Belgium
Under the General Regulations for Health and Safety at Work, Articles 148 and 183, it is necessary to keep the concentration of pollutants in the air at workplaces below the permissible limit values set for a number of chemical substances. Up to 1974 the Commissariat Général à la Promotion du Travail published the ACGIH TLVs for the current year.

Denmark
The legal status of control limits is based on Act No. 681 of 23 December 1975. Recognised standards important to health and safety have to be observed. A list of limits, revised annually, is issued by the Working Environment Council, a governmental body.

France
Act No. 76-1106 of 6 December 1976 is an enabling Act which provides the means to establish upper limits for concentrations of dangerous substances in workplace air. There are specific regulations for benzene, carbon monoxide and free silica. Limits for asbestos in air are specified in Decree 77-949, 17 August 1977.

Italy
Exposure limits are not specified in legal statutes except for carbon dioxide, sulphur dioxide, hydrogen sulphide, nitrogen oxides and carbon monoxide in mines and quarries. Measures to prevent or to reduce as far as possible the spread of air contaminants in workplaces are required by General Regulations, DPR No. 303, 19 March 1955.

Sweden
Under the Workers Protection Act 1949, as amended 31 May 1974, the National Swedish Board of Occupational Safety and Health (NSBOSH) issues recommended TLVs which, in most but not all cases, agree with the ACGIH values. Although these Swedish TLVs are recommended as guidelines, NSBOSH may require employers to institute additional protective measures to reduce worker exposure as much as possible.

* Maximum Allowable Concentration.
† Technical Reference Concentration.

Table 5.28 (continued)

Switzerland
The Caisse Nationale Suisse d'Assurances en cas d'Accidents (CNA) promulgates national MAC values annually which are used as industrial hygiene criteria by CNA for prescribing appropriate health protection measures in industry. In the preamble to the official list, the acronym MAC is not actually explained, but, confusingly, is defined as an 'average value' and not as a 'maximum allowable concentration'. In fact, CNA use of the term MAC is a misnomer because the values relate to and comply with ACGIH TLVs, although permissible excursions are not referred to.

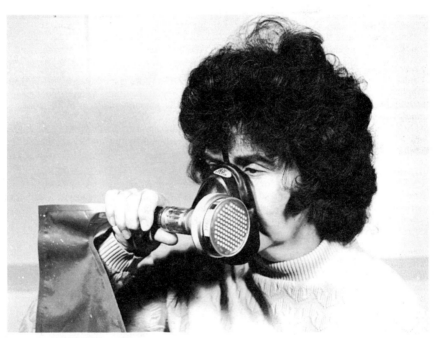

Fig. 5.9 Collection of exhaled breath for analysis (Courtesy Environmental Monitoring Systems)

define lung function include forced vital capacity (FVC), which is the maximum volume of air that can be expelled with maximum effort after full inspiration and timed vital capacity otherwise termed 'forced expiratory volume' (FEV_1), which is the volume of air expelled in 1 second with maximum effort after inspiring maximally. The ratio of FEV_1/FVC (normally expressed as a percentage) for an individual will be affected by various factors such as height, age, sex and exposure to certain chemicals.

Complications arise because of fluctuations in exposure and body burden, effects of strenuous activity, changes induced by environmental factors (heat, altitude, etc.) and water intake, changes in physiological function caused by pre-existing disease or congenital variation of metabolic pathways, and changes in metabolic pathways caused by simultaneous exposure to other chemicals.

An advantage of biological monitoring over environmental analysis is, that whereas the latter measures the composition of the worker's immediate environment, biological monitoring measures the actual amount of substance absorbed into the body and takes into account the worker's individual responses and overall exposure.

However, major problems associated with biological monitoring, which have limited its exploitation, surround the wide variation in individual responses to a particular substance together with the wide spectrum of 'normal' to consider, the lack of sufficiently sensitive analytical methods and the unpopularity of invasive techniques. Compared with environmental standards, therefore, there are few biologic standards of wide acceptability.

The ACGIH, however, has recently issued[82] a list of Biological Exposure Indices (BEIs) for carbon monoxide, ethyl benzene, styrene, toluene, trichloroethylene and xylenes. The publication includes advice on the timing of sample collection within a shift. BEIs represent warning levels of biological responses to the chemical or warning levels of chemical or its metabolites in body tissue, fluid or exhaled air of exposed workers, regardless of whether the chemical was inhaled, ingested or absorbed via the skin.

References

(*All places of publication are London unless otherwise stated*)
1. Burgess, W. A., 'Potential exposures in industry – their recognition and control', in G. D. Clayton & F. E Clayton (eds) *Patty's Industrial Hygiene and Toxicology*, vol. 1 (3rd edn). Wiley-Interscience 1978.
2. Casarett, L. J. & Doull, J., *Toxicology – The Basic Science of Poisons* (2nd edn). Macmillan 1980.
3. Birmingham, D. J., 'Dermatoses' in *Occupational Diseases: A Guide to Their Recognition*. National Institute of Occupational Safety and Health, US Dept of Health Education and Welfare, Washington 1977.
4a. Health and Safety Executive, *Guidance Note EH 26: Occupational Skin Diseases: Health and Safety Precautions*. HMSO 1981.
4b. Cullen, E. J. *HSC News Release*, 1986 (20 Oct.), 1.
5. Health and Safety Commission, *Consultative Document on Control of Substances Hazardous to Health*, 1984.
7. McNaughton, K. J., *Chem. Engr*, 1982, **89** (6), 147.
8. Government Statistical Services, *DHSS Social Security Statistics 1981*. HMSO.

9. *1980 Mortality Statistics: Cause*, OPCS Series DH2 No. 7. HMSO.
10. Sax, N..I., *Dangerous Properties of Industrial Materials* (6th edn). Van Nostrand Reinhold, New York 1984.
11. *IRS Industrial Health and Safety Bulletin*, 1985 (114), 16.
12. Doll, R, & Peto, R., *The Causes of Cancer*. Oxford University Press 1982.
13. Weiss, G., *Hazardous Chemicals Data Book*. Noyes Data Corporation: New Jersey, 1980.
14. *Toxic and Hazardous Industrial Chemicals Safety Manual*. The International Technical Information Institute: Tokyo, Japan 1976.
15. *Handling Chemicals Safely*. Dutch Chemical Industry Association 1980.
16. National Institute of Occupational Safety and Health, *Registry of Toxic Effects of Chemicals*. NIOSH Cincinnati, USA 1978.
17. Anon., *Tamworth Herald*, 4 Feb. 1972.
18. Stokinger, H. E., 'Routes of entry and modes of attack', in *Occupational Diseases: A Guide to their Recognition*. NIOSH, US Dept of Health Education and Welfare, Washington 1977.
19. *Annual Report of H.M. Chief Inspector of Factories*. HMSO 1967.
20. British Chemical Industry Safety Council, *Quarterly Safety Summary*, **38** (151), 1967, 28.
21. *Annual Report of H.M. Chief Inspector of Factories*. HMSO 1970.
22. Health and Safety Executive, *Guidance Note EH 40: Occupational Exposure Limits*, 1984.
23. Kobayashi, J., 'Pollution by cadmium and the Itai-Itai disease in Japan' in F. W. Oehme (ed.), *Toxicity of Heavy Metals in the Environment*, Pt 1. Marcel Dekker, New York 1978.
24. Haradi, M., 'Methyl mercury poisoning due to environmental contamination (Minimata disease)', in F. W. Oehme (ed.) *Toxicity of Heavy Metals in the Environment*, Pt 1. Marcel Dekker, New York 1978.
25. Cambridge, G. W. & Goodwin, B. F. J., *Allergy to Chemicals and Organic Substances in the Workplace*, Occupational Hygiene Monograph No. 12, 1984. Science Reviews.
26. Barlow, S. M. & Sullivan, F. M., *Reproductive Hazards of Industrial Chemicals*. Academic Press 1982.
27. Le Serve, A. W., Vose, C., Wigley, C. & Bennett, T., Nelson 1980.
28. 'Identification and Control of Chemical Carcinogens in the Workplace' Oyez Conference, London, July 1981.
29. Searle, C. E., *Chemical Carcinogens*, A. C. S. Monograph No. 173, 1976.
30. *The Prevention of Occupational Cancer*: An ASTMS Policy 1980.
31. *IARC Monographs on the Evaluation of Carcinogenic Risk of Chemicals to Humans* (various volumes and supplements). World Health Organisation: Geneva, Switzerland.
32. Sax, N. I., *Cancer-Causing Chemicals*. Van Nostrand Reinhold Co. 1981.
33. American Conference of Governmental Industrial Hygienists, *Threshold Limit Values and Biological Exposure Indices for 1986–87*. ACGIH, Cincinnati 1986.
34. Sittig, M., *Handbook of Toxic and Hazardous Chemicals*. Noyes Data Publications: New Jersey 1981.

35. Chemical Industry's Safety and Health Council, *A Guide to the Evaluation and Control of Toxic Substances in the Work Environment*. Chemical Industry's Association 1980.
36. Royal Society for the Prevention of Accidents, *RoSPA Bull* 1980 (Jan.), 3.
37. Chemical Industry's Safety and Health Council, *Chemical Safety Summary*, 1983, **54** (213), 401.
38. *The Health and Safety at Work*, 1982 (Sept.), 48
39. Department of Health and Social Security, *Pneumonconiosis and Byssinosis*. HMSO, 1973.
40. *Isocyanates in Industry – Operating and Medical Codes of Practice for Safe Working with Isocyanates*. BRMA Health Advisory Committee, 1978.
41. *Annual Report of H.M. Chief Inspector of Factories*. HMSO 1968.
42. *American Review of Respiratory Disease*, 1983, **128**, 226.
43. *Thorax 34*, 1979 (1st Feb.), 13.
44. Health and Safety Commission, *Asbestos*, Vol. 1., *Final Report of the Advisory Committee*. HSC 1979.
45. Rankine, A. D., Universities Safety Association, *Safety News*, **17** (May 1983), 11.
46. *Gardner* v *Motherwell Machinery* 1961, 3 All E.R. 831, 314.
47. Mathias, C. G. T., *Archives of Dermatology*, 1982, **118**, 420.
48. Waldron, H. A., *Lecture Notes on Occupational Medicine*. Blackwell Scientific Publications 1976.
49. Fawcett, H. H. & Wood, W. S., *Safety and Accident Prevention in Chemical Operations*. Interscience 1965.
50. *Annual Report of H.M. Chief Inspector of Factories, 1967*, 105–19. HMSO.
51. *Dean Forest Mercury*, 4 Nov. 1977.
52. Grant, W. M., *Toxicology of the Eye* (2nd edn). C. Thomas Charles, Springfield: Illinois 1974.
53. Health and Safety Executive, *Guidance Note MS 11: Eyes*. HMSO 1978.
54. Seitz, B., *Soc. Med. Hyg. du Travail*, 1972 (Apr. 10)
55. Velasquez, *J. Med. del Trabajo*, 1964, **1**, 43.
56. Hewitt, F., Universities Safety Association, *Safety News*, 1983 (17 May), 12.
57. Elkins, H. B., *The Chemistry of Industrial Toxicology*. Wiley, New York.
58. Henderson, Y. & Haggard, H. W., *Noxious Gas* (2nd edn). Reinhold: New Jersey, 1943.
59. National Safety Council of America, *National Safety News*, 1976 (Feb. and Apr.).
60. *Eastern Daily Press*, 28 Oct. 1979.
61. Chemical Industry's Safety and Health Council, *Chemical Safety Summary*, 1977, **48** (190), 22.
62. Bond, J., *Loss Prevention Bulletin*, 1985.
63. Morton, F., *Report of the Inquiry into the Safety of Natural Gas as a Fuel*. Ministry of Technology, 1970–71.
64. Gutierrez, G., 'Carbon monoxide toxicity' in J. J. McGrath & C. D. Barnes (eds) *Air Pollution – Physiological Effects*. Academic Press 1982, 127–45.
65. Atherley, G. R. C., *Occupational Health and Safety Concepts*. Applied Science Publications Ltd. 1978.

66. Royal Society for the Prevention of Accidents, *RoSPA Bull.* 1981 (July), 2.
67. Fagin, J., Bradley, J. & Williams D., *Brit. Med. J.*, 1980, (Nov. 29) 1461.
68. The Carcinogenic Substances Regulations, 1967.
69. Northfield, R. R., *J. Soc. Occup. Med.*, 1981, **31**, 164.
70. Hunter, D., *The Diseases of Occupations* (5th edn). English Universities Press, 1975, 640.
71. Morley, R. & Silk, S. J., *Annals Occup. Hyg.*, 1970, **13**, 101.
72. *Dry Cleaning Plant: Precautions Against Solvent Risks*, Safety, Health & Welfare, No. 15. HMSO.
73. Kelly, W. D., 'Protection of the sensitive individual', American Conference of Government Industrial Hygienists, Ohio 1982.
74. *The Times*, 13 Oct. 1983, 13.
75. *OECD Guidelines For Testing of Chemicals*. OECD: Paris 1981
76. Sellers, J. G., 'Quantification of toxic gas emission hazards', in *Proceedings of Symposium of Process Industry Hazards*, 14–15 Sept. 1976, West Lothian. Institution of Chemical Engineers Symp. Series 1976, No. 47.
77. Amoore, J. E. & Hautala, E., *J. Applied Toxicology*, 1983, **3** (6), 272
78. Verschueren, K., *Handbook of Environmental Data on Organic Chemicals*. Van Nostrand Reinhold Co. 1977.
79. *Occupational Exposure Limits for Airborne Toxic Substances*, Occupational Safety and Health Series No. 37, 1977. International Labor Office, Geneva, Switzerland.
80. Olishifski, J. B., 'Toxicology' in J. B. Olishifski & F. E. McElroy (eds) *Fundamentals of Occupational Hygiene* (2nd edn). National Safety Council, Chicago: Illinois 1979
81. Anon., *Safety*. The British Safety Council 1982 (Sept.), 14.
82. American Conference of Governmental Industrial Hygienists, *Biological Exposure Indices Proposed for 1986/87*. ACGIH: Cincinnati 1986.

CHAPTER 6

Control measures for handling toxic substances

A major problem with risk assessment is the paucity of toxicity data for many chemicals. One reason for this inadequacy stems from the rapid development of the chemical and allied industries over a relatively short time period.[1] Some 4 million new chemicals were prepared and identified during the past decade or so, though 75% are cited but once in the literature. Even so new substances enter the market place at an average of 14 per day. However, of the 60,000 chemicals currently in use at work, hygiene standards exist for only about 500 substances. (Moreover, industrial use of chemicals can result in exposure to mixtures of substances, the individual components of which may have unknown additive or synergistic toxic effects.) Indeed, according to a recent study[2] in the United States no toxicity data exist for nearly 80% of the chemicals used in commercial products or processes. All this hampers meaningful assessment of risk and emphasises the need to minimise exposure.

Sources of exposure

Since it is rarely practicable to isolate workers completely from chemicals under all circumstances, an awareness of likely sources and levels of exposure is crucial in order to assess risk and institute the best control strategy. Notwithstanding the need for ambient air analysis, sufficient knowledge and experience often exist to provide guidance on prediction of sources, and approximate levels, of exposure as indicated below.

For materials toxic by ingestion key sources of exposure stem from poor personal hygiene or contamination of food and drinks in the work area, e.g. as a result of a low standard of housekeeping. Since there is often significant subconscious hand-to-mouth contact, material can also enter the mouth from contaminated hands, finger nails or protective gloves. Inhaled particulate material can also be swallowed. Precautions must also protect against accidental or careless contamination of the food chain.

Sources of exposure can be classed as:
- *Periodic emissions* which arise from the need to open or enter the 'system' ocasionally, for example, during sampling, cleaning, batch additions, bulk tank-car loading, line-breaking, etc. Periodic emissions tend to be large and, as discussed on page 325 include both anticipated events and unplanned releases, in which human error may be a factor.
- *Fugitive emissions* which tend to be small, but continuous, escapes from normally closed points. They occur from dynamic seals, such as valve stems and pump or agitator shafts, or from static seals such as flange gaskets. Leakage rates from a variety of sources have been reviewed[3-5] and a selection is included in Table 6.1 for volatiles and in Table 6.2 for dusts, although exact leakage rates will be influenced

Table 6.1 Emission rates from process equipment and fittings (after ref. 3)

Equipment	Emission rate (mg s^{-1})
Pump shaft seals	
Regular packing without external lube sealant	140
Regular packing with lantern ring oil-injection	14
Grafoil packing	14
Single mechanical seal	1.7
High temperature	1.7
Vapour phase seal	1.7
Single mechanical seal plus auxiliary packing	1.5
Tandem mechanical seal (barrier fluid at lower pressure than the stuffing box)	0.21
Double mechanical seal	0.006
Bellows seal (must include auxiliary packing)	Nil
Diaphragm pump (double)	Nil
Canned pump	Nil
Valve stems (excluding pressure relief valves)	
Rising stem valves and regular packing	
Rating ≤ 300 lb	1.7
Rating > 300 lb	0.03
Rising stem valves and Grafoil packing	
Rating ≤ 300 lb	0.2
Rating > 300 lb	0.004
Non-rising stem valves	0.005
Pressure relief valves	
Average release from a single pressure relief valve	2.8*
Piping/equipment connections	
Open-ended	0.63
Snap-valve end	0.50
Screw-cap	0.009
Blind flange	
Asbestos Gasket	0.009
Grafoil Gasket	0.0009

* Assuming that all valves vented to a closed system and also all valves protected by an upstream rupture disc have zero leakage.

Table 6.1 (continued)

Equipment	Emission rate (mg s^{-1})
Compressors	
Reciprocating	
Rod packing	45
Single	3.6
Double	45
Labrynth seal	Use pump seal data
Mechanical seal	
Liquid film seal	0.006
Agitators	Pump seal emission rate $\times \dfrac{\text{Agitator shaft vel. (m s}^{-1})}{1.91}$
Flanges, bolting and fittings	mg s^{-1} m^{-1}†
Flange and asbestos gasket	
150–300 lb	0.056
>300 lb	0.003
Flange and grafoil gasket	
150–300 lb	0.006
> 300 lb	0.0003
Threaded connection only	0.056
Threaded and welded	Nil
Welded only	Nil

† Flange leakage is influenced by outer circumference of the flange itself therefore unit of emission rate quoted as mg per second per metre.

Table 6.2 Emission rates from powder handling[3–5]
The emission rates refer to the total dust level of release. To *estimate* the respirable level, these numbers should be halved.

Type of equipment/handling	Emission rate (mg/s)
Vibratory screens	
Open top	5.5 × top surface area (m^2)
Closed top with open access port	
15 cm dia. port	0.11
20 cm dia. port	0.21
30 cm dia. port	0.44
Closed cover – no ports	Zero
Bag dumping	
Manual slitting and dumping	3
Semi-automatic (enclosed dumping but manual bag entry/removal)	0.2
Fully automatic (includes negative pressure)	Zero
Bagging machines (filling)	
No ventilation	1.5
Local ventilation	0.01
Total enclosure and negative pressure	Zero

by inherent properties of the substance (e.g. vapour pressure, 'dustiness') and process conditions (e.g. temperature) and the maintenance situation. Leak rates may also vary with time, e.g. as shown for a valve packing in Fig. 6.1.[4]

As an illustration of potential sources of exposure to chemicals, Tables 6.3 and 6.4 (after refs 6, 7, 8) qualitatively identify problem points for a range of specific operations and specific processes, respectively, if control is inadequate. Such exposures may arise due to poor process control, operator and maintenance worker malpractice, inadequate maintenance, incomplete understanding of the process or other system-of-work failures.

In the past insidious exposure to hazardous materials may have arisen due to contaminated clothing being worn or taken outside the factory.

> Mesothelioma cases have been recorded in which asbestos exposure was due to a close relative working in the industry. Since contaminated clothing may also prolong worker exposures, with highly toxic materials measures are generally required to segregate personal and protective clothing.

Having decided that control measures are necessary, the main general strategies available for reducing risk are described below. Clearly the strategies and detailed measures adopted depend upon the nature of the hazard as discussed in Chapter 5, e.g. toxic, corrosive or dermatitic, and the level of risk.

In many situations a combination of techniques is required. Here emphasis is on 'handling', but the measures may be equally applicable to storage, transportation, etc.

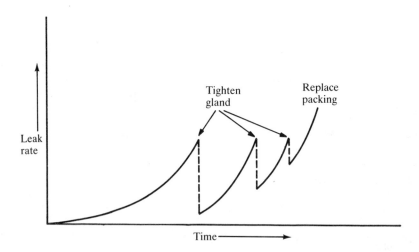

Fig. 6.1 Typical valve leakage patterns with time/maintenance[4]

Table 6.3 Sources of exposure to chemicals from selected specific operations

Operation	Chemical stressor	Potential key sources of exposure
Electroplating	Acid/alkali mists	• Mists released from electrolyte by gassing in the bath. An added danger stems from impurities present in the electrolyte. Gassing is a particular feature associated with chromium plating.
	Dusts	• From the handling and use of powdered, granular or flake materials used as additives or electrolyte compounds. • Resulting from the evaporation of water from mists liberated from tanks. • From secondary dispersal or resuspension of settled dust.
	Cyanides	• Salt handling during weighing and transfer operations. • Hydrogen cyanide or cyanide mist evolved from electroplating baths (control of pH is crucial).
Centrifugation	Vapours	• During charging, washing and spinning cycles when volatile liquids are processed. • During unloading, because of residual vapour or gas in the filtered material. • During transfer of product to containers.
	Spray/mist/wet solids	• During filtration and wash cycles and unloading/transfer of wet solids.
Filtration	Solvent vapours	• Emissions from filter boxes and plate and frame presses during cake removal. • During cleaning and setting up enclosed filters and filter presses. • Emissions from rotary drum filters unless well ventilated.
Metal forming and working	Combustion products	• Toxic gases liberated from holding or stock furnaces due to incomplete fuel combustion.
	Nuisance dusts	• Released from initial processing of the metal (e.g. scale sloughed off in roughing mills).
	Metal fumes and dusts	• From hot work (e.g. casting, forging, welding).
	Die lubricants	• Overspray during the application of lubricant between die and metal piece impacts. • Oil mist and thermal degradation from oil-based lubricants applied to hot dies or heated work pieces.

Table 6.3 (continued)

Operation	Chemical stressor	Potential key sources of exposure
Mixing and blending	Dusts	• During charging of blender through the charging port and during filling of drums directly from blender.
Rubber compounding	Dusts	• Raw material handling as a result of over-dusted rubber stock, leaks in conveyors or piping systems, from damaged containers, spillage from rail car tankers or trailers during hookup or discharge into hoppers. • During weighing operations and master-batch make-up. • Improper use of pneumatic ram on manual loading mixers. • Poor housekeeping. • During maintenance.
Vacuum drying and freeze drying	Product	• Dust as material is dumped from vacuum drier into drums or hoppers.
	Solvent	• From filtered material during loading into trays for drying.

Table 6.4 Sources of exposure to chemicals from selected specific processes

Process	Chemical stressor	Potential key sourcers of exposure
Acetyl chloride production	Acetic acid) Phosphorous) Trichloride	• During filling of storage tanks from railcar or tank trucks. Spillage.
	Hydrochloric acid	• Hydrolysis of spilled product. • From vent during addition of water to crude phosphoric acid.
	Acetyl chloride	• During drum filling.
Acrylic fibre production	Solvent (DMF and DMAC)	• Venting of slurry-dissolving vessels, changing filters when slurry is hot and system is open. • During routine start-up when fibre tow is threaded and spinnerettes are changed.
Acrylonitrile	Acrylonitrile	• Dock loading arms, columns, process filters, piping and pumps, storage tank drain lines during decontamination of process equipment. • Vapour venting during barge, railcar and tank truck loading. • Liquid leaks and vapour emissions due to pump seal failures during routine surveillance. • Vapour emissions from liquids in process sewers and sumps.

Table 6.4 (continued)

Process	Chemical stressor	Potential key sources of exposure
		• Vapour emissions from storage tanks. • Vapour emissions from process liquids in piping, product storage tanks, barges, railcars and tank trucks during process/product sampling.
Alkylamine production	Ammonia	• Unloading bulk material. • Leaks from feed vaporiser.
	Amines and other organics	• Handling of tars/bottoms from purification stage.
Cellulose acetate production	Polymer	• Dust during dry milling or from discharging chutes.
	Acetic acid	• Wet milling. Leaking valves, pumps and pipes during hydrolysis stage. • Transportation of dope-laden filter elements • During precipitation and washing process (e.g. from viewing ports, tank vents)
Chromic acid (anhydrous) manufacture	Hexavalent chromium	• During cleaning and maintenance of roasting kilns and during replacement of refractory linings. • Chrome-bearing mist generated during the initial quenching for leaching.
Ethyl alcohol production from ethylene	Catalyst support media	• Dust generated during charging of catalyst support into mixing vessel.
	Phosphoric acid	• Vapour emission as freshly prepared catalyst is dumped.
	Diethyl ether	• From pumps unless tandem seals are incorporated.
Halogenated flame-retardent production	Host of raw materials	• Leaks during bulk tank car unloading, drum opening, changing drum pumps and drum cleaning prior to disposal or return. • Dust and vapour emissions during manual addition of dry powders into blending tank.
	Product	• Cleaning basket centrifuges, packaging.
	Product and solvent	• Charging and loading of driers.
	Dust	• Cleaning dust collection units.
Iron and steel industry	Ore and coal dust	• Mining.
	Iron oxide dust	• Ore sintering and pelletising.
	Silica dust	• Refractory handling.

Table 6.4 (continued)

Process	Chemical stressor	Potential key sources of exposure
Iron and steel industry (cont'd)	Silica sand	• Foundries (casting, knock-out and fettling operations).
	Iron oxide fume	• Furnaces, scarfing operations, welding.
	Lead fume	• Scrap preparation.
	Flux fume, zinc	• Galvanising.
	Lead and manganese fume	• Leaded and ferromanganese steels.
	Fluorides, carbon monoxide	• Blast furnace.
	Carbon monoxide sulphur dioxide and hydrogen sulphide	• Coking operations.
	Ozone, oxides of nitrogen	• Welding.
	Solvent vapours	• Maintenance and cleaning motors.
	Sulphuric acid mist	• Pickling.
	Lead mist	• Spray painting with lead paint.
Maleic anhydride production	Maleic anhydride	• Maintenance activities (especially cooler wash out). • Bulk loading. • Bagging of briquettes or pastilles. • Leaking pumps.
	Benzene	• From feedstock storage tank during transfer.
	Vanadium	• Removal of spent catalyst from reactor.
Nitric acid manufacture	Nitric acid	• Mist and fume during loading of tank trucks, rail tank cars and drums.
	Hydrochloric acid	• During catalyst pickling.
Nylon 6 fibre production	'Dowtherm'	• Vapour emission from pump packing valves or other parts of the fluid handling system used in chip melting.
	Monomer of oligomers	• As molten fibre leaves spinnerettes if ventilation is poor.
	Fibre finish	• Propelled from yarn into ambient air by high-speed spinning.
Oxo reactions	Carbon monoxide	• Process leaks or during maintenance.
	Monoethanolamine	• Sampling, charging and breaking into process lines/equipment.
	Cobalt oxide and nickel catalyst	• Dust from catalyst charging. • Poor personal hygiene.
Petroleum refining	Hydrocarbon vapours	• Transfer and loading operations. • Flares. • Boilers.

Table 6.4 (continued)

Process	Chemical stressor	Potential key sources of exposure
		• Cooling towers. • Storage tanks. • Cracking unit regeneration. • Pump valves. • Treating operations.
	Sulphur dioxide	• Boilers. • Flares. • Cracking unit regeneration. • Treating operations.
	Carbon monoxide	• Cracking unit regeneration. • Boilers. • Flares.
	Nitrogen dioxide	• Flares. • Boilers.
	Hydrogen sulphide	• Sour crudes. • Pumps. • Hydrogenation. • Liquid waste. • Hydrocracker.
	Particulates	• Cracking unit regeneration.
Refactory manufacture	Raw materials and impurities	• Dust generated during receipt and storage of dry bulk material and during transport by front-end loaders. • During crushing, grinding and screening operations. • Transfer of material in bucket elevators to and from screening equipment and silos. • Dust from batching of raw material to mixers. • Poor housekeeping/spillages
	Graphite, carbon black, amorphous silica, sodium silicate, talc, hydrated lime, chromite ore	• Opening of packaged or bagged materials.
	Coal-tar pitch volatiles	• During blending and mixing of either tar-bonded dolomite or tar-bonded basic brick. • During hot pressing of bricks and transport of material. • From drying and curing of tar-bonded dolomite.
	Silica	• During drying of shaped refractories and during firing.
	Carbon monoxide	• Leaking furnaces for firing of brick • Exhausts from fork-lift trucks and front-end loaders. • From driers due to incomplete combustion.

Table 6.4 (continued)

Process	Chemical stressor	Potential key sources of exposure
Styrene-acrylonitrile and acrylonitrile – butadiene-styrene polymerisations	Acrylonitrile or styrene	• Unloading of raw material. • Tank vents. • Emissions from centrifuges, during maintenance of filters and strainers (especially leaf- or cartridge-type units). • During transfer of slurry from reactor, wet cake from centrifuge and product from drier, sample traps. • During kettle cleaning. • From off-spec batches.

Substitution

It is often possible to replace relatively hazardous materials with safer alternatives, i.e. toxic substances can be replaced by less toxic, or less volatile, or less easily dispersed substances (flammable solvents can similarly be substituted by less flammable or non-flammable materials). Examples of industrial applications of this approach include:

- White phosphorus replaced by sesqui-sulphide in matches.
- Beryllium replaced by halophosphates in fluorescent lighting.
- Benzene replaced by white spirits in the print industry.
- Detergents and water to replace organic solvents.
- Trichloroethylene replaced by 1,1,1-trichloroethane in metal degreasing applications.
- Replacement of asbestos-based insulation by man-made mineral fibre based insulation.
- Use of leadless glazes in pottery manufacture.
- Replacement of toluene di-isocyanate by less volatile isocyanates or by pre-polymers in polyurethane applications.
- Replacement of chromic acid by surfactant solutions for cleaning laboratory glassware.
- Replacement of suds (oil-in-water emulsions) or neat cutting oils by properly constituted synthetic machine coolants.

In the present context the first criterion in seeking a substitute is toxicity, for example solvents with a Threshold Limit Value – Time Weighted Average of 100 ppm or less should if practicable be avoided in formulations. However, evaporation rate is also important since this determines the rate at which a toxic, or flammable, concentration may build up in an unventilated enclosure (as explained in Ch. 3).

Substitution may also include altering the synthetic routes for the chemical substance to avoid the use, or production, of toxic intermedi-

Control measures for handling toxic substances

ates. For example in the dyestuffs industry a change was made in the manufacture of tobias acid (a β-naphthylamine sulphonic acid used in the manufacture of dyes) to the sulphonation of β-naphthol followed by amidation.⁹

Consideration must be given not only to the toxicity of reactants, normal intermediates and products but to any chemicals likely to be generated by side reactions or as impurities. A classical example of the danger of impurities arising in synthesis is the production of the hyperpoison dioxin during the manufacture of 2,4-D and 2, 4, 5-T; dioxin formed accidentally during trichlorophenol production in the Seveso incident (see Ch. 10, page 510).

Minimisation of inventory/concentration

An important principle is to reduce, so far as is reasonably practicable, the inventory of toxic materials (and indeed of flammable or unstable materials) stored or in process. However, continuity of production or supply generally necessitates a significant storage of raw materials and products, and in some cases of intermediates. Therefore, these materials (including quantities held because they are off-specification, or contaminated, or have been temporarily drained from equipment) must be kept in properly designed and labelled containers.

There may often be an advantage in handling chemicals in the dilutest practicable concentration. Examples are:
- Addition of toxic reactants at a rate corresponding to consumption in batch reactors.
- Supply of proprietary weed-killer, herbicides, etc., made up in aqueous solutions.
- Purchase of the most dilute acid, or alkali, suitable.

Maintenance of the correct dilution, and freedom from contaminants, is often an important factor in the prevention of occupational dermatitis. For example:
- In machine shops control is required over the concentration of aqueous-based coolants, or emulsions of mineral oil, used for lubrication of cutting tools and they should be changed on a regular basis. The supplier's instructions should be followed precisely and no additional materials, e.g. biocides, are normally needed. Neat cutting oils should also be properly maintained. Swarf in either particulate, ribbon or chip form should be filtered out and removed to avoid it causing abrasions to the operator's hands.
- Over-concentration of bleach solutions, for cleaning or disinfectant purposes, should be avoided.

- Over-concentration of detergent solutions for manual washing operations should be avoided;

A former chief cook contracted chronic dermatitis on his hands when working in the galley of a ship. The complaint was caused by excessive exposure to the detergent he used to wash pots and pans. Although the detergent was mild and would not cause damage if appropriate precautions were taken, because he was given inadequate instruction and warning by his employers the cook tended to use too much detergent when washing up.
No separate basin for washing hands was provided in the galley and staff had to dry hands on their aprons or oven cloths. As a result of his skin condition the cook is registered as disabled and was awarded £46,829 against his former employer.[10] (This case illustrates the importance of the dose–response concept and of the need for adequate labelling.)

Also batch size may be adjusted to require that hazardous material be used in full bag or drum quantities supplied by the manufacturer in agreed pre-weighed quantities. This removes the necessity for careful weighing with consequential exposure.

Mechanical handling

Dispersion, and hence contact of personnel with toxic chemicals, can obviously be reduced by the use of mechanical handling and transfer equipment in preference to manual handling operations.
Examples include:
- Arranging for in-plant transfer by pipeline, otherwise use specially designed vessels, e.g. lined portable containers with provision for mechanical lifting. Pipelines should be sloped to assist free-draining and consideration should be given at the design stage to provisions to reduce blockages (e.g. hotwater tracing of phenol lines) and to procedures for the safe freeing of blockages (e.g. by steam injection).
- Enclosed transfer of liquids or gases using pressurisation or vacuum with balance lines, or vents provided with knock-out and/or scrubbing facilities or filters.
- Enclosed transfer of solids by pneumatic conveyors, belt conveyors or chutes.
- Transfer of bulk solids, e.g. toxic waste, in skips or using dumper trucks (sealed plastic bags may be used, e.g. for asbestos).
- Enclosed transfer of powders by screw feeders.
- Use of overhead hoists instead of manual handling at solvent degreasing vats (a hold time is required for drainage).
- Use of jigs or hangers for components to be spray painted.
- Provision of portable 'drum pumps' for emptying small containers to avoid manual lifting.

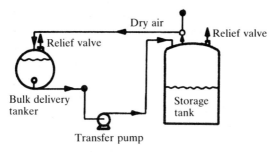

Fig. 6.2 Bulk delivery to storage tank[11]

Hence the filling of TDI storage tanks from bulk delivery vehicles is engineered as in Fig. 6.2 so that a completely closed circuit is formed between the delivery tanker and storage tank.[11] Transfer is usually by pumping; dry air in the storage tank is displaced into the tanker thus excluding moisture ingress.

All vents, even after 'treatment', are directed to a safe place away from the working area. Given correct design and maintenance of the handling facility it is then less of a problem to ensure that inadvertant exposure does not occur due to maloperation or during cleaning operations. Provision is necessary to ensure that flushing out of all vessels, equipment and pipelines is possible prior to cleaning or maintenance.

It may be possible to design for vessel or equipment cleaning *in situ* and without personnel entering it, e.g. high-pressure water sprays in polymerisation reactors for PVC production and in spray drier applications.

Change of process

A change in the process, or purification of materials can often bring about a reduction in the level of risk. For example:
- Removal of water from white lead with linseed oil rather than by filter presses ensures product is dry of water but dampened by oil and hence not dusty.
- Vacuum cleaning dust rather than brushing.
- Brush painting in place of spray techniques.
- Hydroblasting in place of sandblasting in foundries.
- Wetting of dusty materials resulting in agglomeration.
- Incorporation of toxic substances in master-batches.
- Feeding of toxic substances into a process in pre-packed containers or in plastic bags.
- Processing with solutions, pellets or granules rather than fine powders.
- Use of solvent-refined mineral oils in coolant formulations.

In general emission prevention is easier with steady state, continuous processes than with batch processes.[12]

The production of master-batches, or ready-made mixtures of concentrates, in a convenient physical form in one segregated area of a factory helps to reduce the dispersion in other areas. This is common in rubber processing when various rubber chemicals, e.g. accelerators, antioxidants, dyes and pigments are pre-weighed and milled into small batches of rubber compound; these batches are then transferred for mixing with bulk supplies of rubber and fillers.

Rubber manufacture is also an example of an industry to which certain materials are now available as pellets or granules which were previously supplied as powders. This concept is slowly being applied in other industries but quality assessment by the supplier and by the user is desirable.

Pre-weighing into plastic bags, which can be milled into a compound, may be applicable in plastics compounding.

Clearly, with highly toxic chemicals transportation between plants at different sites should be avoided whenever possible. Thus one user of phosgene in the UK manufactures it on the site and feeds it directly into the process for which it is required. This also enables a smaller inventory to be stored, although it still requires the shipment of chlorine. There may also be scope for transporting the active chemical agent required in a safer form, for example shipping sodium hypochlorite instead of chlorine, or sodium bromide instead of bromine.

Suppression

Release of potentially toxic or flammable liquids as mists or sprays, as vapours, or as dusts can sometimes be minimised by suppression methods. Clearly with liquids this includes lowering the operating temperature and hence the vapour pressure.

One strategy with open vessels is to cover the liquid interface with a partial seal. For example a layer of floating inert spheres may serve to reduce vaporisation from a liquid in an open tank. Floating roof tanks are commonly used for the storage of petroleum hydrocarbons. The evolution of mist from electrolytic metal plating vats, e.g. containing chromic acid solution, may be suppressed by addition of a foaming agent which under the action of the bubbles of gas generated at the electrodes forms a foam blanket; however, local exhaust ventilation is still required, together if possible with a cover when not in use.

In the making up of resin bonded components a 'non-touch' technique is commonly adopted; the components are not handled until after the resin has cured. Generally, with the exception of epoxy resins, the 'cured' resins are considered inert but some unreacted chemicals will be present; this must be taken into account when subsequently machining the components, e.g. sawing or drilling which produces fine dust.

Dust generation can often be reduced by pre-wetting of particulate or fibrous solids resulting in agglomeration. (Thus, in the absence of vacuum cleaning, wet-sweeping is preferable to dry-sweeping.) 'Dust-free' powders, e.g. dyestuffs, may be produced by the addition of small proportions of oils to particulate products. A method introduced to minimise exposure to the dust from asbestos–magnesia powder when mixing with water to produce a paste for thermal insulation involved inserting a hosepipe into a hole in the plastic bag in which the powder was delivered and pre-wetting it. Similarly when stripping old, dried-out, insulation an approved practice is to thoroughly pre-wet it and to bag it up for disposal while still wet.[13]

Continuous monitoring and automation

Continuous monitoring of processes and equipment may be used to detect/indicate and give warning of either process problems or venting/leakage of materials. Examples include:
- Temperature or pressure excursions.
- Reduction, increase, or failure, of liquid/solid or gas flows.
- High or low levels of materials in storage vessels or hoppers or process vessels.
- Increase or decrease in pressure drop.
- Deviations in the composition of gas or liquid phases in process equipment or transfer lines.

These can be linked to automatic control systems or, in the extreme, to emergency shut-down systems. This 'on-line' monitoring and control is separate from the workplace/environment testing and monitoring discussed in Chapter 7.

In commercial production of chemicals, particularly in large-scale continuous processes, it is common for automatic control to predominate; thus except for intermittent operations (e.g. sampling, cleaning or maintenance operations), or in an emergency, the majority of the operators work at some distance from the process units (e.g. in a control room). However, in jobbing or batch production, and in the use of various chemicals, this may not be the case and the other strategies then become more important.

Although in the present context 'monitoring' refers to operating parameters and environmental conditions, it is a prerequisite for safety that the systems of work (see Ch. 17) and possibly the health of the workforce are also monitored.

> Automatic, computer-based, fixed-point, vinyl chloride monomer monitoring systems have been installed by most US companies engaged in the production of polyvinyl chloride resin. A sample is collected at each probe location in sequence, analysed automatically, the results recorded in a computer, and the data analysed statistically. Area concentrations which

exceed set limits are indicated by warning lights and alarms. To confirm the effectiveness of vinyl chloride monomer control and area monitoring, personnel often wear active or passive solid sorbent samplers or detector tubes.[14]

Monitoring techniques are described in detail in Chapter 7. It is relevant to note that, since the object of atmospheric monitoring is normally to obtain a measure of the levels of particular pollutants to which workers are exposed, the preferred strategy is to take measurements at head-height in the positions in which people actually work. The activities monitored should be those during normal operation, abnormal operation, cleaning, maintenance, etc. Because of the difficulty of catering for movement of workers there is, as discussed later, a trend towards the use of personal sampling equipment.

Although it might give a useful early warning with some chemicals, odour should not in general be considered a reliable method of monitoring as explained in Chapter 5. For example, ammonia has a TLV-TWA of 25 ppm; since it would be intolerable to work at the TLV concentration, detection by smell is in this case a fairly reliable warning. Conversely, hydrogen sulphide can have its distinctive smell, with an odour threshold of only 0.2 ppm, masked by other vapours or may result in olfactory paralysis; smell is not therefore a reliable warning. Similarly toluene di-isocyanate (TDI) has a TLV-TWA value below the normal odour threshold so that reliance on detection by smell is dangerous.

(Laboratory workers do resort to deliberately smelling the vapour above small flasks or boiling tubes on occasions. Though not generally an operation to be recommended, when it is required the technique then to be used is to waft the vapour towards the nose by gently flapping a card and not to inhale directly above the outlet.)

Segregation

Toxic risks may be controlled by segregation, e.g. by distance, time, age, sex, physiological criteria or physical barriers.[15]

Segregation by distance is common in the chemical process industry; where possible the site is laid out so that potentially hazardous operations (e.g. involving highly reactive chemicals, high pressure or high temperature) or large inventories of process materials are situated as far as is practicable from other plant areas, and in particular from highly populated buildings, e.g. offices. This is discussed further in Chapter 15. Processes with an acknowledged explosion potential, e.g. explosives manufacture and high-pressure hydrogenations, may be segregated behind blast walls. It is common to provide bund walls to limit spillage and to protect drainage systems in the event of liquid leakages.

Similarly in labour-intensive operations involving fixed work stations spacing may provide an additional precaution so that any toxic materials evolved at one station do not spread to another. This is common in large open welding bays. (It is also relevant to note that associated metal degreasing operations using chlorinated solvent vats must be completely segregated from electric arc-welding bays to avoid the conversion of vapours into traces of gases of high toxicity; see Ch. 5.)

In asbestos-stripping the operations are sheeted off and 'sealed' so that non-protected personnel are kept well away from the stripping area; access is restricted to adequately trained, supervised, protected workers.[13] Routine atmospheric monitoring is then used to check the adequacy of the 'sealing' and the local exhaust ventilation provisions.

The principle of segregation by time is exemplified by some foundries in which the dustiest 'knocking out' operations, in which sand is tipped/knocked/vibrated off castings, are performed on the night-shift when the factory is sparsely populated. In more modern foundries 'knocking-out' is segregated at a point, or points, provided with very efficient local exhaust ventilation and to which castings are transferred on conveyors. The sand is removed by covered/underground conveyors. Cleaning operations involving potentially harmful dusts may similarly be done outside normal working hours. Limitation of exposure by the allowance of adequate breaks spent in a non-contaminated environment, or by rotation of work, is an example of 'partial' segregation.

The concept that young workers are especially vulnerable to risk is questionable nowadays. Nevertheless, segregation by age is a tactic imposed on work with certain substances or processes, e.g. asbestos,[16] lead compounds,[17] chromium plating,[18] ionising radiation sources,[19] etc. On occasions it may also be necessary either to exclude women of child-bearing age and workers who are particularly sensitive, or to apply stricter standards of control to ensure their safety.[15,20]

Ventilation

The principles of control by containment or enclosure are discussed below. *Complete* enclosure may be physically impossible, for example in the following situations:

- At manual solid/liquid charging points.
- At drum or small container filling points; at bagging-off points.
- At sampling points.
- At waste or effluent discharge points.
- During cleaning or maintenance operations; during manual welding.
- During removal and disposal of empty portable containers or bags.
- During manufacturing or assembly operations involving the application, reaction, curing or removal of chemicals.
- During machining, cutting, sawing or abrading operations, (e.g. of synthetic resin-bonded fabrications) or grinding (e.g. fettling).
- During operations involving dipping into open vats/tanks (e.g. metal plating or treatment – nitriding, carburising).

286 *The Safe Handling of Chemicals in Industry*

Local exhaust ventilation designed to remove contaminants, located as near as practicable to their source and before they reach the operator's breathing zone, may then be an effective control measure.
Examples include:
- Provision of lip extraction on acid pickling and solvent degreasing vats; provision of combined blowing and rear extraction.
- Provision of slatted benches with extraction from beneath and from the rear for manipulation, assembly and finishing of synthetic resin components or where synthetic adhesives are applied.
- Provision of open-fronted, extracted booths for repetitive epoxy resin handling and use.
- Provision of small extract hoods to the rear of the manual soldering operations involving the use of colophony-based resin fluid.
- Provision of efficient local exhaust ventilation at pedestal grinders and, more recently, on portable grinders.
- Laboratory fume cupboards.
- Provision of portable extractors on welding operations.

Dilution by general ventilation is a secondary measure. Often chemical plants are located outside. Toxic gas storage facilities should be housed outdoors and supplies piped in wherever possible. Good design and maintenance are of paramount importance and are discussed in Chapter 12.

However, atmospheric dispersion does not necessarily ensure that operatives working in the open air will not be exposed to harmful concentrations of toxic chemicals. For example, hazardous exposures to gas, vapour or particulate matter may be encountered at solid-waste-disposal landfill sites, at quarries, stonemasons' yards and construction sites, during the loading of chemical tank wagons, in agriculture, etc.

Containment

Clearly effective complete enclosure will prevent any contamination of the workplace atmosphere. In the simplest case this may involve putting a lid on an open vessel. In more complex situations the plant and process may be completely segregated from personnel, e.g. by physical separation and remote handling techniques. This approach is particularly relevant for potent poisons, radioactive isotopes and carcinogens.

Laboratory operations

The precautions for handling toxic materials on a laboratory scale are discussed in Chapter 11.

Generally small quantities of toxic dusty or volatile materials may be handled in glove boxes. Alternatively, provided the level of hazard and

Control measures for handling toxic substances 287

Fig. 6.3 Pilot-scale liquid–liquid extraction plant using liquid ammonia as solvent. Equipment enclosed with exhaust ventilation canopy over and sliding access door (shown open). Provision for local alarm, and general alarm with emergency water sprays along front side of canopy

location of the vent renders it safe to do so, they may be handled in properly designed and maintained fume cupboards with use of appropriate personal protection.

Normally all reactions involving toxic substances in laboratories should, at least, be conducted in the back of a fume cupboard with the sash pulled down. Similar arrangements can be made by means of ventilated canopies on a semi-pilot plant scale, e.g. as illustrated in Fig. 6.3.

Toxic gases and volatile liquids should be handled in an efficient fume cupboard; experiments should be designed for containment, e.g. waste gases should be scrubbed, condensed or filtered through an High Efficiency Particle Absorption (HEPA) system. Operating procedures should be such as to prevent aerosol generation, i.e. avoiding dropping of liquids from a height or blowing-out of pipettes. (Obviously the use of mouth-pipetting should be prohibited.) Experiments should be undertaken in spill trays, or the fume-cupboard base should be lined to contain spillage, and the drainage points sealed off.

Hyper-poisons or toxic, dusty solids should be manipulated in a glove-box. (The draught in a fume cupboard may otherwise be sufficient to encourage fine powders to become airborne; contamination of the fume cupboard is then likely, since transport velocities are possibly inadequate to prevent particulate material settling in the ventilation system.)

Glove-boxes under negative pressure should be ventilated at a rate of at least two complete air changes per hour with an internal pressure of at least 0.5 cm water gauge below the ambient atmosphere; effluent should be scrubbed or passed through a HEPA filter.

Plant operations

On a plant scale the technology for containment of gases and liquids, which can be relatively easily transferred in pipes, is well developed. Solids, pastes and slurries present problems because they tend to be difficult to transport.[12] Thus solids handling needs to be minimised particularly if they are not free flowing; therefore toxic solids handling is considered separately later.

All equipment and pipework should be in accordance with good chemical engineering practice and conform to established codes for design, fabrication, inspection and testing. The following are general guidelines:

- Use all-welded pipelines wherever practicable. Avoid mere screwed joints; minimise joints and flanges.
- Where possible transfer lines should be self-draining with provision for back-flushing.
- With high melting-point liquids provision should be made to avoid blockages due to chilling (e.g. steam, hot water, or electric heating tape tracing).
- Provide a double block and bleed arrangement, e.g. as illustrated in Fig. 6.4, for isolation purposes.
- Minimise the use of flexible hoses or piping which have to be dismantled and re-made.
- All flexible hoses should be securely fastened (i.e. avoiding simple jubilee clips).
- Consider tank drainage arrangements and provide adequate instrumentation on storage and reactor vessels (e.g. temperature, pressure and level indicators and alarms) and well thought-out pressure relief systems (bursting discs and/or relief valves).
- Adopt in-line analysis to avoid manual sampling except when it is unavoidable (e.g. with pastes and solids).
- Consider with separation processes the inclusion of as much of the system as possible in one volume, e.g. distillation with the reboiler, column and condenser all in one shell.[12]

Emergency relief systems are conventionally employed to reduce fire hazards, explosion hazards and over-pressurisation of equipment. With toxic relief systems the best approach may be to adopt a policy of containment, i.e. to design process systems to withstand the maximum pressures and temperatures likely to occur so that relief is unnecessary.[12] Flaring and quenching and absorption have to be resorted to in some cases.

The degree of exposure coupled with the degree of hazard associated with the toxic materials dictates the extent to which total containment will be strived for. Important sources of leaks include seals, pumps and valves as summarised in Table 6.1.[3-5]

For hazardous fluids

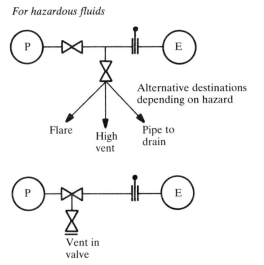

(P = process side; E = engineering access required)

For high pressures (600 psi) and/or high temperatures or for fluids with isolation problems

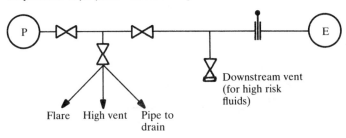

Fig. 6.4 Double block and bleed arrangements

With enclosed handling and transfer of materials on an industrial scale particular attention is required to any intermittent operations which result in 'uncontainment'. Sampling of process materials and drainage or venting of process equipment and pipelines are common examples. Methods of sampling which reduce potential exposure to process fluids are shown diagrammatically in Fig. 6.5.

Whenever practicable, transfer points, pipes, sample points and drain points should not be located over access ways. In addition to possibly shielding sample points as in Fig. 6.5(b), and indeed drain points, it is important that discharge from them does not result in splashing from sinks or gulleys.

Continuous filling of drums, bottles or tanks with liquid may result in vapour emissions in the displaced air. The quantities can be simply esti-

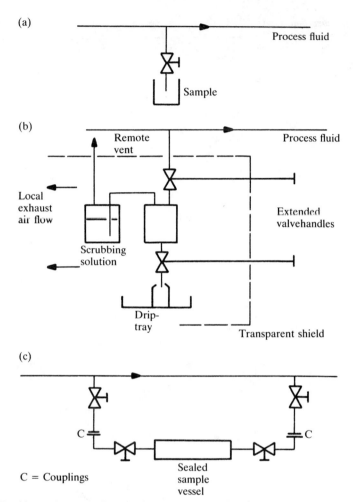

Fig. 6.5 Alternative sampling methods
(a) Simple sample valve only; (b) Shielded and exhausted sampling station; (c) Sealed sampling – precautions needed on uncoupling

mated from data on the filling rate, filling temperature and liquid vapour pressure. For example[3–5]:

Solvent X filled at 25 °C at a rate of 80 litres/minute. Saturated vapour concentration at 25 °C is 22,000 mg/m³.

Emission rate is calculated from 80 litres of air leaving the drum per minute with 22,000 mg/m³ of solvent, i.e.

$$\frac{80 \times 10^{-3} \times 22{,}000}{60} \text{ mg/second}$$

$$= 29 \text{ mg/second}$$

Control would involve a local ventilation system attached to the filling arm and withdrawing expelled air. The reduction in emission rate expected from this control would be 99.9% based on a properly designed air-moving system.

Consideration also needs to be given to precautions in the event of accidental loss of containment, e.g. by the provision of bunds to contain spillages, gas detection systems, emergency drenching facilities, and alarms. Important features are:
- Critical examination of the operations involved to identify, and if possible quantify, the problems of hazards and operability.[21]
- Plan for different, credible, leakage scenarios.
- Avoid sumps, pits and other low points where liquids or heavier-than-air vapours can accumulate if released.
- Provide conveniently accessible escape rooms which are clearly identified and provided with telephones.

Stocks of neutralising substances or adsorbents should be maintained to deal with spillages and splashes.

When it is necessary to break a pipeline which has contained corrosive chemicals, or indeed hazardous chemicals of any description, then an appropriate procedure for draining, flushing and pressure reduction should be followed (i.e. permit to work system). Eye protection, at least equivalent to chemical goggles, should be worn together with gloves and other protective clothing. Flanges should be broken by slackening the nuts on the side away from the worker who should also, when practicable, be positioned with his face above the work and not beneath it.

Temporary enclosure

In some operations, such as the stripping of asbestos-containing insulation, temporary enclosure of the working area may be used to prevent the transfer of asbestos dust to other areas. This may involve the use of existing partitions, impervious sheeting and sealing to prevent escape of dust via external openings, e.g. doors and windows.[13]

Partial enclosure

The installation and proper use of spray and splash guards may serve to reduce exposure to liquids. Examples are:
- Provision of fixed, sliding or hinged splash guards on lubricated machine tools, e.g. lathes, grinding wheels.
- Manipulation of laboratory chemicals within a fume cupboard with the front screen partially raised.
- Shielding of all glass vessels and pipework, including level gauges.

Storage

Bulk transport by road or rail in specially designed tanks or tankers is preferable to the use of small containers, but a balance is required having regard to the inventory.

Any small containers used should be of the correct design with regard to free space, pressure rating, corrosion resistance and mechanical strength. Any chemicals stored in glassware should be transported in baskets or other special carriers. Chemicals, e.g. mineral acids, which are shipped in 10 gallon polythene plastic carboys should not be handled without any outer plastic-covered metal cage provided for mechanical protection. (Some thicker gauge carboys do not require cages.) Before they are handled the cradle handles, or handles on the carboy, should be checked as a matter of routine to see that they are sound and clean.

With small containers separate 'full' and 'empty' areas should be provided. Incompatible chemicals must be stored in segregated areas. The storage area should be adequately bunded or provided with kerbs to restrict spread in the event of spillage.

Areas in which corrosive chemicals are handled, other than in completely enclosed vessels and pipes, should be marked with clear warning notices. Kerbs should be provided where appropriate to retain spillages. Good access should be provided and under no circumstances should it be necessary to stand on top of containers to lift down other containers or to reach valves, etc. Access may be restricted by means of barriers; in any event instructions regarding eye protection must be clearly displayed.

All small containers, particularly if as is common they are returnable, should be subjected to regular inspection prior to filling, movement and emptying. Plastic containers should be stored out of direct sunlight (to slow down UV initiated degradation). It is preferable to minimise manual handling of such containers; palletisation and mechanical handling result in the operator being relatively remote from the containers during lifting. Particular care is required with portable containers which may, for some reason, contain pressure slightly above atmospheric pressure.

With 45-gallon drums the bowing out of the ends is a clear indication of over-pressure but such deformation is not always present: plastic containers or Winchester bottles do not provide any indication. Over-pressure may occur if the container is filled at one process, or ambient temperature and the temperature is subsequently raised in transit or in storage, e.g. through radiant heating by sunlight. The pressure must be released by gently cracking open the cap while covered with a cloth or plastic bag; sudden removal may result in ejection of some of the contents as a spray. Transfer should whenever possible be by means of a portable drum pump or a purpose-designed siphon, i.e. avoiding lifting and pouring. Alternatively, an approved tipping device may be used. In any case any liquid which runs outside the mouth of the carboy/drum should be cleaned up regularly.

All storage containers or tanks/tankers used for transportation of chemicals should be clearly labelled with name, hazard, precautions and first-aid measures. Marking and labelling procedures are described further in Chapter 13. Guidance on the storage or laboratory chemicals is given in Chapter 11.

Health surveillance and biological monitoring

The medical background of workers must be considered for work involving certain chemicals to screen out, for example pregnant women from work with radio-active materials, hypersensitive atopics from work with sensitisers, etc.

Advice on general requirements for health surveillance by routine procedures has been published.[22] Medical surveillance, biological monitoring and pre-employment examination have been identified either as a legal requirement or as a recommendation in the UK for only a few materials or diseases. These include isocyanates,[23] mercury,[24] organophosphorus pesticides,[25] platinum salts,[26] sodium or potassium chromate and bichromate,[27] aniline and related amino and nitro compounds,[28] trichloroethylene,[29] chromic acids,[30] arsenic and its compounds,[31] beryllium,[32] antimony,[33] lead,[34] dermatitis[35] and dust diseases.[36]

Restriction of occupational exposure

The potential for exposure may be limited by minimising the number of persons permitted access to the process, and the duration of exposure. The degree of exposure of such essential personnel must still, of course, be limited to within the appropriate hygiene standard.

General hygiene

Further protection can be provided by the application of 'good housekeeping', e.g. segregation of substances, replacement of caps or seals on containers not in use, prompt disposal of waste materials and a high standard of workplace cleanliness. Where solids are involved the latter is assisted by the provision and use of vacuum equipment for general cleaning and for the removal of spillages.

Approved vacuum cleaners are available for asbestos. In some factories ducts are provided to which flexible hoses can be coupled up. Dry-sweeping should be avoided since it tends to redisperse dusts into the workplace. Pre-wetting to encourage agglomeration is an old-established technique.

The provision of good washing facilities and the encouragement of personal hygiene are discussed later. Where the minimum standards are not specified by statutory legislation, not less than one wash-basin or 0.6 m of trough should be provided for every five persons handling toxic materials. In certain cases it may be desirable to provide the washing facilities within the workrooms so that every opportunity is offered for frequent washing.

Clearly, depending on the nature of the hazard, the consumption of food or drink may need to be prohibited in factories handling toxic materials except in specially designated areas, e.g. mess-rooms or canteens. An adequate mess-room or canteen is specifically required when certain chemical processes are performed in the UK.[37] Strict personal hygiene and the discarding of overalls or laboratory coats, etc., before entry are important.

Smoking should also be prohibited except in designated areas. This not only minimises the possibility of accidental ingestion following hand contamination, but avoids any chance of toxic thermal degradation products being inhaled, e.g. highly toxic fluorine compounds from the combustion of polytetrafluoroethylene dust; toxic chlorine compounds, such as phosgene, from the thermal oxidation of chlorinated hydrocarbon vapours. There is also a proven synergistic effect of smoke with some toxic gases, e.g. sulphur dioxide, and fibres, e.g. asbestos.

Personal protection

General

Preference should be given to safe-place rather than safe-person policies. This is achieved by the foregoing handling strategies. Use of personal protection represents the safe-person approach and only offers protection for the wearer. Furthermore, personal protection must be worn continuously throughout the duration of exposure to the hazard. For example, it has been demonstrated that hearing protectors with infinite attenuation worn for 50% of the exposure time confer not more than 3 dB(A) protection. Furthermore, the defenders must be worn for 99.9% of the time in order to afford 30 dB(A) protection.[38] Similarly, respiratory protection keeping all contamination out of the operative's air supply must be worn for 99% of the time to be effective, and if worn for only 95% of the exposure period then effective protection drops to about 25%.

There is little point in providing protection which is highly efficient if the wearer is reluctant to wear it for most of the exposure period. Since most highly efficient systems tend to be cumbersome and uncomfortable, the role of personal protection must be considered a last line of defence and be restricted to hazardous situations of short duration (e.g. emergencies, maintenance, or temporary arrangements while engineering control measures are being introduced).

In summary the provision and use of personal protective equipment should normally be regarded as a back-up for other strategies which attempt to control the risk at source rather than as a first line of defence. The degree of protection afforded depends largely on selection, correct use and condition of equipment. In certain circumstances personal protection is the only reasonably practicable measure and indeed may be required by law as indicated by Table 6.5. For example, the requirements for protective clothing under the Poisonous Substances in Agriculture Regulations 1984 are summarised by Table 6.6 (see also Ch. 16).

Protection should conform to accepted standards of manufacture and performance with careful consideration of anthropometrics. In the UK the British Standards Institution has published a number of standards on personal protective equipment as shown in Table 6.7. The National Safety Council of Chicago (see Ch. 18), for example, provides details of some American equipment.

If reliance is placed on personal protection the provision of equipment must be supplemented with training, supervision, inspection and maintenance schedules and operatives must co-operate by using the equipment as instructed.

Respiratory protection

Since inhalation is the prime route of entry into the body by toxic chemicals a consideration of respiratory protection is crucial when handling such substances. Provision of clean uncontaminated air should be accomplished by engineering controls whenever practicable. However, respiratory protection may be required in the following circumstances:
- To protect against materials in the atmosphere for which no exposure is permissible (by regulations) or desirable.
- To protect against known toxic chemicals in the atmosphere where engineering controls are not yet sufficient to reduce exposure to acceptable levels.

Table 6.5 Selection of UK legislation with requirements for use of personal protection

The Aerated Water Regulations 1921.
The Asbestos Regulations 1969.
The Blasting (Castings and Other Articles) Special Regulations 1949.
The Bronzing Regulations 1912.
The Cement Works Welfare Order 1930.
The Chemical Works Regulations 1922.

Table 6.5 (continued)

The Chromium Plating Regulations 1931.
The Clay Works (Welfare) Special Regulations 1948.
The Construction (General Provisions) Regulations 1961.
The Control of Lead at Work Regulations 1980.
The Docks and Harbours Act 1966.
The Dyeing (Use of Bichromate of Potassium or Sodium) Welfare Order 1918.
The East India Wool Regulations 1908.
The Electric Accumulator Regulation 1925.
The Electricity Regulations 1908.
The File-Cutting By Hands Regulations 1903.
The Flax and Tow Spinning and Weaving Regulations 1906.
The Foundries (Protective Footwear and Gaiters) Regulations 1971.
The Fruit Preserving Welfare Order 1919.
The Grinding of Metals (Miscellaneous Industries) (Amendment) Special Regulations 1950.
The Grinding of Metals etc. (Metrication) Regulations 1981.
The Grinding of Metals (Miscellaneous Industries) Regulations 1925.
The Gut Scraping, Tripe Dressing etc., Welfare Order 1920.
The Factories Act 1961.
The Health and Safety (Foundries etc.) (Metrication) Regulations 1981.
The Hollow-ware and Galvanising Welfare Order 1921.
The Ionising Radiations (Unsealed Radioactive Substances) Regulations 1968.
The India Rubber Regulations 1922.
The Iron and Steel Foundries Regulations 1953.
The Jute (Safety, Health and Welfare) Regulations 1948.
The Laundries Welfare Order 1920.
The Magnesium (Grinding of Castings and Other Articles) Special Regulations.
The Non-Ferrous Metals (Melting and Founding) Regulations 1962.
The Offshore Installations (Operational Safety, Health and Welfare) Regulations 1976.
The Oil Cake Welfare Order 1929.
Order for Securing the Welfare of the Workers Employed in the Bevelling of Glass 1921.
The Paints and Colours Manufacture Regulations 1907.
The Pottery (Health and Welfare) Special Regulations 1950.
The Protection of Eyes Regulations 1974.
The Refractory Materials Regulations 1931.
The Regulations for the Use of Horsehair from China, Siberia or Russia 1907.
The Ship Building and Ship-Repair Regulations 1960.
The Tanning (Two-Bath Process) Welfare Order 1918.
The Tanning Welfare Order 1930.
The Tin or Terne Plate Manufacture Welfare Order 1917.
The Vehicles Painting Regulations 1926 (amended 1973).
The Vitreous Enamelling Regulations 1908.

Some of these may be repealed with the introduction of COSHH Regulations (see Ch. 16)

Table 6.6 Protective clothing required by the Poisonous Substances in Agriculture Regulations 1984. (*See Key at end of the table*)

Operations for which protective clothing must be worn	Part 1 Chemicals	Part 2 Chemicals	Part 3 Chemicals	Part 4 Chemicals
General applications				
1. Opening containers, diluting, mixing or transferring contents, adjusting apparatus, washing out containers clearing spillages, etc.				
Liquids and wettable powders	A,F,G,K,L	B,F,G,K,L	B,G,L	—
DNOC or dinoseb (used as an insecticide)	—	B,G,L	—	—
Granules	B,G,I,L *or* A,C,G,I,L	B,G,I,L *or* A,C,G,I,L	B,G,I,L *or* A,C,G,I,L	B,G,I,L *or* A,C,G,I,L
2. Spraying crops (see below if spraying climbing plants, bushes or trees or using aerosols/smoke)	(except chloropicrin) A,D,F,G,L	B,D,F,G,L	—	—
Spraying climbing plants, bushes or trees	(except chloropicrin) A,D,F,G,J	B,D,F,G,J	—	—
3. Soil application				
(i) in a greenhouse	A,F,G,K,L	F,G,K,L	—	—
(ii) other than in a greenhouse by				
(a) driver of tractor mounted apparatus				
(b) unaccompanied driver of tractor-drawn apparatus; or	F,G,L	F,G,L	—	—
(c) any operator on foot (including tractor driver while not driving)	F,G,K,L	F,G,K,L	—	—
4. Using aerosols/smoke Working in a greenhouse or livestock house when an aerosol, smoke-generator or smoke shreds are being used	A,D,G,L	A,D,G,L	A,D,G,L	A,D,G,L
5. Granule placement				
By hand or hand-operated apparatus	—	I,L	—	—
By apparatus not operated by hand, or when operating any apparatus mounted on or drawn by a tractor engaged in granule placement	—	L	—	—
6. Washing or cleansing apparatus Washing or cleansing spraying, soil-application, soil-injection or granule placement apparatus, including aircraft and their apparatus	(except chloropicrin) B,F,G,K,L	B,F,G,K,L	—	—

298 The Safe Handling of Chemicals in Industry

Table 6.6 (continued)

Operations for which protective clothing must be worn	Part 1 Chemicals	Part 2 Chemicals	Part 3 Chemicals	Part 4 Chemicals
Special applications				
7. Application from aircraft Acting as a ground-marker				
(a) When spraying liquids	—	—	L (or B,D,F,L)*	—
(b) During granule placement	—	E,L (and A+C or B)	—	—
8. Using chloropicrin Opening containers, diluting, mixing or transferring contents, adjusting apparatus, washing out containers etc	(chloropicrin) A,F,G,K,L	—	—	—
Soil injection in a greenhouse	(chloropicrin) A,F,G,L	—	—	—
Soil injection (other than in a greenhouse) by:				
(a) Driver of tractor-mounted apparatus (b) Unaccompanied driver of tractor-drawn apparatus, or	(chloropicrin) F,L	—	—	—
(c) Any operator on foot (including the tractor driver whilst he is not driving)	(chloropicrin) F,G,L	—	—	—
Washing and cleansing soil-injection apparatus	(chloropicrin) A,F,G,K,L	—	—	—
Removing the sheeting after oil injections				
(a) in a greenhouse	(chloropicrin) A,F,G,L	—	—	—
(b) Out-of-doors	(chloropicrin) F,G,L	—	—	—
9. Using nicotine Opening a container of smoke shreds, or a smoke generator (containing not more than 40% by weight nicotine) or transferring contents	—	—	(nicotine) G	—
Application to roosts, perches and other surfaces in a livestock house	—	—	B,G,L or B,D,G,L	B,G,L or B,D,G,L
10. Bulbs and root dipping Dipping, steeping and handling wet bulbs or plants, disposing of the solution and washing the apparatus	—	F,H,K,L	F,H,K,L	—

* With fentin acetate or fentin hydroxide.

Control measures for handling toxic substances 299

Table 6.6 (continued)

Operations for which protective clothing must be worn	Part 1 Chemicals	Part 2 Chemicals	Part 3 Chemicals	Part 4 Chemicals
11. Handling hops if sprayed: (a) within the previous 24 hr	—	mevinphos *or* TEPP G	—	—
(b) within the previous 4 days	G	except mevinphos *or* TEPP G	—	—
12. Livestock applications e.g. opening a container of sheep dip, warble fly dressing; diluting, mixing or transferring contents, adjusting apparatus, washing out containers, etc.	—	—	B,G,L	—

Key: *Precautions*

- A = respirator
- B = face-shield
- C = goggles
- D = hood
- E = head protection
- F = rubber boots
- G = rubber gloves
- H = rubber gauntlet gloves
- I = sleeves worn over rubber gauntlet gloves
- J = rubber coat
- K = rubber apron
- L = overall

Key: *Specified substances*

Part 1
Chloropicrin
Demeton
Dimefox
Mazidox

Part 2
Aldicarb
Amiton
Carbofuran
Carbosulfan
Cycloheximide
Dialifos
Dinoseb and its salts
Dinoterb and its salts
Disulfoton
DNOC and its salts
Endosulfan
Endothal and its salts
Endrin
Fluoroacetamide
Fonofos
Medinoterb and its salts
Mephosfolan
Methomyl
Mevinphos
Mipafox
Oxamyl
Parathion
Phorate
Potassium arsenite
Schradan
Sodium arsenite
Sulfotep
TEPP
Thiofanox
Thionazin

Part 3
Amitraz
Azinphos-ethyl
Azinphos-methyl
Chlorfenvinphos
Deltamethrin
Demephion
Demetonmethyl
Demeton-S-methyl
Demeton-S-methyl sulphone
Dichlorvos
Dioxathion
Drazoxolon
Ethion
Fenaminosulf
Fenazaflor
Fentin acetate
Fentin hydroxide
Formetanate
Mecarbam
Methidathion
Nicotine and its salts
Omethoate
Oxydemeton-methyl
Phenkapton
Phosphamidon
Pirimiphos-ethyl
Quinalphos
Thiometon
Triazophos
Vamidothion

Part 4
Organo-mercury Compounds

This table is for guidance only. It is not intended as an authoritative interpretation of the Regulations.

Table 6.7 British Standards for protective clothing and equipment

Part protected	Short title or description	BS NO.
Eyes	Industrial – general (non-radiation)	2092
	Filters for welding, etc.	679
	Green, for steel workers	1729
	Filters for intense sun	2724
	X-ray protection	4031
Respiratory system	Guide to selection	4275
	Breathing apparatus (air line and self-contained – in 4 parts)	4667
	Respirators, general dusts and chemical	2091
	Dust respirators, high efficiency	4555
	Dust respirators, powered	4558
Face and eyes, etc.	Radiation protection, welding	1542
	Dust hoods and blouses, powered	4771
Head	Industrial safety helmets	5240
	Scalp protection, light duty	4033
	Welders' helmets	1542
Hands	Industrial gloves	1651
	Rubber gloves – for electricians	697
Feet	Safety footwear (3 parts)	1870
	Womens safety footwear	4972
	Antistatic footwear	2506
	Conducting footwear	3825
	Footwear and gaiters, for foundries	4676
Body (clothing)	Air and liquid impermeable	4724
	For construction workers	4679
	For intense heat	3791
	Waterproof	4170
	Welders	2653
	X-ray protecting aprons	3783

- To protect against toxic chemicals during repair or maintenance work.
- As standby equipment for emergency use in case of accidental release of toxic materials.
- For use in occupational situations where inhalation hazards clearly exist, but their precise nature may be qualitatively, or quantitatively, indeterminate in any given situation and subject to unpredictable variation (e.g. fire-fighting).
- For use in oxygen-deficient atmospheres (see Ch. 5).
- For use in situations where there is a combination of an oxygen-deficient atmosphere and the presence of toxic chemicals.
- For use in situations where materials are present in the atmosphere which are not necessarily toxic at the concentrations likely to be encountered, but which may cause other effects which are detrimental to a safe and efficient working environment, e.g. the presence of peripheral sensory irritant materials which may be objectionable or

predispose to accidents by producing incapacitating effects during exposure.

The earliest record of respiratory protection is attributed to Pliny (AD 23–79) who described the use of animal bladders that minium refiners wore over their faces as protection against dust inhalation (minimum, the Latin for cinnobar, i.e. red mercuric sulphide). However, most significant advances in respirator design have occurred since the First World War. The main types of respirator are described below.

Air purifying respirators

Normally these comprise a face piece (quarter, half or full face) and a filter through which local air is drawn via a non-return valve to remove contaminant. Expired air is eliminated from the face piece to the environment via a non-return exhalation valve. These devices (one type is shown in Fig. 6.6a) may be designed to remove gases, vapour or particulate matter. They are of no value in oxygen-deficient atmospheres.

Vapour- and gas-removing filters These devices are designed to remove specific substances, or classes of substance, by sorption in a bed of particulate chemical housed in a cartridge or canister. Combinations of filters may be required and indeed prefilters to remove dust or aerosol are available. Although canisters and cartridges are coded there is no internationally acceptable system. The life of the canister depends upon its loading, which is determined by atmospheric concentrations of contaminent plus its duration of use. Generally no indication is given of failure of the filter. Complications arise from previous usage and unsealed storage. Inspired air not only enters via the canister but the face piece also; even when perfectly fitted this can account for one-tenth of the total inspired air, rising to half in the case of poorly fitted devices or when the user has a beard or wears spectacles.

> A senior operator worked for most of his shift cleaning a troublesome batch of crude intermediate products from a plant. Fumes of sulphur dioxide were being evolved from the product so the operator wore a respirator.
>
> The following morning he had a particularly tight chest as a result of constant inhalation of low concentrations of sulphur dioxide. This arose because he was bearded which prevented the respirator from forming a good seal with his face thus allowing gas to pass.[39]

Particulate removing filters These devices protect the wearer from airborne particulate matter such as dust, fumes or aerosols. Filters are usually of loose fibre of cotton, wool, synthetics or a paper of these.

Efficiency increases as contaminants are deposited though increased resistance to breathing is also associated with such loadings. Frame- and pad-dust respirators are generally only suitable for use with nuisance particulates.

For very toxic substances of small aerodynamic diameter (e.g. below 1 μm) high efficiency filters are recommended such as encapsulated filter cartridges of resin impregnated merino wool (electrostatically charged). Where dust levels are high, e.g. at or above 10 × TLV then breathing apparatus is required.

Disposable face pieces such as that shown in Fig. 6.6b are available with high user acceptance. For certain well-known hazards it has received

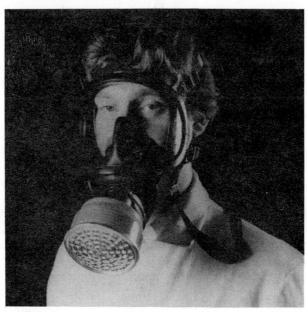

Fig. 6.6a A lightweight full-face cartridge respirator. (Courtesy Arco Ltd): A comprehensive range of DIN-approved filter cannisters is available for a host of gases, dusts or combinations of both.

approval by the Health and Safety Executive and the British Standards Institution: International Committee Européan de Normalisation (CEN) standards are in draft.

Positive pressure, powered dust respirators are available. These consist of a full face-piece mask supplied with filtered air from a battery-operated power pack pulling through a high-efficiency filter carried in a haversack. These can be used in atmospheres containing up to 500 times the TLV. An alternative version is a protective helmet into which is incorporated an electrically operated fan and filter unit and complete with face visor and provision for ear muffs. With clean filters and a fully charged battery an air flow rate of 200 l/min clean air is achieved, more than adequate for heavy manual work. The slight positive pressure generated within the visor reduces inward leakage. Various grades are available for dusts,

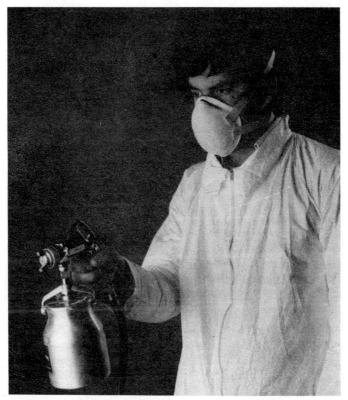

Fig. 6.6b Disposable 3M filtering face piece (Courtesy Arco Ltd). A range of models are available including a version approved by the Health & Safety Executive for use against asbestos exposure.

fumes and vapours. A fuller description is given in reference 40 and an example is shown in Fig. 6.7.

Atmosphere-supplied respirators

Three basic types exist:
1. *Short-distance fresh-air breathing apparatus* These respirators consist of face mask linked by a flexible hose to uncontaminated air which is made available either:
    (a) by being drawn directly by the wearer's inspiration effort (exhaled air leaves via a non-return valve in the mask); or
    (b) by a hand-operated rotary blower or bellows located in uncontaminated air.

Neither relies on a compressor. Problems are associated with the possibility of contamination of air supply, e.g. owing to winds, the kinking of hose, and inward leakage into the face piece owing to negative internal pressure.

Fig. 6.7 Airstream helmet (Courtesy Arco Ltd). A range of models is available to guard against a variety of hazards including nuisance dusts, toxic pollutants, welding hazards and an intrinsically safe version for use in flammable environments.

2. *Compressed airline breathing apparatus* These comprise a face piece or hood linked to a filter and hand-operated valve which is connected by tubing to an independent remote source of compressed air. In some designs the hood is an extension of a plastic blouse with an elasticated waist and cuffs or part of a complete plastic suit.

   The air is supplied from a compressor, compressed air cylinders or via a ring main system. The air inlet line to the compressor must extend into an area where contaminants will not be sucked in and be provided with a filter; rainwater must be excluded. The compressor itself should be of low temperature, lubricant-free design unless linked to special purification facilities to remove contaminants such as oil mist, carbon monoxide, moisture, etc. To account for periods of heavy exertion a capacity of 150 l/min per mask is required, even though normal consumption is below 60 l/min. For the loose hood variety about 140 l/min is required to prevent inward leakage of contaminant during mask movement.

3. *Self-contained breathing apparatus* The two main types available are:
   (a) *Regenerative oxygen respirators* which consist of an appropriate face piece connected by flexible tubing to a reservoir breathing bag and absorbent for exhaled carbon dioxide. Free air returns to the reservoir and oxygen adjustment is provided by a high-pressure cylinder via reduction and demand valves.

      Generally this version of aspirator is not recommended. Because the carbon dioxide absorbing reaction is exothermic the temperatures of the canister can reach 100 °C. Care is required in flammable atmospheres of vapour with low auto-ignition temperature (e.g. carbon disulphide).

   (b) *Open circuit compressed air type* Air is supplied from cylinders on the user's back via reduction and demand valves to the face piece. (See Fig. 6.8.) Compared with the canister and dust respir-

Control measures for handling toxic substances

Fig. 6.8 Self-contained breathing apparatus in use (Courtesy Sabre Gas Detection Ltd)

ator this system operates at lower face piece suction pressure resulting in lower face piece leakage. Exhaled air is vented via a non-return valve. Cylinder pressure must be readily checked by the wearer using a pressure gauge attached to an extension tube. Some are equipped with a low-pressure audible alarm. Cylinders of 1200 l capacity pressurised to 132 atmospheres provide protection for about 75 minutes dropping to 25 minutes for heavy exertion. Smaller cylinders of 400 l capacity last approximately 20 minutes.

Since stresses may be imposed as a result of respiratory protective gear, medical examination of workers may be required where there occupations demand wearing respiratory protective equipment for significant periods of time.

Any user of respiratory protection requires frequent training and the manufacturer's instructions must be followed. Training should cover potential hazards, the need for respiratory protection, the type chosen and why, how it works, fitting, care and maintenance, how to recognise faults and leaks and the limitations of the equipment. Equipment should be stored out of direct light and maintained monthly, details logged and defects reported immediately. Whenever fitted the face piece should be tested for air-tightness by kinking the supply tube, disconnecting the air supply line from valve or closing the cylinder outlet valve. If the face piece collapses on the face the face-piece outlet non-return valve and head harness should be checked. After use the face piece should be cleaned with mild disinfectant (e.g. 5% solution) rinsed and dried thoroughly.

> Two operators had been called to change a filter on a plant. One donned an air-fed hood and the other self-contained breathing apparatus. Both felt the air supply to be only just adequate. Shortly the operator wearing the hood felt uncomfortable and removed it. He immediately inhaled a very high concentration of methylene chloride/methanol vapour and collapsed. His companion removed his own breathing apparatus to drag him away. A third man was summoned and first aid was given. The affected man recovered consciousness and was given medical attention.
>
> On investigation it was revealed that the hood had not received the normal cleaning treatment after use, had not been disinfected, and had not been inspected and placed in a protective polythene bag. The silencer in the air supply was blocked with rust and there was a defective regulator on the compressed air system.[41]

Thus prior to selecting respiratory protection the identity of airborne contaminant must be known together with the type of work, duration of task, ease of access and the oxygen content of the air. Consideration needs to be given to hazards generated in the operation to be undertaken (e.g. cleaning, hot work).

> When welding is necessary on any pipe which passes hydrogen from the decomposer of a mercury cell (used for the electrolysis of brine) the pipe should be thoroughly cleaned initially and self-contained breathing apparatus should be worn by the welding crew. The reason for this is that mild steel absorbs mercury, volatilised in the hydrogen, and although a pipe appears clean, mercury vapour will be released on heating.[42]

Additional criteria surround ergonomics such as limited restriction of use or vision, weight, size, tightness and fit of face piece, quality of inhalation and exhalation valves, comfort, draining of sweat, headbands, resistance to breathing, ease of cleaning and sterilisation, portability and strength. Respirators, where possible, should be approved for use. Relevant British Standards are given in Table 6.8. In the UK some

Table 6.8 British Standards relevant to respiratory protection

BS Number	Title
BS 2091/1969	Specification for respirators for protection against harmful dusts, gases and scheduled agricultural chemicals.
BS 4275/1974	Recommendations for the selection, use and maintenance of respiratory protective equipment.
BS 4400/1969	Method for sodium chloride particulate test for respirator filters.
BS 4555/1970	Specification for high efficiency dust respirators.
BS 4558/1970	Specification for positive pressure, powered dust respirators.
BS 4667/1974	Specification for breathing apparatus Pt 1 Closed circuit breathing apparatus. Pt 2 Open circuit breathing apparatus. Pt 3 Fresh air hose and compressed air-line breathing apparatus.
1982	Pt 4 Escape breathing apparatus.
BS 4771/1971	Specification for positive pressure powered dust hoods and blouses.
BS 6061/1980	Specification for filtering face-piece dust respirators.

respirators have approval by the Health and Safety Executive.[43] In the US approval of respirators meeting minimum performance standards is given by the Mines Safety and Health Administration and by the National Institute of Occupational Safety and Health.

Figure 6.9 illustrates a first approach to respiratory selection. An indication of the atmospheric concentration of contaminant is needed in order to determine the nominal protection factor (NPF) required. The latter describes the level of protection afforded by a specific respirator and is defined thus:

$$\text{NPF} = \frac{\text{concentration of contaminant in air}}{\text{concentration of contaminant inside face piece}}$$

Table 6.9 summarises the NPF achievable with different respiratory protection devices and Table 6.10 provides guidance on choice of NPF in various environments.[44]

Skin (hand and arm) protection

Two approaches are available for providing a barrier between hand-skin and chemicals, viz. gloves and barrier creams.

Gloves

A wide selection of gloves, mitts and gauntlets of different styles and specifications is available. As a consequence the wrong type is often

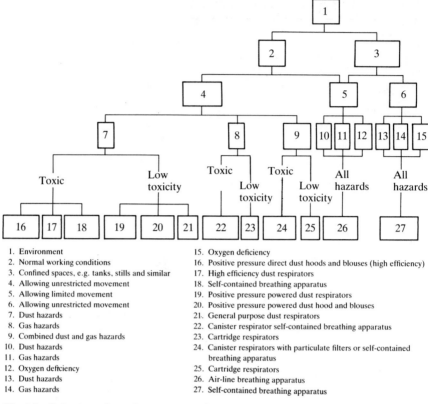

Fig. 6.9 Selection of respiratory protective equipment

1. Environment
2. Normal working conditions
3. Confined spaces, e.g. tanks, stills and similar
4. Allowing unrestricted movement
5. Allowing limited movement
6. Allowing unrestricted movement
7. Dust hazards
8. Gas hazards
9. Combined dust and gas hazards
10. Dust hazards
11. Gas hazards
12. Oxygen deficiency
13. Dust hazards
14. Gas hazards
15. Oxygen deficiency
16. Positive pressure direct dust hoods and blouses (high efficiency)
17. High efficiency dust respirators
18. Self-contained breathing apparatus
19. Positive pressure powered dust respirators
20. Positive pressure powered dust hood and blouses
21. General purpose dust respirators
22. Canister respirator self-contained breathing apparatus
23. Cartridge respirators
24. Canister respirators with particulate filters or self-contained breathing apparatus
25. Cartridge respirators
26. Air-line breathing apparatus
27. Self-contained breathing apparatus

chosen out of ignorance and inadequate protection afforded. Clearly for wet chemicals they must be impervious since otherwise liquid penetration or saturation of the material will rapidly occur, which may prolong exposure.

The factors to be considered when selecting gloves are the combination of hazards to be protected against (e.g. corrosive chemicals, irritants, heat and abrasion), the degree of resistance required, the extent of sensitivity required, and the area to be protected.

In the UK BS 1651 provides a classification system (Table 6.11) together with recommendations for specific hazards (Table 6.12). The types of rubber used in rubber gloves are designated by the letters A, B and C where A = natural rubber latex solution, B = chloroprene latex and C = nitrile latex. R refers to reinforcement. BS 3120 should be consulted if flame resistance is required.

Leading glove manufacturers also publish guidance on the suitability and degree of resistance of glove materials to different chemicals, as

Table 6.9 Nominal protection factors

Type of equipment	Protection factor
General purpose dust respirators	7 to 10
Positive pressure, powered dust respirators	20 to 500
Positive pressure, powered dust hoods and blouses	20 to 500
High efficiency dust respirators	1,000
Cartridge-type gas respirators	20 (gas only)
Canister-type gas respirators	400 (gas only)
Escape respirators	BS in course of preparation
Fresh air hose apparatus without blower	2,000
Fresh air hose apparatus with blower	2,000
Compressed air line apparatus: demand valve type	2,000
Compressed air line apparatus: constant flow type	1,000 to 2,000
Self-contained breathing apparatus: closed-circuit type	2,000
Self-contained breathing apparatus: open-circuit type	2,000
Escape breathing apparatus	BS in course of preparation
Combination self-contained and air-line breathing apparatus	According to use

Note The figures are based on the test procedures and requirements detailed in the relevant British Standards to which reference should be made.

Table 6.10 Protection factor formula

$$\text{Protection factor (PF)} = \frac{\text{Ambient air concentration}}{\text{Concentration inside face-piece or enclosure}}$$

Protection factor	Respirator efficiency (%)	Face-piece penetration (%)	Selection guide for maximum use concentration ($\times$ TLV)
5	80	20	5
10	90	10	10
20	95	5	20
50	98	2	50
100	99	1	100
200	99.5	0.5	200
500	99.8	0.2	500
1,000	99.9	0.1	1,000
2,000	99.95	0.05	2,000
5,000	99.98	0.02	5,000
10,000	99.99	0.01	10,000

exemplified by the list given in Table 6.13 (e.g. by James North & Sons Ltd), or as a wallchart (e.g. from Edmont Europe – a division of Becton Dickinson, Benelux). Certain glove materials enable chemicals to penetrate, a property facilitated by the presence of defects in the glove such as holes, perishing, etc. Gloves should be inspected carefully before use, and thoroughly washed prior to removal.

Table 6.11 Classification of gloves

Type 1.	Chrome leather wrist gloves
Type 2.	Chrome leather inseam mitts and one finger mitts, wrist and gauntlet style
Type 3.	Chrome leather gauntlet gloves
Type 4.	Chrome leather inseam gauntlet gloves, with canvas or leather cuffs, with or without rein for cement between thumb and forefinger
Type 5.	Discarded
Type 6.	Chrome leather stapled double palm wrist gloves
Type 7.	Horse hide or cattle hide inseam gauntlet gloves, with vein patches and aprons covering palm to first joints of fingers
Type 8.	Horse hide or cattle hide palm, chrome leather back, inseam wrist gloves.
Type 9.	Felt mitts, palms faced with canvas, mole or chrome leather, wrist and gauntlet style
Type 10.	Polyvinyl chloride wrist gloves and gauntlet gloves, lined
Type 11.	Polyvinyl chloride wrist gloves and gauntlet gloves, unlined
Type 12.	Lightweight natural and synthetic rubber wrist gloves and guantlet gloves, unlined
Type 13.	Medium weight, as type 12
Type 14.	Heavy weight, as type 12
Type 15.	Handguards, rubber, leather or fabricated with rubber
Type 16.	Cotton drill gloves
Type 17.	Cotton drill gloves with chrome leather palms
Type 18.	Lightweight natural and synthetic rubber wrist gloves and gauntlet gloves, lined
Type 19.	Medium weight, as type 18
Type 20.	Heavy weight, as type 18

Table 6.12 Classification of hand hazards and recommended glove types (BS 1651)

Hazard group	Protection against	Typical operations	Recommended types of gloves*
A	Heat, but wear not serious and no irritant substances present	Furnace work drop-stamping, casting and forging, handling hot tyres and similar operations	1, 2, 3, 4, 9
B	Heat, when wear more serious or irritant substances present	Stoking gas retorts, riveting, holding-up, hot chipping	2, 3, 9
C	Heat, when fair degree of sensitivity is required, and splashes or spatters of molten metal may occur	Welding, case-hardening in cyanide bath	2, 3, 4
D	Sharp materials or objects	Swarf, metal after guillotining, blanking or machining	2, 4, 7, 9, 10, 14A, 14A(R), 15, 20A
E	Sharp materials or objects in an alkaline de-greasing bath		10, 11, 14A, 14A(R), 19BC, 20BC

Table 6.12 Continued

Hazard group	Protection against	Typical operations	Recommended types of gloves*
F	Glass or timber with splintered edges		15, 17
G	Abrasion	Handling cold castings or forgings precast concrete, bags of cement and bricks	2, 6, 8, 10, 12A(R), 13, 13AC, 14AC, 15, 17, 18AC 19AC, 20AC
H	Gross abrasion	Shot blasting	2, 9, 12A(R), 13A(R), 14A
J	Light abrasion	Light handling operations	1, 2., 9, 10, 12ABC, 13ABC, 15, 16, 17
K	Chemicals	Acids, alkalis, dyes and general chemical hazards not involving contact with solvent or oils	10, 11, 12ABC, 13ABC, 14ABC, 19ABC, 20ABC, 12A(R), 13A(R), 14A(R)
L	Solvents, oils and grease	General chemical hazards involving contact with solvents or oils	10, 11, 12BC, 13BC, 14BC, 18BC, 19BC, 20BC
M	Electrolytic deposition	Plating and subsequent operations	10, 11, 12A(R), 12ABC, 13A(R), 13ABC, 14A(R), 14ABC, 19ABC, 20ABC
N	Hot alkaline cleaning baths		10, 11, 13(R), 13ABC, 14ABC, 19ABC, 20ABC
O	Spraying paints or cellulose lacquers		1, 9, 11, 12BC, 13BC, 18BC, 19BC
P	Special hazards: lead tetraethyl, mercury, lead salts		10, 13ABC, 14ABC
Q	Classification discarded		
R	Electric shock	See BS 697 Rubber gloves for electrical purposes	

* The key to the code is given in BS 1651.

The barrier function of the skin itself can be reduced by increased hydration, which can be the result of wearing impervious gloves, which do not allow sweat to evaporate. The added use of cotton liners can be of benefit. These must be changed fairly often. This build-up of moisture otherwise promotes bacterial and fungal growth, enhances the damaging effects of irritants, and facilitates penetration of allergens. These chemicals can come into contact with skin as a result of: (a) re-exposed hands

Table 6.13 Glove resistance ratings for various glove materials (by courtesy of James North & Sons Ltd)

Resistance rating code: E, Excellent; F, Fair; G, Good; NR, not recommended

Chemical	Natural rubber	Neoprene	Nitrile	Normal PVC	High grade PVC
Organic acids					
Acetic acid	E	E	E	E	E
Citric acid	E	E	E	E	E
Formic acid	E	E	E	E	E
Lactic acid	E	E	E	E	E
Lauric acid	E	E	E	E	E
Maleic acid	E	E	E	E	E
Oleic acid	E	E	E	E	E
Oxalic acid	E	E	E	E	E
Palmitic acid	E	E	E	E	E
Phenol	E	E	G	E	E
Propionic acid	E	E	E	E	E
Stearic acid	E	E	E	E	E
Tannic acid	E	E	E	E	E
Inorganic acids					
Arsenic acid	G	G	G	E	E
Carbonic acid	G	G	G	E	E
Chromic acid (up to 50%)	G	F	F	E*	G
Fluorosilicic acid	G	G	G	E	G
Hydrochloric acid (up to 40%)	G	G	G	E	G
Hydrofluoric acid	G	G	G	E*	G
Hydrogen sulphide (acid)	F	F	G	E	E
Hydrogen peroxide	G	G	G	E	E
Nitric acid (up to 50%)	NR	NR	NR	G*	F*
Perchloric acid	F	G	F	E*	G
Phosphoric acid	G	G	G	E	G
Sulphuric acid (up to 50%)	G	G	F	E*	G
Sulphurous acid	G	G	G	E	E
Saturated salt solutions					
Ammonium acetate	E	E	E	E	E
Ammonium carbonate	E	E	E	E	E
Ammonium lactate	E	E	E	E	E
Ammonium nitrate	E	E	E	E	E
Ammonium nitrite	E	E	E	E	E
Ammonium phosphate	E	E	E	E	E
Calcium hypochlorite	NR	G	G	E	E
Ferric chloride	E	E	E	E	E
Magnesium chloride	E	E	E	E	E
Mercuric chloride	G	G	G	E	E

* Resistance not absolute, but the best available.

Table 6.13 (continued)

Resistance rating code: E, Excellent; F, Fair; G, Good; NR, not recommended

Chemical	Natural rubber	Neoprene	Nitrile	Normal PVC	High grade PVC
Potassium chromate	E	E	E	E	E
Potassium cyanide	E	E	E	E	E
Potassium dichromate	E	E	E	E	E
Potassium halides	E	E	E	E	E
Potassium permanganate	E	E	E	E	E
Sodium carbonate	E	E	E	E	E
Sodium chloride	E	E	E	E	E
Sodium hypochlorite	NR	F	F	E	E
Sodium nitrate	E	E	E	E	E
Solutions of copper salts	G	G	G	E	E
Stannous chloride	E	E	E	E	E
Zinc chloride	E	E	E	E	E
Alkalis					
Ammonium hydroxide	E	E	E	E	E
Calcium hydroxide	E	E	E	E	E
Potassium hydroxide	E	G	G	E	E
Sodium hydroxide	E	G	G	E	E
Aliphatic hydrocarbons					
Hydraulic oil	F	G	F	G	E
Paraffins	F	G	E	G	E
Petroleum ether	F	G	E	F	G
Pine oil	G	G	E	G	E
Aromatic hydrocarbons[†]					
Benzene	NR	F	G	F	G
Naphtha	NR	F	F	F	G
Naphthalene	G	G	E	G	E
Toluene	NR	F	G	F	G
Turpentine	F	G	R	F	G
Xylene	NR	F	G	F	G*
Halogenated hydrocarbons[†]					
Benzyl chloride	F	F	G	F	G
Carbon tetrachloride	F	F	G	F	G
Chloroform	F	F	G	F	G
Ethylene dichloride	F	F	G	F	G
Methylene chloride	F	F	G	F	G
Perchloroethylene	F	F	G	F	G
Trichloroethylene	F	F	G	F	G

[†] Aromatic and halogenated hydrocarbons will attack all types of natural and synthetic gloves. Should swelling occur, switch to another pair, allowing swollen gloves to dry and return to normal.

Table 6.13 (continued)

Resistance rating code: E, Excellent; F, Fair; G, Good; NR, not recommended

Chemical	Natural rubber	Neoprene	Nitrile	Normal PVC	High grade PVC
Esters					
Amyl acetate	F	G	G	F	G
Butyl acetate	F	G	G	F	G
Ethyl acetate	F	G	G	F	G
Ethyl butyrate	F	G	G	F	G
Methyl butyrate	F	G	G	F	G
Ethers					
Diethyl ether	F	G	E	F	G
Aldehydes					
Acethaldehyde	G	E	E	E	E
Benzaldehyde	F	F	E	G	E
Formaldehyde	G	E	E	E	E
Ketones					
Acetone	G	G	G	F	G
Diethyl ketone	G	G	G	F	G
Methyl ethyl ketone	G	G	G	F	G
Alcohols					
Amyl alcohol	E	E	E	E	E
Butyl alcohol	E	E	E	E	E
Ethyl alcohol	E	E	E	E	E
Ethylene glycol	G	G	E	E	E
Glycerol	G	G	E	E	E
Isopropyl alcohol	E	E	E	E	E
Methyl alcohol	E	E	E	E	E
Amines					
Aniline	F	G	E	E	E
Butylamine	G	G	E	E	E
Ethylamine	G	G	E	E	E
Ethylaniline	F	G	E	E	E
Methylamine	G	G	E	E	E
Methylaniline	F	G	E	E	E
Triethanolamine	G	E	E	E	E
Miscellaneous					
Animal fats	F	G	G	G	E
Bleaches	NR	G	G	G	E
Carbon disulphide	NR	F	G	F	G
Degreasing solution	F	F	G	F	G
Diesel fuel	NR	F	G	F	E
Hydraulic fluids	F	G	G	G	E
Mineral oils	F	G	E	G	E

Table 6.13 (continued)

Resistance rating code: E, Excellent; F, Fair; G, Good; NR, not recommended

Chemical	Natural rubber	Neoprene	Nitrile	Normal PVC	High grade PVC
Ozone resistance	F	E	G	E	E
Paint and varnish removers	F	G	G	F	G
Petrol	NR	G	G	F	G
Photographic solutions	G	E	E	G	E
Plasticisers	F	G	E	G	E
Printing inks	G	G	E	G	E
Refrigerant solutions	G	G	E	F	G
Resin oil	F	G	G	G	E
Vegetable oils	F	G	G	G	E
Weed killers	G	E	E	G	E
White spirit	F	G	G	F	G
Wood preservatives	NR	G	G	F	G

once gloves are removed; (b) penetration under cuffs or direct penetration through glove membrane; or (c) by contaminated hands being gloved.

Gloves must be of sufficient length to avoid ingress of chemicals via the open tops. Gloves with knitted cotton wristbands are inadvisable on wet work since impregnation of these bands may cause chaffing and irritation of the wrists. If long gauntlets are impracticable the gloves may be sealed at the wrists by the use of elasticated armlets. The most dexterity is retained with surgeons'-type rubber gloves or with thin disposable plastic gloves but neither have good abrasion resistance.

In some operations in factories the use of gloves may be impracticable, e.g. assembly of very small components, or dangerous, e.g. when working with the hands close to moving machinery. (In such cases the other measures to limit contact with potential dermatitic agents become predominant.)

Any gloves provided should be changed whenever necessary. This is essential since exposure to chemicals together with flexing and abrasion will eventually bring about deterioration of all types of gloves. The use of faulty gloves, with holes or tears in them should be avoided since if ingress of contaminant occurs exposure to it may be prolonged.

Certain individuals may develop an allergic reaction towards contact with rubber gloves themselves.

Creams

Basically two types of cream are available, viz. cleansing cream and barrier cream. Each has a different function and the two must not be confused. Cleansing creams are used at the end of the work period to aid removal of dirt or to condition skin with humectant. These do not provide protection from chemicals. Barrier creams are applied before working with chemicals to provide a protective film. Unrestricted use of barrier creams must be prevented otherwise the risk of injury may actually increase. Limitations of barrier creams include the following:
- They can become a reservoir of harmful chemical.
- Unless properly chosen and applied they can lead to a false sense of security.
- Water-soluble preparations intended for dry work must not be used for aqueous situations or where perspiration is excessive. Water-repellant creams are marketed for these duties.
- There is a limit to the quantity of chemical a film of cream can repel.
- Their effectiveness diminishes as a result of massage (e.g. rubbing on cuffs, tools, etc.), which forces chemical into hair follicles.
- Some barrier creams contain mild irritants such as soaps and their regular use may cause chapping of the skin.

Barrier creams are usually only applied to the hand and forearms, and are not generally used beneath gloves. However, suitable ointment (e.g. Vaseline) for application inside the nose is provided for chrome platers[45] and the provision of barrier cream, generally applied to the face, is a requirement for workers using pitch as a binder.[46]

Eye protection

In the UK the proper bases for the provision of eye protection are the Protection of Eyes Regulations 1974 and British Standard 2092.

The main requirements of the Regulations are, in summary:
- Employees shall be provided with eye protectors and/or with fixed shields, according to the nature of the process (35 different processes are listed in the Schedule of the Regulations).
- Eye protectors shall be given into the possession of the worker, and sufficient eye protectors should be provided for use by occasional workers.
- The employer shall replace lost, defective or unsuitable eye protectors, and shall keep available a sufficient number to enable him to do this.
- Eye protectors and shields must be suitable for the person for whose use they are provided.
- Eye protectors and shields shall be made in comformity with an appropriate and approved specification and shall be marked to indicate the use for which they are intended.

Persons provided with eye protectors and shields shall use them while engaged in a specified process or in any place where there is a reasonably foreseeable risk of eye injury.

The user shall take reasonable care of eye protectors and shields and report any loss or defect.

The selection of the type of personal eye protection depends upon the nature of the hazard. (Protection against UV radiation and laser light is discussed in Ch. 2.) The normal range comprises:

1. *Safety spectacles* – a variety of styles and colours are available with different side-shields to protect against lateral hazards. Attention must be given to both the lens (available toughened, tinted or even to personal prescription for wearers requiring corrected vision) and the frame (since the best safety lens is only as effective as the frame which holds it).

    Eye protection should be provided for visitors and a wide selection of clip-ons or visitors' spectacles exists.

2. *Cup and wide-vision goggles* – these tend to be more versatile and cheaper and offer more protection and, according to design, can protect against fine dust, fumes, liquid splashes and impact from flying particles. They tend to be less comfortable than safety glasses and usually cannot be worn over ordinary spectacles. They often mist up and as a rule are not fitted with prescription lenses.

3. *Face shields* – these are intended to protect from the forehead to the neck. Some are attached to head gear and equipped with a chin-guard to prevent upward splashing of acids or alkalis. Often the transparent screen is made of polycarbonate though heavier special versions are available for, e.g. welding.

Fixed shields are also available. They may be constructed of polycarbonate plastic, to provide protection against splashing and projectiles, or of toughened glass or perspex for protection against splashing only. Design should allow for 'wrap-around' of the hazard and if the need for access cannot be eliminated, e.g. by extending valve handles or switches outside the screen, then personal protection is also required.

Head protection

The most obvious hazard against which head gear is required is that of falling objects, but it may also protect against heat, chemical splashes, entanglement of hair in machinery parts etc. Specifications are given in BS 2095 and 2826. Head gear may also protect against extremes of weather and provide shade. Hoods and suits designed for fire fighting or protection against toxic chemicals cover head and neck and provide protective head gear in their design.

Footwear

The range of protective footwear includes shoes, boots with steel toe-caps, and full boots of the Wellington type. The material of construction varies and choice dictates durability, acid resistance, oil resistance, non-electrically conducting, heat resistance, non-slip and impact resistance. Special non-conducting footwear is required for work on electrical equipment; conversely conducting rubber boots may be provided to reduce static electricity generation/accumulation. The arrangements for issue of protective footwear vary between companies from free-issue to repayments through wage-deduction schemes.

Protective clothing

Types of garment available include overalls, bibs, duffle coats, aprons, spats and leggings and are selected with a view to giving protection against chemicals, oils, extremes of temperature, or mechanical hazards. The nature of the hazard dictates the choice of material but user comfort must be considered (e.g. cotton and terylene overalls allow the body to 'breathe' more than pure synthetic garments). Table 6.14 lists the properties of a range of protective clothing materials.[47] The UK Chemical Industries Association recently prepared guidelines on the selection, use, care, training, management control aspects and comments on materials of manufacture of protective garments.[48] A recent development is disposable clothing made from lightweight paper or non-woven material which also have benefits from the point of view of personal hygiene.

The importance of using impervious protective clothing when handling corrosive chemicals, or chemicals toxic by skin absorption, cannot be over-emphasised.

> A petroleum refinery worker was splashed in the face with anhydrous hydrofluoric acid while wearing a visor, neoprene boots, gloves and jacket. Some ten minutes after the accident he reappeared in the control room with wet clothes. Water washing and treatment with magnesium oxide was begun immediately but an ambulance arrived and he was taken to hospital. Despite extensive treatment and monitoring he died the following morning.

Protective clothing should be properly maintained, i.e. in a sound condition, cleaned and/or washed regularly as appropriate, and be stored in a place apart from everyday clothing. With toxic or dermatitic materials a double locker system is advisable.

Washing facilities

Personal hygiene is important in the prevention of occupational diseases. Therefore, adequate washing facilities should be provided in convenient areas and time must be allowed for their use.

Table 6.14 Properties of materials of protective clothing (after ref. 47)

Material	Advantages	Disadvantages	Notes
Cotton	Lightweight, reasonably hard-wearing, no static generation. Resists penetration of direct splashes of corrosive, etc. Unaffected by oils.	Liable to shrinkage unless treated. More flammable than wool. Vulnerable to hot splashes.	Suitable for under-gloves
Wool	Resists rapid penetration of direct splashes (more effective than cotton). Resists penetration of dust. High absorption and porosity absorbs perspiration.	Not resistant to hot splashes. Takes up water and dirt, therefore, difficult to wash.	
Artificial fibres (nylon, terylene)	Hardwearing. Terylene has good acid resistance, etc.	Not resistant to hot splashes. High initial cost. Can allow dust to pass through. Static electricity can cause rapid soiling.	Ceramic coating of fibres can render dust-proof.
PVC	Impervious. Non-flammable. Chemically and biologically resistant. Abrasion resistant.	Not resistant to hot splashes. Can cause sweating unless well ventilated.	Can be used alone or impregnated on fabric. For gloves armlets, hats, bibs, spats, suits.
Paper	Hygienic (disposable). Fairly resistant to chemicals if treated with polythene film.	Liable to wetting if not treated with polythene film. No strength. Flammable. Not resistant to hot splashes.	Possibly used for disposable underwear or for clothing for visitors. Used under headgear (disposable).
Polythene	Hygienic. May be disposable. Impervious.	Low abrasion resistance. Not resistant to hot splashes and low melting point could cause adhesion to skin.	Suitable for disposable gloves or headgear.

The facilities comprise wash-basins or troughs with a plentiful supply of hot and cold running water, or running warm water, soap or liquid hand cleanser, and clean towels or disposable paper towels or hot-air driers, at least one wash-basin, or the equivalent of 0.6 m of trough. As an example, UK legislation requires provision of one wash-basin for every ten persons employed at any time. For the handling of toxic or highly irritant materials this should be increased to one wash-basin for every five persons.[49,50]

For hot and dirty work, e.g. boiler cleaning, where clothing has to be changed it is desirable for bathing facilities, e.g. showers to be provided.

In the UK the provision of suitable bath accommodation is a statutory requirement in certain chemical works, viz. for the use of persons employed in any nitro or amido process, certain parts of a chrome process, the process of distilling gas or coal tar or processes in which this tar is used, and in the refining of shale oil. It is also required in patent fuel manufacture and under other special regulations.[49]

These facilities should be conveniently situated with respect to the working positions, so that frequent washing of hands or other exposed areas is encouraged. In some cases, e.g. repetitive use of epoxy-resins, it is advisable for some facilities to be installed at the benches so that irritants can be removed rapidly.[51]

The use of solvents, thinners or home-made chemical formulations for skin cleansing should be strongly discouraged. De-greasing of the skin by such materials renders it more susceptible to attack by other irritants.

Supervision

Where there is a known toxic hazard it may be advisable to pre-select workers. Medical advice should be sought regarding any with a previous history of dermatitis, since one attack may predispose to another. It is also important that the worker should be capable of understanding the nature of the hazard and the measures necessary to minimise it.

Clearly all exposed workers, even if – as with cleaners or some maintenance workers – the contact is only on a casual basis, should be instructed of the risks and precautions. This may form part of an induction course for new recruits but should also be repeated periodically. Leaflets, such as those reproduced as Figs 6.10 and 6.11, may be issued to workers. Cautionary notices, such as those listed below may be displayed at convenient positions, e.g. near the clocking-in point or in the wash-rooms.

Cautionary notices:

Effects of chrome on the skin.	SHW 398
Effects of mineral oil on the skin	SHW 397
Dermatitis from flour, dough or sugar	SHW 355
Dermatitis from synthetic resins	SHW 366
Effects of lemon- and orange-peeling on the skin	SHW 396
Dermatitis	SHW 367

However, these need to be reinforced by formal instruction and verbal advice and adequate supervision. It is common experience that where precautions slow up a job, or make it harder to perform, and the risk is not immediately apparent, they tend over a period to be neglected.

An important measure in the prevention of occupational dermatitis is for workers to be encouraged to report any skin irritation or rashes for

Employment Medical Advisory Service

DERMATITIS AND WORK (OCCUPATIONAL DERMATITIS)

What is occupational dermatitis?

Occupational dermatitis is a rash caused by substances used at work. It can look like some common rashes not connected with work. Dermatitis is not catching.

How is it caused?

Too much contact with irritant substances can cause dermatitis in anyone, although some people are worse affected than others. The rash comes out on skin exposed to the irritant, most commonly on the fingers, hands, forearms or legs; when the irritant is dust or fumes the face, neck and chest may also be affected. Some common irritants are:
- mineral oil and 'soluble' cutting oils;
- chemicals such as alkalis and acids, pitch and tar;
- solvents such as paraffin, trichloroethylene and white spirit.

There are some substances that do not cause a rash to begin with. Further contact after weeks, months or even years, however, may cause dermatitis. This is because the body has become allergic to the substance. Sometimes this type of dermatitis spreads beyond obvious sites of skin contact. There can be swelling around the eyes. Substances that can cause skin allergy include:
- chrome and nickel compounds;
- some resins and their hardeners;
- some woods and plants;
- many individual chemicals.

How do you prevent it?

The best ways to prevent occupational dermatitis are:
- avoid or cut down contact with irritant and allergy-causing substances;
- keep your skin and clothes clean.

If you work with a substance that can cause dermatitis, follow these simple rules:

1 Your skin needs protection, so wear gloves, overalls or other protective clothing when you work with an irritant substance.
2 Get first-aid treatment for all cuts and grazes, however slight, and see that such injuries are protected at work by a dressing.
3 When using a synthetic resin, don't let it harden on the skin; remove it immediately with a damp swab.
4 Whenever you stop work, wash your hands and face and any other exposed area of skin carefully with soap and water, then rinse and dry the skin thoroughly. This is particularly important before meals and at the end of the day.
5 Do not use abrasives to clean the skin.
6 Look after your working clothes. Remember that clothing soaked with oil or other substances can cause skin irritation or make it worse.
7 Report any skin trouble. Delay in treatemnt may prolong the attack and increase the risk of your getting it again.
8 If you suffer from eczema or allergic rashes, avoid work where there is a known risk to the skin if you can.

Note: Further advice may be obtained from your local employment Medical Adviser. He or she can be reached at the Area Office of the Health and Safety Executive, which is listed in your local telephone directory.

MS (B) 6

December 1983: Health and Safety Executive

Fig. 6.10 Dermatitis and work (occupational dermatitis) (Crown copyright; courtesy of HMSO)

Employment Medical Advisory Service

Skin Cancer caused by Pitch and Tar

People who work, or have worked, with pitch and tar sometimes get warts on their faces, necks, hands, arms, or scrotum (the bag which contains the testicles). You might get a wart after only a few months in the industry, but it usually takes years for a wart to develop.

There are several kinds of tar wart, but one of them is **cancer** and will not go away unless it is treated. It is, however, easily cured by prompt treatment.

If you work in contact with pitch or tar and develop a wart or a little sore that does not heal, go to your doctor. If it occurs on the scrotum it is a serious danger and **delay is dangerous.**

The scrotum is at special risk. Examine it each time you have a bath. If you feel a patch of hard skin or a little lump, it could be a dangerous wart.

Go at once to your doctor and take this leaflet with you.

If you work with pitch or tar, **watch out for warts.**

If you used to work with pitch or tar, **watch out for warts.**

You could work with pitch or tar for many years without getting a wart, and then get one years after you leave the work, so **watch out for warts.**

General Precautions

Keep pitch and tar off your skin.
Wear protective clothing.
If you have a protective device, such as a dust extractor, use it.
Change your underclothes often.
Change your outer working clothes often, because tar works its way in.
Do not put dirty rags, tools or other materials contaminated with pitch or tar in your trouser pockets.
Wash your hands before going to the lavatory to make water.
Have a bath after work.
Watch out for warts.

MS (B) 4
20M Reprinted 7/86/Health and Safety Executive

Fig. 6.11 Skin Cancer caused by pitch and tar (Crown copyright; Courtesy of HMSO)

medical examination; similarly all illness potentially due to exposure to chemicals should be reported. Inevitably, a proportion of ailments will initially be wrongly attributed to working conditions but to discourage reporting may result in significant 'early warnings' being lost.

All complaints regarding working conditions, or reports of conditions believed to have been implicated in an outbreak of illness, should be properly investigated on the shop floor. If more than one person employed on similar work is affected then, of course, the probability of the illness being occupational in origin is greater. In any event enquiries should be directed towards any changes in chemical formulations, process, operating procedures, ventilation provisions, etc. In isolated cases enquiries should also be made regarding chemicals used at home or in hobbies.

Emergency equipment

In general with accidental exposure of the skin or eyes to chemicals the sooner the contaminants are removed or neutralised the less the damage. Therefore, emergency equipment comprising showers (or baths full of water), eye-wash bottles or fountains, and first-aid kits should be readily accessible from all areas in which toxic chemicals are handled. With regards to eye-wash bottles, the possibility of microbiological contamination may be a serious issue.

First aid necessitates the training of first aiders and that they are always available to render assistance at the scene of any accident. Furthermore, it is clearly better if lone working is avoided in areas where toxic chemicals are handled without complete containment. In many cases in the past immediate assistance from colleagues, in removing contaminated clothing and applying water to the affected areas of the body or in assisting with eye-irrigation, has mitigated the effects of potentially serious exposures.

Appropriate advice on first-aid measures should be made available in an appropriate, easily read form. Figure 6.12 provides an ideal example,[52] although recommended first-aid treatment and antidotes may differ between countries.

Solids handling

The design and operation of equipment for toxic 'solids', whether as solid, slurries or pastes, may involve special problems because of the inherent materials-handling characteristics. It is also necessary, to recognise and assess the potential fire and dust explosion hazards; to specify appropriate systems, equipment and operating procedures; and to provide appropriate protection.[53,54]

Plant for use with toxic solids will generally involve:
- total enclosure of the process, or,
- enclosure of the process, so far as is practicable, and prevention of the escape of dust by keeping the enclosure under negative pressure, or,
- application of local exhaust ventilation as near as possible to the point of origin of the dust.

However, as for gaseous and liquid chemicals, it is also necessary to envisage the potential for accidental 'uncontainment'. This involves consideration of all the credible events and scenarios which may result in unplanned release of material. These will generally include the conditions listed in Table 6.15 (after ref. 55). Transfer points generally represent the main sources of leakage.

The next step is to attempt to minimise the probability of these hazardous conditions arising, by process and plant design and by the use

Cyanides (nitriles)

General: these chemicals are extremely poisonous. Medical aid should be obtained as soon as possible and the casualty transferred urgently to hospital. The onset of symptoms may be delayed for 1–12 hours.

Skin Contact

Symptoms
There may be irritation, redness and pain at the site of contact. Burns can occur. These chemicals are absorbed through the skin causing symptoms similar to those of inhalation and ingestion.

Action
Immediate first aid. Remove all contaminated clothing; wash affected areas with soap and copious amounts of water.

Eye Contact

Symptoms
There may be irritation, redness and pain. Corneal burns can occur.

Action
Immediate first aid. Rinse with flowing water.

Ingestion

Symptoms
There will be a burning sensation in the throat with nausea and vomiting. Severe poisoning causes dizziness, rapid breathing, unconsciousness and convulsions.

Action
Immediate First Aid. Convulsions: see FA-21. If oxygen is available, give 100% by mask, with the casualty sitting quietly. If he is not breathing, *do not* use mouth to mouth or mouth to nose ventilation because of the danger to the rescuer. Use a resuscitation bag and mask instead. If the casualty is breathing break two amyl nitrite capsules open under the casualty's nose so that he inhales the vapour. If Kelo-cyanor (dicobalt edetate) and personnel trained in its use are available and you are certain that the casualty has been poisoned with cyanide, give it immediately. Note that Kelo-cyanor is *extremely dangerous* when given to anyone not suffering from cyanide poisoning.
NB: amyl nitrite and Kelo-cyanor have a limited shelf life. Check the expiry dates before use.

Inhalation

Symptoms
Mild poisoning may produce shortness of breath with chest pains and agitation. Severe poisoning causes dizziness, shortness of breath and a rapid onset of unconsciousness, with convulsions.

Action
As for ingestion (above).

Fig. 6.12 Desirable quality of first-aid advice[52] (FA – 21 is further advice on medical complications of chemical poisoning) (Courtesy of Croner Publications Ltd, Croner House 173 Kingston Rd, New Malden, Surrey KT3 3SS)

Table 6.15 Conditions causing unplanned release of material (after ref. 55)

Spillage due to:
  Overflow, backing-up, blowback, air lock, vapour lock
  Excess pressure, wrong routing, loss of vacuum
  Vessel being damaged, tilted, collapsed, vibrated, overstirred
  Overloading of an open flow/conveyor
  Poor isolation, drains or doors open, flanges uncovered
  Failure of control or major services
  Surging, priming, foaming, puking, spraying-over of vessels
  Condensation products present in vapour, spitting
  Change in normal discharge.
Leakage due to:
  Broken, damaged or badly fitted pipe, vessel, instrument, glass, spade, gasket, gland, seal, flange, joint or seam weld
  Internal leaks
  Overpressure of pipe or vessel.
Venting caused by:
  Evaporation through open line, drain, cover
  Relief valves leaking, bursting discs blown, lutes blown or drawn, failure of pads
  Valve stuck, scrubber overloaded, loss of vacuum, ejector failure
  Failure of control or major service, wrong routeing, excess vapour
  Dust formation, escape and collection
  Change in normal discharge
Failure of item at normal working pressure due to:
  Inadequacy of design, materials, construction, support, operation, inspection or maintenance
  Deterioration due to corrosion, erosion or fatigue.
Failure of item due to excessive pressure resulting from:
  Overfilling or overpressurising
  Overheating or undercooling
  Internal release of chemical energy
  Exposure to fire or other source of external heating (e.g. radiation)

---

of control and safety instrumentation, and to reduce the number of locations from which releases are possible. The maximum quantity which could be released can also be limited, e.g. by the control of inventory, by segregation (e.g. by chokes and shut-off devices) and by the use of smaller equipment items (albeit with some cost penalty).

Good operating practice can also minimise dust generation. For example gentle removal of lids from portable containers will tend to reduce the draught induced on opening; avoidance of empty bags being flattened will also minimise dispersion.

Equipment design

Consideration has to be given at the design stage to the need for cleaning and maintenance. Equipment which can be operated with a minimum of manual charging, connection/disconnection of piping, opening for cleaning/charging/discharging will result in less solids release and dispersal problems. Thus in the pharmaceutical industries *in situ* cleaning and sterilisation should be used, with internal surfaces suitably designed for this purpose, wherever possible. Similarly, production by continuous reaction may be advantageous, e.g. in pharmaceuticals processing[56]:

> High yields and rapid reactions are usually most economical and, depending on the side reactions, they may also be chosen to reduce emission problems. When high yields are required using slow reactions, this usually involves batch reaction, necessitating the opening up of the reactor for cleaning or charging, which increases the emission problem. Continuous low yield reactions can be carried out economically by employing a recycle. The build-up of by-products and inerts in the recycle frequently presents an emission problem when purging becomes necessary. Oxidising reactions using air in this situation, for example, may cause a nitrogen build-up. The use of oxygen instead of air would remove the need for purging, although process hazards may be increased.

Equipment which is not subjected to vibration, load cycling or high speed rotation is less likely to result in leaks. The extent, and quality, of joints seals and packed glands are also significant. Equipment, pipework and ducting should be fabricated with smooth interior surfaces to facilitate cleaning.

Materials transfer

Slurries and solid-in-gas dispersions can be transported in pipes to provide total containment. However, system integrity may be weakened by the need to provide for regular cleaning and maintenance.[56] Since solids handling presents problems – particularly with cohesive solids – it may be preferable to convey them as slurries or pneumatically and to separate them off as late as possible in the process. The simplest fluid motive mechanism is recommended,[56] i.e. gravity followed by slight pressurisation by inert gas (or air), and then pumps.

Free-flowing solids can be discharged from hoppers under gravity; cohesive solids may require the application of vibration or aeration devices. However, manual 'scooping' or shovelling may have to be resorted to at times; the appropriate suppression, ventilation and personal protective precautions described earlier then require consideration.

Estimates of emission rates from conventional solids handling operations are summarised in Tables 6.1 and 6.2.[3-5]

A preferred arrangement for sack emptying is for a sack tipper, which fits into the dust extraction system, e.g. into a hooded canopy with a hinged door or curtained front, so that the solids can either enter the process directly or be transported to a closed hopper and then to a closed weighing system.

A typical enclosure designed for dust-free bag opening, asbestos handling and empty-bag disposal in asbestos-cement pipe manufacture is shown in Fig. 6.13.[57] It is essential to ensure that the height of the

Control measures for handling toxic substances 327

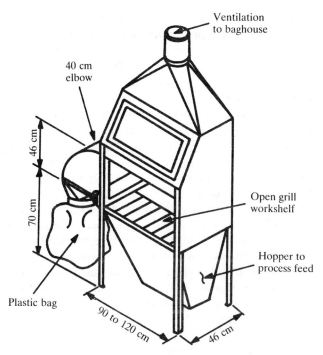

Fig. 6.13 Bag-opening station for asbestos fibres[57]

opening is such that the operator is discouraged from leaning into the hood in order to handle the bags.

If only one solid, or mixture of solids, is transferred via a specific route then a screw conveyor or pneumatic conveyor can provide containment with minimal operator involvement and exposure.

Solids blending using a tumbler blender necessitates the making and breaking of a pipe connection for filling and emptying purposes; this may result in some dust emission. The use of a static blender, e.g. a ribbon or orbital-screw blender, enables the feed and discharge pipes to be located in fixed positions; they can hence be contained, or be provided with efficient local exhaust ventilation.[56]

Separation operations

In the separation of solids from liquids, i.e. filtration or centrifugation, emissions from the liquid phase may be significant. It has been shown that, in terms of a 'special process hazard factor' the choice between equipment for containment is in descending order of preference: pressure filter, peeler centrifuge, belt filter, 'Nutsche' filter.[56] The latter suffers in this respect because of open surfaces and the need for manual handling.

The selection of equipment for the separation of solids from gases depends on the gas and solid properties, the minimum acceptable separation efficiency and whether the solid is required in a dry condition. (For the gas the important properties are temperature and composition, and for the solid particle size and 'stickiness'.) Typical efficiencies of dust collection equipment are shown in Table 6.16.[58]

A major problem may arise with blockages in certain designs, e.g. packed tower wet collectors or small diameter high-efficiency cyclones. Equipment deterioration due to erosion or mechanical damage, e.g. perforation of a filter or damage to ductwork, may result in reduced separation efficiency after a period of satisfactory performance. (Hence the need for routine inspection, testing and scheduled maintenance.)

Inertial collectors are generally preferred for high-temperature operation and wet de-dusters for the removal of 'sticky' dusts. Electrostatic precipitators can be used with high-temperature gases; these, or fabric filters, give good performances with fine dusts. Loss of containment may arise during maintenance and in the removal of dust from the system. For example, filter bags may give rise to emission problems during dust collection from the system; if practicable this can be prevented by removal as a wetted sludge.

Drying of solids may result in emissions from operations because of:
- Purging of the gas recycle stream.

Table 6.16 Typical efficiencies of dust arrestment equipment[58]

Equipment	Percentage efficiency at		
	50 μm	5 μm	1 μm
Inertial collector	95	16	3
Medium-efficiency cyclone	94	27	8
Cellular cyclone	98	42	13
High-efficiency cyclone	96	73	27
Tubular cyclones	100	89	40
Jet-impingement scrubber	98	83	40
Irrigated cyclone	100	87	42
Self-induced spray deduster	100	94	48
Spray tower	99	94	55
Fluidised-bed scrubber	> 99	98	58
Irrigated-target scrubber	100	97	80
Electrostatic precipitator	> 99	99	86
Disintegrator	100	98	91
Irrigated electrostatic precipitator	> 99	98	92
Annular-throat scrubber – low energy	100	> 99	96
Venturi-scrubber – medium energy	100	> 99	97
Annular-throat scrubber – medium energy	100	> 99	97
Venturi-scrubber – high energy	100	> 99	98
Shaker-type fabric filter	> 99	> 99	99
Low-velocity bag filter	100	> 99	99
Reverse-jet fabric filter	100	> 99	99

- Drier feed and discharge arrangements, e.g. during traying-up of batch tray driers; or during manual filling and discharge of removable bowl, batch-operated fluid-bed driers.
- Particulates dispersal at the end of the drying cycle.
- Vapours emitted at the start of the drying cycle.
- Cleaning-out of residues (e.g. material from the walls of spray driers).

The range of driers available is summarised in Table 6.17.[58] In general, continuous driers present fewer problems in controlling emissions; however, these depend upon the specific design and ancillary operations. Spray-drying is efficient and can often produce a granular, dust-free product; alternatively, gas recycle loops via cyclone separators, or the use of slurry-sprayed fluid-bed after-driers, may be used.[59]

Dust dispersion may be reduced by a combination of measures, e.g.
- Automation of batch driers.
- Operation under vacuum, so that any air leaks are inwards.
- Discharge by sucking the product out into a hopper under vacuum.
- Exhausting of gas or air via efficient dust arrestment equipment (Table 6.17).
- Limitation of drier size so that cleaning and sealing are easier.

Similar considerations to those discussed above apply to other solids-processing operations, e.g. leaching, comminution, crystallisation, agglomeration and heat transfer.

Table 6.17 Drying equipment

Material form	Batch operation	Continuous operation
Granules	Tray drier Shelf drier	Conveyor drier Filter drier Rotary drier Turbo drier
Paste	Agitated pan drier	Flash drier Screw conveyor drier Tower drier
Preformed cakes	Tray drier Shelf drier	Tunnel drier
Solution or slurry	Agitated pan drier	Spray drier Rotating drum drier

Housekeeping

Application of sophisticated dust-control devices to dusty processes will not result in safe working conditions if continual contamination occurs due to redispersion of dust deposits through the movement of personnel or machinery.

Therefore, deposits, whether deliberate or unintentional, should be minimised by 'good housekeeping'. This involves consideration of the following precautions.[60]

- All spillage of dust should be prevented. Methods of handling and transport, and design of workbenches, should have this in view. It is a major risk if toxic dust is trodden around the working area and contaminates floors and gangways.
- The construction of all buildings and workrooms used for potentially dusty processes should be such as to permit easy and frequent cleaning. Floors should be smooth and impervious without slats or grids, walls and doors smooth without ledges or mouldings, and rooms should have plain ceilings without projecting beams or rafters. Machinery and plant should be designed for easy cleaning. Floors should at all times be kept as free as possible from obstructions.
- Cleaning should be regular and systematic and should include not only floors and workbenches, but also ledges, lighting fittings and all places where dust can lodge. Special attention should be given to overhead plant and equipment. The best method of cleaning is by a built-in vacuum system or, failing this, by a portable vacuum cleaner with a high-efficiency filter. Other methods of cleaning tend to have the immediate effect of stirring up dust, and should not be done at times, or in ways, that will increase danger to other workers.
- Piles or heaps of dust should be kept in containers with covers (e.g. in a bin or ark with a lid), especially where there may be air currents. Deposits of toxic dust on, or near, vibrating machinery are especially dangerous.
- Damaged bags or kegs should be patch-repaired immediately defects are detected.
- Paste or wet suspensions of dust may become dangerous when they dry out. All scraps, drips and spillages should, therefore, be kept wet and cleaned up or removed before they can dry.

Mitigatory measures

The effects of the release scenarios listed in Table 6.15 should be assessed, so that mitigatory measures can be planned. These may include:
- Provision for easy observation, or rapid detection by measurement, of instantaneous or prolonged leaks or spills.
- Provision for rapid detection of the source of any leaks or spills.
- Methods of isolation of contaminated equipment or work areas.
- Provision for covering, neutralisation, removal and disposal of spilled material. Efficient cleaning methods.
- Provision for dealing with airborne contamination, e.g. emergency ventilation.

Normally, occupational exposure to airborne toxic chemicals arises from accidental releases (Table 6.15) or from fugitive emissions (Tables 6.1 and 6.2). However, in the horticultural and agricultural industries herbicides, fungicides, insecticides, fertilisers, field dressings, etc., are purpose-generated in the form of dusts and aerosols by pneumatic or hydraulic propulsion. Dispensing may be from operator-carried packs, from tanks mounted on tractors, or by aerial application. Table 6.18 lists accidents in agriculture from accidental poisoning, dust disease and dermatitis.[53,61-69] Guidance has been published on the use of toxic chemicals in farming and horticulture[61-69] and legislation has recently been updated by introduction of the Poisonous Substances in Agriculture Regulations 1984 (see Ch. 16). A summary of advice for operators using chemicals on the land, on livestock, and in greenhouses, etc., is given in Table 6.19.

Table 6.18 Fatal and non-fatal accidents in agriculture from accidental poisonings, dust disease and dermatitis (Great Britain* 1973–1975)

	Non-fatal			Fatal†		
	1973	1974	1975	1973	1974	1975
Scheduled pesticides and veterinary products	1	3	5	—	—	—
Other pesticides and veterinary products	15	15	13	1	—	—
Other chemicals	25	13	4	—	—	—
Gases	5	2	1	3	—	1
Total	46	33	23	4	—	1
Farmer's lung	4	10	17	1	3	—
Non-infectious dermatitis	27	20	26	—	—	—
Total	31	30	43	1	3	—

* Great Britain includes England, Wales and Scotland.
† Includes children under 16.

Conclusion

Clearly the safe handling and use of toxic chemicals are tasks requiring thought by the designers and commitment from management. In some ways it is more difficult than dealing with highly flammable substances (Ch. 4) since the concentrations to be maintained are lower and the hazard arises with no secondary initiator (i.e. an ignition source) being required. It is for management to select the appropriate combination of

Table 6.19 Precautions for use of toxic chemicals in farming and agriculture

1. Read the label on the container to ascertain if material is on the Approved List and follow the manufacturer's instructions.
2. Wear appropriate personal protection (e.g. see page 297).
3. Wash overalls frequently and if badly contaminated remove them instantly.
4. Store overalls in a 'clean' area.
5. Use only approved respirators and dust masks, replacing filters as necessary.
6. In preparing sprays, etc., guard against splashes.
7. Spray with care and avoid drift. Do not spray in windy weather.
8. Switch off spray prior to attending to nozzles: never clear nozzle by 'blowing' or 'sucking'.
9. Remove protective clothing (except boots and overalls) and wash hands before eating, drinking or smoking.
10. Wash gloves and boots before removing them; wash inside of gloves before they are put away.
11. Keep all equipment and outsides of containers free of contamination.
12. Limit hours of exposure to chemicals.
13. Never transfer pesticides, etc., into other containers particularly empty lemonade or beer bottles.
14. Keep containers closed and store in a safe place away from children.
15. Dispose safely of all used containers (do not contaminate ponds, waterways and ditches with used containers). Wash out before disposal.

control measures. To be confident that the system of work is safe, however, worker co-operation is essential.

All the techniques and strategies discussed rely upon there being a sound 'system of work' to reinforce them, e.g. operator training and supervision, inspection and maintenance, control over process and plant modifications. There should also be a procedure for monitoring the environment and management procedures for monitoring the 'system of work', e.g. safety surveys and/or safety audits.

> A steelworks was fined £250 for having an unsafe system of work which led to the death of an assistant chief chemist working at the firm.
>
> The man collapsed while trying to carry out tests with a colleague on gas-cleaning equipment. The smelting furnaces where he was working gave off a gas which could be used as fuel if its 30% carbon monoxide content was removed. An electrostatic precipitator was used to clean the gas. The equipment had been out of action for some time and the gas was being pumped into it under pressure to remove all the air.
>
> The Health and Safety Executive's inspector said that when the gas sampling had been requested the two men climbed up to the sampling platform where there was a bleeder pipe discharging carbon monoxide, as it was designed to do, about 1.2 m above their heads. Both men began to feel

unwell. As they were getting down, the man collapsed due to the inhalation of toxic fumes.

None of the men, said the inspector, had been told to wear respirators. It appeared that while the gas was going straight up into the air some was diffusing and billowing back at them. The inspector pointed out that the steel industry had produced a document advising that certain operations, including the purging of gas, be done under strict safety standards. There were respirators at hand but no procedures to use them.

Ethyleneimine (aziridine), used in the synthesis of various chemotherapeutic agents, is a highly volatile (b.p. 56 °C) and toxic substance (OEL 0.5 ppm)[70] which is capable of being absorbed through the skin. The lack of adequate enclosure and local exhaust ventilation caused a fatal accident to an operator who was dispensing ethyleneimine from a large cylinder into smaller containers. The liquid was forced by nitrogen from the cylinder into a jug from which it was poured into bottles placed on scales on the workbench. A canopy was situated 1.2 m above the bench and a fan was mounted in the wall below this canopy. Because of the inadequate enclosure, the air speed at bench level was only about 6–9 linear m min^{-1} and although the operator wore a full-face respirator this did not provide sufficient protection against the high concentration of vapour in the breathing zone. For such highly toxic volatile chemicals an air speed of at least 45 linear m min^{-1} is necessary to ensure adequate control.[71]

This case demonstrates that although management had provided control by a combination of ventilation and personal protection they had failed to discharge their duties fully by thinking through the details of the precautions and checking the adequacy of the arrangements.

A material was required of which the physiological effects were unknown. From its chemical structure it was suspected that it might be carcinogenic. Positive test data could not be developed in the time available. Therefore, the operation was carried out in completely enclosed equipment. Room ventilation was provided at three air changes per minute. Operators were placed in suits providing complete body and breathing protection for all work. Suits were decontaminated with sodium bisulphite and checked for fluorescence with UV light after work. This is an illustration of the level of consideration which may be required when handling materials when not all the safety data have been determined (e.g. at the research stage of development) and when there is reason to suspect the material of being highly toxic.

It is important that workers are made aware of the toxic hazards, particularly with the undiluted chemicals, and warned against taking excess or unused quantities for use at home, etc. This practice and the use of unlabelled bottles referred to in Table 6.19 has unfortunately resulted in numerous cases of accidental ingestion – some with fatal consequences.

A 15-year-old boy ingested a mouthful of Paraquat (1,1-dimethyl-4,4-dipyridium dichloride) from a bottle and developed severe respiratory distress. Treatment included the transplantation of one lung but he died two weeks after the operation. (Swallowing one mouthful is in fact nearly always fatal.)

A child drank from a Coca Cola bottle in which his parent had put some Paraquat illicitly obtained from a market garden. The child appeared to recover from the acute effects but died two weeks later.[73]

References

(*All places of publication are London unless otherwise stated*)
1. Langley, E., *Chem. Ind.*, 1978 (14), 504.
2. Ember, L., *Chem. Eng. News*, 1984 (12 Mar.), 12.
3. Jones, A. L., 'Industrial hygiene input ot the design process' presented at Conference on Industrial Hygiene, Toxicology and the Design Engineer, Royal Lancaster Hotel, London, Mar. 1982.
4. The BOHS Technology Committee, *Fugitive Emissions of Vapours From Process Equipment*, British Occupational Hygiene Society Technical Guide No. 3. Science Reviews Ltd. Leeds, 1984.
5. Jones, A. L., *Annals Occup. Hyg.*, 1984, **28** (2), 211.
6. Cralley, L. V. & Cralley, L. J. (eds), *Industrial Hygiene Aspects of Plant Operations*, Vol. 1 *Process Flows*. Macmillan New York, 1983.
7. Cralley, L. V. & Cralley, L. J. (eds) op. cit., Vol. 2 *Unit Operations and Product Fabrication*. Macmillan, New York, 1984.
8. Burges, W. A., 'Potential exposures in industry – their recognition and control', in *Patty's Industrial Hygiene and Toxicology* (3rd edn) Vol. 1. Wiley 1978.
9. Gadien, T., *Instn Chem. Engrs Symp. Series*, No. 34, 1972, 98–102.
10. *Sandwell Evening Mail*, 23 Feb. 1979.
11. Corbett, E., 2nd Symposium on Chemical Process Hazards, Institution Chemical Engineers, 1963, 41–5.
12. Gillet, J. E. *Annals Occup. Hyg.*, 1976, **19**, 301.
13. Health and Safety Commission, *Approved Code of Practice* (COP 3) and Health and Safety Executive, *Guidance Note on Work with Asbestos Insulation and Asbestos Coating*.
14. Lynch, J., 'Polyvinyl chloride', in L. V. Cralley & L. J. Cralley (eds), *Industrial Hygiene Aspects of Plant Operations*, Vol. 1. *Process Flows*. Macmillan, New York, 1982, 440–2.
15. *Handbook of Occupational Hygiene*. Kluwer Pulo Ltd 1980.
16. Asbestos Regulations 1969, Reg. 20.
17. Factories Act 1961, S.74. HMSO.
18. Chromium Plating Regulations 1931, Reg. 9. HMSO.
19. The Ionising Radiation (Unsealed Radioactive Substances) Regulations, 1968; The Ionising Radiation (Sealed Source) Regulations, 1969.

20. Kelly, W. D., 'Protection of the sensitive individual', American Conference of Governmental Industrial Hygienists, Ohio, 1982.
21. Lihou, D. A., 'Operability studies and hazard analysis', in *Hazard Identification and Control in the Process Industries, 1981*. Oyez 1981, Ch. 7.
22. Health and Safety Executive 1981, *Guidance Note MS 18: Health Surveillance by Positive Procedures*; Health and Safety Executive 1982, *Guidance Note MS 20: Pre-Employment Health Screening*; Health and Safety Executive 1977, *Guidance Note MS 5: Lung Function*. HMSO.
23. Health and Safety Executive 1982, *Guidance Note MS 8: Isocyanates – Medical Surveillance*; Health and Safety Executive Rev. 1984, *Guidance Note EH 16: Isocyanates: Toxic Hazards and Precautions*. HMSO.
24. Health and Safety Executive 1979, *Guidance Note MS 12: Mercury – Medical Surveillance*. HMSO.
25. Health and Safety Executive 1980, *Guidance Note MS 17: Biological Monitoring of Workers Exposed to Organophosphorus Pesticides*. HMSO.
26. Health and Safety Executive 1983, *Guidance Note MS 22: The Monitoring of Workers Exposed to Platinum Salts*. HMSO.
27. Health and Safety Executive 1976, *Guidance Note EH 2: Chromium – Health and Safety Precautions*. HMSO.
28. Health and Safety Executive 1977, *Guidance Note EH 4: Aniline – Health and Safety Precautions*. HMSO.
29. Health and Safety Executive 1976, *Guidance Note EH 5: Trichloroethylene – Health and Safety Precautions*. HMSO.
30. Health and Safety Executive 1976, *Guidance Note EH 6: Chromic Acid Concentrations in Air*. HMSO.
31. Health and Safety Executive 1977, *Guidance Note EH 8: Arsenic – Health and Safety Precautions*. HMSO.
32. Health and Safety Executive 1977, *Guidance Note EH 13: Beryllium – Health and Safety Precautions*. HMSO.
33. Health and Safety Executive 1978, *Guidance Note EH 19: Antimony – Health and Safety Precautions*. HMSO.
34. Health and Safety Commission, *Approved Code of Practice – Control of Lead at Work*. HMSO 1985.
35. Health and Safety Executive 1977, *Guidance Note MS 3: Skin Tests in Dermatitis and Occupational Chest Disease*. HMSO 1977.
36. Health and Safety Executive 1977, *Guidance Note MS 5: Chest X-Rays in Dust Disease*. HMSO.
37. Chemical Works Regulations 1922. HMSO.
38. Else, D., *Annals Occup. Hyg.*, 1973, **16**, 81.
39. Chemical Industries Safety and Health Council, *Chem. Safety Summary*, 1978, **49** (194), 30.
40. 'Respiratory protection – principles and applications' in B. Ballentyne & P. H. Schwabe. Chapman and Hall 1981.
41. Chemical Industries Safety and Health Council, *Chem. Safety Summary*, 1980, **50** (201), 11.
42. Pennell, A. *2nd Symposium on Chemical Process Hazards*, Institution of Chemical Engineers, 1963, 35–40.

43. Certificate of Approval (Respiratory Protective Equipment) 1985. Health and Safety Executive.
44. Hyatt, E. C., *Respirator Protection Factors*, Los Alamos Scientific Laboratory Report No. LA-6084-Ms, 1976.
45. Chrome Plating Regulations 1931, Reg. 5. HMSO.
46. The Patent Fuel Manufacture (Health and Welfare), Special Regulations 1946. HMSO.
47. Department of Employment, *Lead – Code of Practice for Health Precautions*, August 1973. HMSO.
48. *Safety in the Use of Chemical Protective Garments*. Chemical Industries Assoc. 1985.
49. *Cloakroom Accommodation and Washing Facilities*, Safety, Health & Welfare, No. 5, 1968. HMSO.
50. *Industrial Dermatitis – Precautionary measures*, Health & Safety at Work No. 18, 1972. HMSO.
51. *Memorandum on the Prevention of Industrial Dermatitis from Synthetic Resins*, HMFI SHW 331.
52. Houston, A. (ed.), *Dangerous Chemicals Emergency First Aid Guide* (2nd edn). Walters Samson (UK) Ltd 1986.
53. Ministry of Agriculture, Fisheries and Food, *Approved Products for Farmers and Growers, 1983* (Agricultural Chemicals Approvals Scheme). HMSO.
54. Working Party of the Engineering Practice Committee of the Institution of Chemical Engineers, *User Guide to Fire and Explosion Hazards in the Drying of Particulate Materials*. Institution of Chemical Engineers 1977.
55. Wells, F. L., *Safety in Process Plant Design*. George Godwin 1980.
56. Gillet, J. E., *Annals Occup. Hyg.*, 1970, **19**, 301–8.
57. Swallow, G. L., 'Asbestos – cement pipe manufacture', in L. V. Cralley & L. J. Cralley (eds), *Industrial Hygiene Aspects of Plants Operations*, Vol. 1. Process Flows. Macmillan, New York, 1982.
58. Bridgwater, A. V. & Mumford, C. J., *Waste Recycling and Pollution Control Handbook*. George Godwin 1979, 282.
59. Masters, K., *Spray Drying Handbook*. George Godwin 1979.
60. HM Factory Inspectorate, *Health, Dust in Industry*, Technical Data Note 14. HMSO 1970.
61. Health and Safety Executive, *Crop Spraying*, AS6. HMSO 1982.
62. Health and Safety Executive, *Poisonous Chemicals on the Farm* (2nd edn). HMSO 1980.
63. Turner, D. J., *The Safety of the Herbicides 2, 4-D and 2, 4, 5-T*, Forestry Commission Bulletin. HMSO 1977.
64. Forestry Commission Booklet 51, *The Use of Herbicides in the Forest*. 1986.
65. Dept. of Environment, Waste Management Paper No. 21, *Pesticide Wastes*. HMSO 1980.
66. Health and Safety Executive, *Storage of Pesticides on Farms*, AS 18. HMSO 1981.
67. Health and Safety Executive, *Biological Monitoring of Workers Exposed to Organophosphorus Pesticides* MS 17. HMSO 1986.
68. Health and Safety Executive, *Poisoning by Pesticides* MS (B) 7. HMSO 1977.

69. *Safe Use of Pesticides*, Occupational Health and Safety Booklet 38. International Labor Organisation, Geneva 1977.
70. Health and Safety Executive, *Guidance Note EH 40: Occupational Exposure Limits*, 1984.
71. *Annual Report of HM. Chief Inspector of Factories, 1974*, 61.
72. Anon., *Brit. Med. J.*, 16 Sept. 1967, 690.

CHAPTER 7

Ambient air analysis for hazardous substances

Introduction

Previous chapters have referred to the importance of environmental monitoring when handling chemicals, and in certain cases this may be a legal requirement.

Substances can become airborne as a result of their volatility (e.g. gases, vapours) or as a result of the process (e.g. aerosols, dusts, mists and fumes) because of:
- inadequate process control (typical fugitive emission rates from process equipment are given in Tables 6.1 and 6.2, and causes of unplanned releases are listed in Table 6.15);
- operation and maintenance malpractice;
- inadequate maintenance;
- incomplete understanding of the process.

Accumulation of a specific substance or mixtures of substances in the working environment, particularly in confined spaces, can produce health risks or be of nuisance value at concentrations as low as ppm, as described in Chapter 5. At higher concentrations flammable gases, vapours and dusts produce mixtures in air which may ignite or explode on reaching an ignition source, as discussed in Chapter 4. Similarly, hazards may arise in atmospheres in which the oxygen content either rises, or falls below its normal concentration of 21% by volume. Oxygen-enriched atmospheres (e.g. leakage of oxygen from oxy-acetylene welding gear in enclosed locations) pose a severe fire risk while oxygen-deficient atmospheres (e.g. in mines, manholes, vessels purged with inert gas) create a risk to health.

Thus air quality tests may be required for a variety of reasons such as:
- detecting smoke for fire prevention;
- assessing risk from atmospheres containing flammable gas or vapour;
- determining oxygen content;
- detecting leak sources of toxic or flammable chemicals;
- identifying unknown pollutants;
- determining employee exposure to known toxic chemicals;

- assessing performance efficiency of engineering control measures;
- testing compliance for legal requirements.

On a routine visit to a factory which used a two-pack polyurethane foam packaging system, a Factory Inspector found that the airborne concentration of isocyanate was in excess of the TLV at a distance of 3 metres from the packaging point, and up to six times the limit in the operators' breathing zone. It transpired that no precautions were taken because the suppliers had made no mention in their product data that high levels of isocyanate vapour could develop. The suppliers were prosecuted under Section 6(1)c of the HASAWA.[1]

Analysis may be on an occasional basis, e.g. as part of the requirements of a permit-to-work system prior to maintenance, or on a continuous basis with more sophisticated systems. Though monitoring technology has advanced rapidly during the past decade, and is continuing to be a dynamic development, clearly there is no universal system applicable to all situations. Each installation or environment poses unique problems to consider (e.g. monitoring a toxic gas may require intrinsically safe or explosion-proof equipment if a flammable hazard coexists). To meet these requirements equipment available ranges from simple portable devices to fixed multipoint units equipped with automatic sampling points, chart-recorders, high- and low-level alarms, sample flow and power failure alarms, etc.

An appreciation of the techniques available is required by those with responsibility for assessing the environmental risk from hazardous chemicals such as occupational hygienists and nurses, safety officers and chemists. Therefore, this chapter introduces the subject of environmental monitoring by discussing, selectively, some of the more common types of monitoring equipment and the principles upon which they are based. A summary is given of techniques for both sample collection and subsequent analysis. The latter is classified according to the type of detection system employed, for gases and then for particulate matter. (The examples of commercial hardware are intended to illustrate the basic principles under discussion and should not be taken as recommendations by the authors, nor as the only suitable apparatus available.) Appendices 7.1 and 7.3 list some appropriate procedures for monitoring several hundred individual substances.[2] Appendix 7.2 lists several hundred substances for which National Institute of Occupational Safety and Health (NIOSH) recommended methods of analysis are available[3]; Appendix 7.3 lists analytical methods published in the UK by the HSE; and Appendix 7.4 lists a selection of relevant British Standards. Detailed handling advice for specific instruments should be obtained from the supplier or from texts included in the references. The chapter concludes with advice on sampling strategies and data interpretation.

General requirements

Desirable characteristics for monitoring equipment include specificity, sensitivity, accuracy, self-calibration, direct reading, simplicity, reliability, light weight, robustness, rapid response, ease of maintenance and low cost. Any device is unlikely to satisfy all requirements and choice will be based on compromise.

The objectives of the monitoring exercise will dictate the sampling strategy and to some extent the hardware specification. Thus leak detection is usually achieved using portable sensors. Increases in concentration are noted as the instrument is moved until the leak source is traced. It follows that these instruments must respond rapidly to fluctuations in concentration, be sensitive over a wide concentration range, and give reproducible readings. Specificity is not usually a requirement and accuracy is not the prime consideration since actual concentration will depend upon position of the probe relative to a leak, air movement and temperature.

Background monitoring is useful for providing a base line of plant conditions and can allow future circumstances to be predicted. Periodic checks by static samplers will reveal whether conditions have changed compared with those monitored during commissioning as a consequence of process modifications, changes in working practices, deterioration of machines, or fluctuations in climatic conditions (air movements, temperature). Concentration profiles can be mapped out for the workplace to highlight areas of unacceptable pollution. The programme should include regular inspections for fugitive emissions with emphasis on pumps, glands, seals, joints, valves, conveyor lines, sampling points, etc. Fixed monitors used in background surveillance of hazardous installations (e.g. indoors on plants processing highly flammable or toxic materials) are generally single or multipoint high-volume sampling systems performing sequential analysis at a range of sampling points located throughout the workplace. Careful design and installation are essential to prevent sorption losses, condensation and contamination, and to ensure sampling points are correctly located taking into account gas density and movement of ambient air. The systems are usually expensive and purpose installed. They are often left unattended for long periods of time and output is usually an alarm with action levels set low. The equipment should be accurate, have stable zero and be safe for the environment in which it is located. Portable background monitors on the other hand are used to check manually for toxic or flammable contaminants, such as in areas usually unmanned or as part of a safety-to-work permit system, e.g. prior to, and during entry into confined spaces for maintenance or inspection purposes (see Ch. 17). The equipment must be accurate with good visual and audible outputs, specific to the contaminant in question, portable, and safe for the environment in which it is located. Since they are sometimes used in emergencies (e.g. spillages) they should have short warm-up times. When sample lines are used in conjunction with these instruments errors can arise from defects in the tubing or poor connections.

Portable or fixed static samplers can be used for routine process control purposes, etc., but data do not necessarily reflect personal exposures.

Approaches adopted for assessing risk of exposure to toxic chemicals depend upon the nature of the substance. For substances which are fast acting, or which have special toxic properties such as those assigned a 'ceiling' TLV,[4] then it may be possible to gauge compliance with background monitors. Alternatively short-term (grab) samples can be collected when exposures are expected to be greatest (e.g. opening-up reactors). Grab sampling is also useful for following phases of a cyclic process and for determining peak airborne concentrations in a new process. However, to accurately assess employees' exposure during a working day, personal dosimetry is required using portable equipment worn by operators to sample air in their breathing zone throughout the working shift. This technique is used in testing compliance with time-weighted-average hygiene standards[4,5] or for accumulating exposure data for epidemiology purposes. Besides providing an average measurement and being safe to use wherever the operator goes, the ergonomics of equipment design are important: it should not be heavy or impede the wearer's movements.

General principles of air monitoring

Impetus for development of flammable gas detectors (explosimeters) came from the coal-mining industry which has experienced many explosions as a result of ignition of mixtures of naturally-occurring methane and air by naked flames or sparks from tools or other equipment. Thus the Davy lamp, designed to provide illumination, also functioned as a primitive methane detector and oxygen-deficiency indicator by the respective effects of those two gases on the lamp's flame-height. Atmospheres containing less than 16% oxygen extinguished the lamp, whereas methane-enriched air caused the flame-height to increase. Nowadays, meters are usually electronically operated and give direct readings (usually expressed for flammable materials as a percentage of the lower explosive limit).

Early monitors for toxic environments relied on 'natural' detectors such as dogs and canaries, and other small animals, or the collection of air samples by scrubbing with subsequent laborious analysis. More recent trends have been to direct read-out instrumental methods.

Whatever the technique adopted, atmospheric monitoring fundamentally involves obtaining representative samples of the air followed by analysis of the sample so collected. In some devices these processes are integrated so as to give direct read-out in the field. Often the stages are completely separate requiring laboratory analysis of the sample. For convenience the stages of sampling and analysis are described separately first for gases then for particulates. More detailed reviews are given in references 6, 7 and 8.

Both the air monitoring equipment and the analytical procedure or instrument being used must be calibrated. The former is usually achieved by checking flow-rate indicators or sample volumes using direct physical methods, e.g. soap bubble meters, dry-gas meters or rotameters. For calibrations of the analytical procedure both the collection efficiency and detection response must be validated. Additionally, techniques such as 'twicing' should be employed, whereby two sample units are arranged such that one samples twice the concentration or for twice the period or at twice the rate of the other, to ensure twice the sample is collected. Repeated discrepancies between the two approaches by 10% or more suggest systematic error. Generally these calibrations necessitate access to 'standard atmospheres' containing known levels of contaminant. Techniques for this are many-fold and guidance should be sought[6,9] (e.g. see Appendix 7.3).

Gas monitoring

Air sampling

The sampling process involves inducement of air to flow into collection vessels. Arrangements include collection of gas in plastic bags, syringes, or by absorption in liquid or on solid adsorbents. A selection of absorbers, used mainly with liquids, is shown in Fig. 7.1.[10] Small, light-weight, unbreakable, non-spillable gas bubblers are available for personal monitoring.

Certain gases can be collected during personal monitoring by absorption in small, commercially-available, lapel-mounted tubes pre-packed with solid adsorbents, and coupled to a personal sampling pump. Figure 7.2 shows a variety of personal pumps and sampling devices, including a charcoal tube. Adsorbents include silica gel, activated charcoal and a range of organic resins. Charcoal is useful for adsorbing non-polar compounds whereas silica-gel attracts polar chemicals. Tubes should remain sealed until ready for use, and care is needed to ensure the tube is not overloaded with sample causing 'break-through' of the pollutant.

The sorbed gas is removed from the sorbent by solvent extraction for subsequent analyses, or preferably by thermal desorption directly into an analyser, using special units of the type shown in Fig. 7.3(a) and 7.3(b).

Ambient air analysis for hazardous substances 343

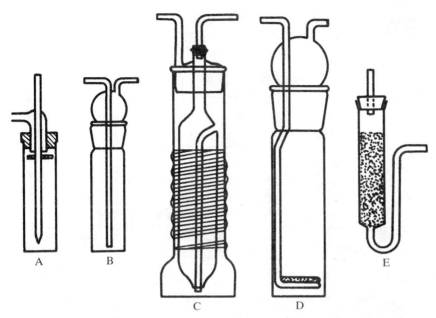

Fig. 7.1 Basic absorbers. Gas washing (A and B), helical (C), fritted bubbler (D), glass bead column (E), provide contact between sampled air and liquid surface for absorption of gaseous contaminants

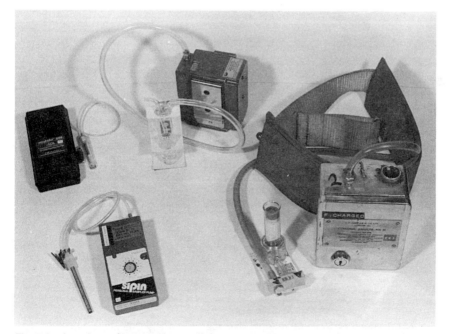

Fig. 7.2 A variety of personal sampling pumps with sampling devices attached (Courtesy Environmental Monitoring Systems)

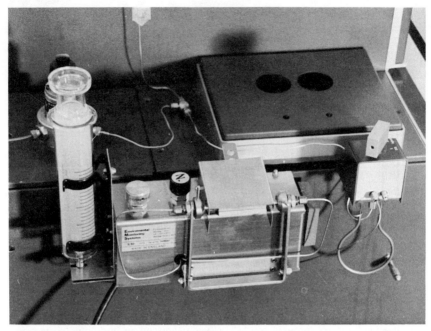

Fig. 7.3(a) Thermal desorption unit connected to gas–liquid chromatography system (Courtesy Environmental Monitoring Systems)

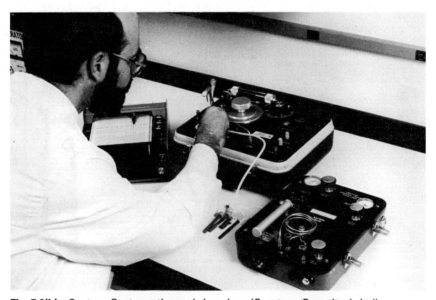

Fig. 7.3(b) Century Systems thermal desorber. (Courtesy Quantitech Ltd)

Advantages of thermal desorption include:
- Elimination of sample preparation and handling of toxic solvents of which carbon disulphide is a common example.
- Absence of solvent simplifies chromatograph.
- Increased sensitivity.
- Sample tubes can be reused.
- Greater range of detection systems to which the desorbed gas can be subjected (e.g. gas chromatography, infrared and ultraviolet spectroscopy, colorimetry).

Limitations
- Certain resins undergo degradation even below 250 °C which limits their use in thermal desorption.
- The test sample may not be thermally stable.
- Not all organic compounds are readily removed from resins by application of heat.
- The entire test sample tends to be used up with no opportunity to repeat the analysis.

There are occasions when solvent extraction is advantageous: the solvent must not react with the polymer and it must possess good solvent properties to ensure quantitative extraction.

Table 7.1 lists a range of compounds which can be collected on charcoal tubes together with some sampling details.[11]

The 'passive sampler' or 'dose badge' is a variant on the above adsorbent tubes. This device is unique in not requiring a pump, relying instead on molecular diffusion or penetration across a membrane into a chamber of known geometry containing adsorbent, normally charcoal. The equation below, based on Fick's law, describes the steady-state relationship for the rate of mass transfer:

$$W = D \frac{A}{L} (C_1 - C_0)$$

Where: W = mass transfer rate; D = diffusion coefficient; A = frontal area of static layer; L = length or depth of static layer, C_1 = ambient concentration, and C_0 = concentration at collection surface. The collection rate is proportional to ambient vapour concentration C_1 when the collection surface has C_0 equal to zero. The units of $D(A/L)$ are volume per unit time, i.e. the same as for the volumetric flow in a pump system. Figure 7.4(a) shows a selection of commercially-available, passive dosimeters and Fig. 7.4(b) depicts one in use. In the main, the sample is retrieved by solvent extraction although thermal techniques can be applied to the tube. Figure 7.4(c) depicts a passive dosimeter in use, which is based on the colour indicator tube (see page 362): exposure is read directly from the scale on the tube.

Table 7.1 Charcoal tube user guide
(Values shown only apply when sampling with 100 mg coconut-shell charcoal tubes based on the NIOSH recommended design) (after ref. 11). (See Notes for guidance at end of the table – pp. 355–6)

Substance	Expected concentration range		Recommended sampling (ml/min)				Recommended maximum tube load (mg)	Approximate desorption efficiency (%)	Eluent
	ppm	mg/m³	2 hr	4 hr	8 hr				
Acetaldehyde	2–50	4–90	100	100	50			
	50–400	90–720	50	25	10			
Acetone	5–200	12–480	100	50	25	9	86±10	CS₂
	200–2000	480–4800	10	5	1			
Acetic acid	0.1–1	0.25–2.5	—	200	200	10.4		11
	1–20	2.5–50	200	100	50			
Acetonitrile	1–10	1.8–17.5	100	100	50	2.7		
	10–80	17.5–140	50	25				
Acrolein	0.02–0.2	0.05–0.5	—	—	200			
Acrylonitrile	0.2–2	0.45–4.5	200	200	100	>2	<80	12
	2–40	4.5–90	100	50	25			
Allyl alcohol	0.1–0.5	0.24–1.20	—	200	200	<0.4	89±5	2
	0.5–4	1.2–9.6	200	100	50			
Allyl chloride	0.05–2	0.15–6	—	—	200	0.75		13
n-Amyl acetate	1–25	5.3–131.3	100	100	50	15	86±5	CS₂
	25–200	131.3–1050	50	25	10			
sec-Amyl acetate	1–25	5.3–131.3	100	100	50	15.5	91±10	CS₂
	25–250	131–1300	50	25	10			
Isoamyl alcohol	1–25	3.6–90	100	100	50	10		
	25–200	90–720	50	25	10			
Benzene	0.1–1	0.31–3.1	—	200	200		96	CS₂
	1–20	3.1–62.6	100	100	50			
Benzyl chloride	0.05–2	0.25–10	—	200	200	>0.4	90±5	CS₂
Bromoform	0.05–1	0.5–10	200	100	50	>0.25		CS₂

Ambient air analysis for hazardous substances 347

Butadiene	5–100	11–220	100	50	25	4		6
	100–2000	220–4400	10	5	1		99±5	
2-Butoxy ethanol	0.5–10	2.4–48	100	100	50	15	95	CS$_2$
	10–100	48–480	100	50	25			
n-butyl acetate	1–25	4.7–118	100	100	50	15	91±5	CS$_2$
	25–300	118–1420	50	25	10			
sec-Butyl acetate	1–50	4.8–237.5	100	100	50	12.5	94±5	CS$_2$
	50–400	237.5–1900	50	25	10			
tert-Butyl acetate	1–50	4.8–237.5	100	100	50	10.5	88±5	3
	50–400	237.5–1900	50	25	25			
Butyl alcohol	1–25	3–75	200	100	50	6	93±5	3
	25–200	75–600	100	50	10			
sec-Butyl alcohol	1–30	3–90	200	100	50	5	90±5	3
	30–300	90–900	50	25	10			
tert-Butyl alcohol	1–25	3–75	200	100	50			
	25–200	75–600	50	25	10			
Butylamine	0.1–1	0.3–3	—	200	200			
	0.5–10	1.5–30	200	100	50	11.5	85±10	CS$_2$
Butyl glycidyl ether	0.5–10	1.4–54	200	100	50			
	10–100	54–540	100	50	25	2.5	100+	CS$_2$
p-tert Butyl toluene	0.1–1	0.6–6	200	200	100			
	1–20	6–120	100	50	25	13.4	98±5	1
Camphor	0.05–0.5	0.63–6.3	200	200	100			
	0.1–4	1.3–25	200	100	50		95	13
Carbon disulphide	0.5–5	1.5–15	200	200	100			
	5–40	15–120	200	100	50	7.5	97±5	CS$_2$
Carbon tetrachloride	0.2–2	1.3–13	200	200	100			
	2–20	13–130	200	100	50	15.5	90±5	CS$_2$
Chlorobenzene	0.75–10	3.5–23.3	100	100	50	9.3	94±5	CS$_2$
	10–150	23.3–700	50	25	10			
Chlorobromomethane	2–20	10.5–105	100	50	25	11	96±5	CS$_2$
	20–400	105–2100	25	10	5			
Chloroform	0.5–10	2.4–48	200	100	50			
	10–100	48–480	100	50	25			

348   The Safe Handling of Chemicals in Industry

Table 7.1 (continued)

Substance	Expected concentration range		Recommended sampling (ml/min)				Recommended maximum tube load (mg)	Approximate desorption efficiency (%)	Eluent
	ppm	mg/m³	2 hr	4 hr	8 hr				
1-Chloro-1-Nitropropane	0.2–5	1–25	200	100	50			
	5–40	25–200	100	50	25			
Chloroprene	0.5–5	1.8–18	200	100	50			
	5–50	18–180	100	50	25			
o-Chlorotoluene	1–10	5–50	100	50	25				CS₂
	10–100	50–500	50	25	10			
Cresol (All isomers)	0.1–1	0.44–4.4	200	200	100	>2.0		
	1–10	4.4–44	100	100	50			
Crotonaldehyde	0.1–1	0.3–3	—	200	200			
	0.5–4	1.5–12	100	100	50			
Cumene	0.5–10	2.5–49	100	100	50	11	100+	CS₂
	10–100	49–490	50	25	10			
Cyclohexane	3–50	10.5–175	100	50	25	6.3	100+	CS₂
	50–600	175–2100	25	10	5			
Cyclohexanol	0.5–10	2–40	200	100	50	10	99±5	2
	10–100	40–400	100	50	25			
Cyclohexanone	0.5–10	2–40	200	100	50	13	78±5	CS₂
	10–100	40–400	100	50	25			
Cyclohexene	3–50	10.2–169.2	100	50	25		100+	CS₂
	50–600	169.2–2030	25	10	5			
Diacetone alcohol	0.5–10	1.4–48	200	100	50	12	77±10	2
	10–100	48–480	100	50	24			
o-Dichlorobenzene	0.5–10	3–60	200	100	50	15	85±5	CS₂
	10–100	60–600	50	25	10			
p-Dichlorobenzene	1–25	6–150	100	100	50		85±5	CS₂
	25–150	150–900	50	25	10			

Ambient air analysis for hazardous substances

Compound	Range 1	Range 2					Recovery (%)	Desorbent
Dichlorodifloromethane (Freon 12)	5–100	24.8–495	100	50	25			
	100–2000	495–9900	10	5	1	7.5	100+	CS_2
1,1-Dichloroethane	1–15	4–60	100	100	50			
1,2-Dichloroethylene	15–200	60–300	50	25	10	5.1	100+	CS_2
	2–25	7.9–99	100	50	25			
1,1-Dichloro-1-Nitroethane	25–400	99–1580	25	10	5	9		CS_2
	0.1–1.5	0.6–9	200	200	100			
Dichloroethyl ether	1.5–20	9–120	100	50	25			
	0.5–5	3–30	200	100	50			
Dichloromonofluoromethane	5–30	30–180	100	50	25			
(Freon 21)	5–100	21–420	100	50	25			
Dichlorotetrafluoroethane	100–2000	420–8400	10	5	1			
	5–100	35–700	50	25	10			
	100–2000	7000–14,000	5	1	1			
Difluorodibromomethane	1–15	8.6–129	100	50	25	15		CS_2
	15–200	129–1720	25	10	5			
Diisobutyl ketone	0.5–7.5	2.9–43.5	200	100	50	12.5		CS_2
	7.5–100	43.5–580	100	50	25			
Dimethylaniline	0.1–1	0.5–5	200	200	100	>1.1	>80	CS_2
	1–10	5–50	100	50	25			
Dimethylformamide	0.5–4	1.5–12	200	100	50			
	2–20	6–60	100	100	50			
f-Dioxane	1–15	3.6–54	200	100	50	13	91±5	CS_2
	15–200	54–720	100	50	25			
Dipropylene glycol methyl ether	1–15	6–90	100	50	5		75±15	CS_2
	15–200	90–1200	25	10	5			
Epichlorohydrin	0.1–1	0.4–4	—	200	100	>1	>80	CS_2
	1–10	4–40	100	50	25			
2-Ethoxyethanol	2–30	7.4–111	200	50	50			
	30–400	111–1480	50	25	10			
2-Ethoxyethyl acetate	1–15	5.4–81	200	100	50	19	74±10	CS_2
	15–200	81–1080	50	25	10			
Ethyl acetate	5–75	17.5–263	100	50	25	12.5	90±10	CS_2
	75–800	263–2800	25	10	5			
Ethyl acrylate	0.5–5	2–20	200	200	100	<5	95±5	CS_2
	5–50	20–200	200	100	50			

Table 7.1 (continued)

Substance	Expected concentration range		Recommended sampling (ml/min)			Recommended maximum tube load (mg)	Approximate desorption efficiency (%)	Eluent
	ppm	mg/m³	2 hr	4 hr	8 hr			
Ethyl alcohol	5–100	9.4–188.5	100	50	25	2.6	77±10	5
	100–2000	188.5–3770	5	1	1			
Ethyl benzene	1–15	4.4–65.3	200	100	50	16	100+	CS$_2$
	15–200	65.3–870	100	50	25			
Ethyl bromide	2–50	8.9–223	100	50	25	7.1	83±5	4
	50–400	223–1780	25	10	5			
Ethyl butyl ketone	0.5–10	2.3–46	200	100	50	>5.5	93±5	1
	10–100	46–460	50	25	10			
Ethyl chloride	10–150	26–390	100	50	25	9.7		CS$_2$
	150–2000	390–5200	10	5	1			
Ethyl ether	5–75	15–227	100	50	25	7.5	98±5	7
	75–800	227–2420	10	5	1			
Ethyl formate	1–15	3–45	200	100	50	4.8	80±10	CS$_2$
	15–200	45–600	50	25	10			
Ethylene chlorohydrin*	0.1–2	0.32–6.4	—	200	200	16.0	92±5	10
	1–10	3.2–32	200	100	50			
Ethylene dibromide	0.2–5	1.6–38.88	200	200	100	>10.7	93±5	CS$_2$
	2–40	15.5–310	100	50	25			
Ethylene dichloride	0.5–10	2–40.5	200	200	100	12	95±5	CS$_2$
	10–100	40.5–405	100	50	25			
Ethylene oxide	0.5–10	0.9–18	200	100	50	1.1		CS$_2$
	10–100	18–180	100	50	25			
Fluorotrichloromethane (Freon 11)	5–100	28–560	100	50	25			CS$_2$
	100–2000	560–11,200	5	1	1			
Furfural	0.1–2	0.4–8	200	200	100		>80	CS$_2$
	1–10	4–40	200	100	50			

Ambient air analysis for hazardous substances

Compound	Range 1	Range 2	C1	C2	C3	Value	Recovery (%)	Solvent
Furfuryl alcohol	0.5–20	2–80	200	100	50			
	10–100	40–400	50	25	10			
Glycidol	0.5–20	1.5–60	200	100	50	22.5	90±5	8
	10–100	30–300	100	50	25			
Heptane	5–100	20–400	100	50	25	12.5	96±5	CS$_2$
	100–1000	400–4000	10	5	1			
Hexachloroethane	0.05–2	0.5–20	200	200	100		98±5	CS$_2$
Hexane	5–100	18–360	100	50	25	11	94±5	CS$_2$
	100–1000	360–3600	10	5	1			
Isoamyl acetate	1–15	5.3–78.8	200	100	50	16.5	90±5	CS$_2$
	15–200	78.8–1050	50	25	10			
Isoamyl alcohol	1–15	3.6–54	200	100	50	10	99±5	2
	15–200	54–720	50	25	10			
Isobutyl acetate	1–20	4.7–93	200	100	50	14	92±5	CS$_2$
	20–300	93–1400	50	25	10			
Isobutyl alcohol	1–15	3–46	200	100	50	10.5	84±10	3
	15–200	46–610	50	25	10			
Isophorone	0.5–5	2.8–28	200	100	50	13	>80	CS$_2$
	5–50	28–280	50	25	10			
Isopropyl acetate	2–50	7.6–190	100	100	50	13	85±5	CS$_2$
	50–500	190–1900	25	10	5			
Isopropyl alcohol	5–75	12.3–185	100	50	25	5.6	94±5	5
	75–800	185–1970	25	10	5			
Isopropyl ether	5–100	21–420	100	50	25			
	100–1000	420–4200	10	5	1			
Isopropyl glycidyl ether	0.5–10	2.4–48	200	100	50	10.5	80±10	CS$_2$
	10–100	48–480	100	50	25			
Mesityl oxide	0.5–10	2–40	200	200	100	4.8	79±5	1
	5–50	20–200	100	50	25			
Methyl acetate	1–50	3.1–152.5	200	100	50	7	88±5	CS$_2$
	50–400	152.5–1220	25	10	5			
Methyl acrylate	0.1–2	0.35–7	—	200	100	>1.5	80±10	CS$_2$
	1–20	3.5–70	200	100	50			
Methyl alcohol	Charcoal tube method not recommended. Use silica gel tubes.							

352 The Safe Handling of Chemicals in Industry

Table 7.1 (continued)

Substance	Expected concentration range		Recommended sampling (ml/min)				Recommended maximum tube load (mg)	Approximate desorption efficiency (%)	Eluent
	ppm	mg/m³	2 hr	4 hr	8 hr				
Methylal (Dimethoxymethane)	5–100	15.5–311	100	50	25		11.5	78±10	9
	100–2000	311–6220	10	5	1				
Methyl amyl ketone	1–15	4.7–70	200	100	50		7.5	80±10	1
	15–200	70–930	50	25	10				
Methyl bromide	0.5–4	2–16	200	100	50		1.5		CS₂
	4–40	16–160	50	25	10				
Methyl butyl ketone (2-Hexanone)	1–15	4.1–61.5	200	100	50		2.0	79±10	CS₂
	15–200	61.5–820	50	25	10				
Methyl isobutyl ketone (Hexanone)	1–15	4.1–61.4	200	100	50		10	81±5	CS₂
	15–200	61.4–818	50	25	10				
Methyl cellosolve	0.3–10	1–32	200	200	100		10	97±5	6
	5–50	16–160	100	50	25				
Methyl cellosolve acetate	0.3–10	1.5–48	200	100	50		5	76±10	CS₂
	5–50	24–240	100	50	25				
Methyl chloride	1–15	2.1–31.5	200	100	50		1		CS₂
	15–200	31.5–420	25	10	5				
Methyl chloroform (1,1,1 Trichloroethane)	1–75	5.4–407	200	100	50		18	98+	CS₂
	75–700	407–3800	25	10	5				
Methyl cyclohexane	5–100	20–400	100	50	25			95±5	CS₂
	100–1000	400–4000	10	5	1				
Methyl ethyl ketone (2-Butanone)	2–50	5.9–148	100	100	50		9.5	89±10	CS₂
	50–400	148–1180	50	25	10				
Methyl formate	1–15	2.5–37.5	200	100	50				
	15–200	37.5–500	50	25	10				
5-Methyl 3-heptanone	0.25–5	1.3–26	200	100	50		>5		1
	5–50	26–260	100	50	25				

Compound								
Methyl iodide	0.1–1	0.56–5.6	200	200	100	7		14
	1–10	5.6–56	100	50	25	5.7	99±5	2
Methyl isobutyl carbinol	0.5–5	2.1–21	200	100	50			
	5–50	21–210	200	100	50	14		CS$_2$
Methyl isoamylacetate	1–10	6–60	200	50	25			
	10–100	60–600	200	100	50	21	91±5	CS$_2$
α-methyl styrene	1–15	4.8–72	200	50	25			
	15–200	72–960	100	50	25	9.3	95±10	CS$_2$
Methylene chloride	5–75	17.4–261	200	100	50			
	75–1000	261–3480	10	5	1	14.8	88±5	CS$_2$
Naphtha (coal tar)	1–15	4–60	200	100	50			
	15–200	60–800	100	50	25			
Naphthalene	0.1–2	0.5–10	200	200	100			
	1–20	5–100	100	50	25			
Nitromethane	1–15	3.1–46.5	200	200	100			
	15–200	46.5–620	50	25	10			
1-Nitropropane	0.5–10	1.8–36	200	100	50			
	5–50	18–180	100	50	25			
2-Nitropropane	0.5–10	1.8–36	200	100	50			
	5–50	18–180	100	50	25			
Nonane	2–30	10.5–157.5	100	50	25			
	30–400	157.5–2100	25	10	5	15	93±5	CS$_2$
n-Octane	5–75	23.5–352	100	50	25			
	75–1000	352–4700	10	5	1	9	96±5	CS$_2$
Pentane	5–100	15–295	200	100	50			
	100–2000	295–5900	10	5	1			CS$_2$
2-Pentanone	1–75	3.5–263	200	100	50		88±5	CS$_2$
	75–400	263–1400	25	10	5			
Perchloroethylene	1–15	6.8–102	200	100	50			
(tetrachloroethylene)	15–200	102–1362	50	25	10	29	95±5	CS$_2$
Petroleum distillates	5–75	20–300	100	50	25	12.3	96±5	CS$_2$
	75–1000	300–4000	10	5	1			
Phenol	0.1–2	0.38–7.6	—	200	100			
	1–10	3.8–38	200	100	50			

Table 7.1 (continued)

Substance	Expected concentration range			Recommended sampling (ml/min)			Recommended maximum tube load (mg)	Approximate desorption efficiency (%)	Eluent
	ppm		mg/m³	2 hr	4 hr	8 hr			
Phenyl ether (vapour)	0.05–2		0.35–14	—	200	200	0.6	90±5	CS₂
Phenyl glycidyl ether	0.5–4		3–24	200	100	50	12.5	97±5	CS₂
	2–20		12–120	100	50	25			
n-Propyl acetate	1–75		4.2–315	200	100	50	14.5	93±5	CS₂
	75–400		315–1680	50	25	10			
n-Propyl alcohol	1–75		2.5–184	200	100	50	9	87±5	3
	75–400		184–980	50	25	10			
Propylene dichloride	1–15		4.6–70	200	100	50	5	97±5	CS₂
	15–150		70–700	50	25	10			
Propylene oxide	1–15		2.4–36	200	100	50	2	90±5	CS₂
	15–200		36–480	25	10	5			
n-Propyl nitrate	0.5–5		2.2–22	200	100	50	12		CS₂
	5–50		22–220	100	50	25			
Pyridine	0.5–2		1.5–6	200	200	100	>7.3	70±10	CS₂
	1–10		3–30	200	100	50			
Stoddard solvent	5–75		29.5–443	100	50	25	13	96±5	CS₂
	75–1000		443–5900	10	5	1			
Styrene (monomer)	1–15		4.3–64	200	100	50	18	87±5	CS₂
	75–1000		64–850	100	50	25			
1,1,1,2-Tetrachloro-2,2-Difluoroethane	5–75		42–625	100	50	25	19.5	100+	CS₂
	75–1000		625–8340	10	5	1			
1,1,2,2-Tetrachloro-1,2-Difluoroethane	5–75		42–625	100	50	25	26	96±5	CS₂
	75–1000		625–8340	10	5	1			
1,1,2,2-Tetrachloroethane*	0.1–2		0.7–14	200	200	100	4.5	85±5	CS₂
	1–10		7–70	100	100	50			
Tetrahydrofuran	1–75		2.5–184	200	100	50	7.5	92±5	CS₂
	75–400		184–1180	25	10	5			

Substance	Range	Sampling rate (ml/min)			% Recovery	Desorbing solvent
Tetramethyl succinonitrile	0.05–1	—	200	200	>0.8	
Toluene	1–15	200	100	50		CS_2
	15–200	3.8–56	50	25		CS_2
		56–750	200	10		
1,1,2-trichloroethane	0.1–2	0.6–11	100	50	5	CS_2
	2–20	11–110	50	25		
Trichloroethylene	1–15	5.4–80	200	50	21	CS_2
	15–200	80–1070	100	25		
1,2,3-Trichloropropane	0.5–10	3–60	200	50	14	CS_2
	10–100	60–600	100	25		
1,1,2-Trichloro-1,2,2-	10–150	77–1150	100	25	20	CS_2
Trifluoroethane (Freon 113)	150–2000	1150–15,300	10	1		
Trifluoromonobromomethane†	10–150	61–914	100	25	25	
	150–2000	914–12,180	10	1		
Turpentine	1–15	5.6–84	200	50	13	CS_2
	15–200	84–1120	50	10		
Vinyl chloride†	0.05–1	0.13–2.5	—	200		100+
	0.5–2	1.3–5.2	200	100		CS_2
Vinyl toluene	1–15	4.8–72	200	50	17	CS_2
	15–200	72–960	100	25		
Xylene	1–15	4.4–65	200	50		CS_2
	15–200	65–870	50	10		

* 100 mg petroleum based charcoal tubes based on the NIOSH recommended design should be used for sample collection of these compounds.

† These compounds migrate rapidly to the back-up section of the charcoal tube. A 400 mg tube should be used for sample collection with a second 100 mg tube in series behind the large tube to determine breakthrough.

Guide to use of Table 7.1

Expected concentration range: There are two concentration ranges given for most substances. The low range is approximately 1%–15% of the TLV and the high range from 15%–200% TLV. The user should select from these two ranges his expected mean concentration.

Recommended sampling rate: A sampling rate is recommended for each concentration range and for a 2-hour, 4-hour, or 8-hour sampling period. Each sampling rate is given in ml/min and has been calculated to provide a minimum tube loading of at least .01 mg at the minimum concentration shown and not to exceed the recommended tube loading at the highest concentration shown for that range. These figures are based on the use of 100 mg coconut-shell charcoal tubes to the NIOSH recommended design except where otherwise noted.

Table 7.1 (continued)

Maximum tube loading: A recovery of 5% of the total sample from the back-up section of charcoal in a sample tube was defined as the breakthrough point: 50% of this value is shown as the recommended maximum tube loading, to allow for high humidity or the presence of other substances which reduce the normal tube capacity.

Approximate desorption efficiency: The figures given are not intended to be used as exact desorption efficiencies and are only given as a guide when carrying out system calibrations. Actual desorption efficiencies should always be determined at the time of analysis. However, these figures represent the best obtainable data from several sources, and any significant deviation should be regarded as a possible indication of a systematic error in the analytical technique. The figure given for desorption efficiency is an average figure. The desorption efficiency for a compound will vary with the amount of substance on the tube. With reduced tube loadings, in most cases, the desorption efficiency will be lower. Significant errors may be introduced when analysing small amounts of substance and an average desorption efficiency factor is used.

Eluent: The desorption efficiencies given are directly related to the eluent used. All data in the desorption efficiency column correspond to the specific eluent listed.

Eluent code:

1 –CS_2 + 1% methanol
2 –CS_2 + 5% 2-propanol
3 –CS_2 + 1% 2-propanol
4 –Isopropanol
5 –CS_2 + 1% butanol
6 –Methylene chloride + 5% methanol
7 –Ethyl acetate 0.5 ml
8 –Tetrahydrofuran 0.5 ml
9 –Hexane 0.5 ml
10 –CS_2 + 5% isoamyl alcohol
11 –Formic acid
12 –Methanol
13 –Benzene
14 –Toluene

Ambient air analysis for hazardous substances 357

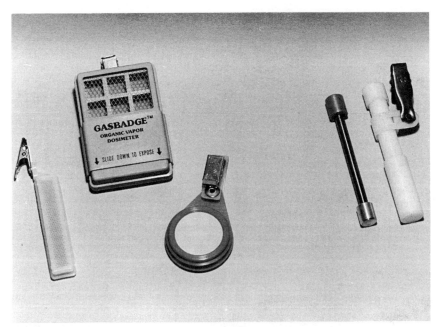

Fig. 7.4(a) A selection of passive samplers for organic vapours (Courtesy Environmental Monitory Systems)

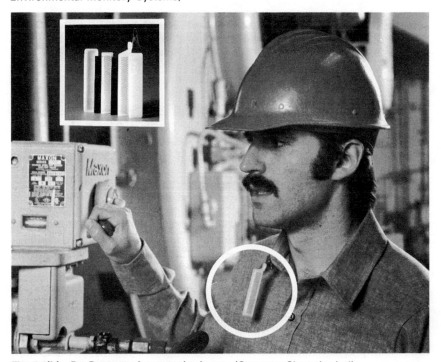

Fig. 7.4(b) Du Pont passive sampler in use (Courtesy Shawcity Ltd)

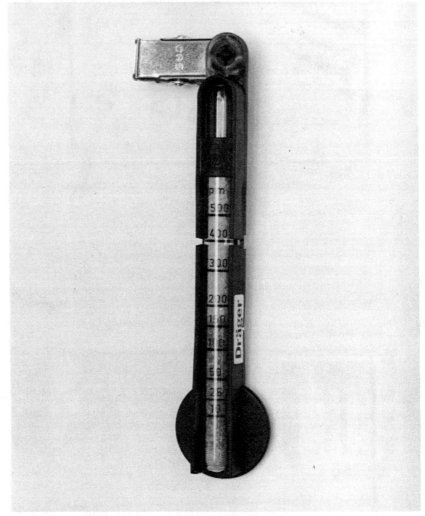

Fig. 7.4(c) Diffusion detector tubes (Courtesy Draeger Safety)

Progress with some passive samplers has been good while others have received a mixed reception. In general, passive systems are considered to be still under development and in some cases, less reliable than pumped systems.[12-15]

A range of techniques for sampling gases, vapours and mists are summarised in Table 7.2.[16] A variety of pumps are used in sampling systems ranging from large mains-driven versions with high sample rates (e.g. 3 m^3/second) to lightweight battery designs as illustrated in Fig. 7.2, or manually operated bellows or pistons as in Figs 7.5(a) 7.5(b) respectively. For personal dosimetry the electric pump is worn in the operator's pocket or attached to his belt. It samples up to 2 litres of air per minute. The rate of air-flow must always be known so that the total volume of air sampled can be calculated to enable the airborne concentration of contaminant to be determined along with operator exposures.

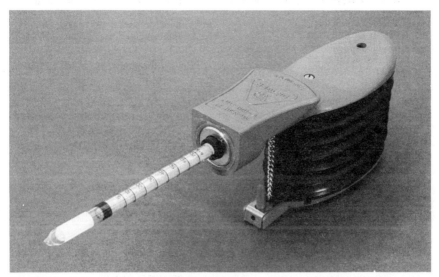

Fig. 7.5(a) Bellows pump and colour indicator tube (Courtesy Draeger Safety)

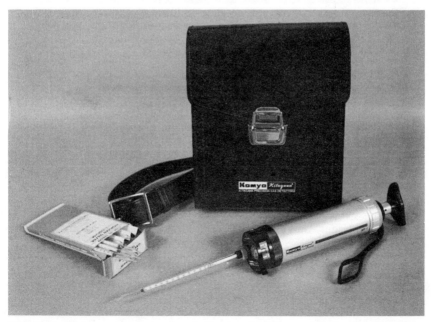

Fig. 7.5(b) Piston pump and colour indicator tube (Courtesy Sabre Gas Detection Ltd)

Table 7.2 Some examples of sample collection methods for air contaminants

Contaminant	Principle	Apparatus	Collecting agent
Gases or vapours that are water soluble or miscible or that are only soluble or highly reactive in other agents.	Absorption with simple washing.	Petri bottle. Wash bottle Drechsel bottle. Impinger.	Water, acid, alkali, or organic solvent.
Gases or vapours of all types. Mists and fumes.	Absorption with multiple contact washing by dispersing gas into fine bubbles of large surface area.	Fritted or sintered glass, alundum, stainless steel, or plastics.	Water, acid, alkali, or organic solvents.
Gases or vapours of all types. Mists.	Absorption with multiple contact provided by wetted surfaces.	Packed colums of glass beads, glass spirals, or fibres. Plastic packing also feasible in some cases. Large surface wall area units also included.	Water, acid, alkali, or organic solvents.
Gases or vapours that are water soluble or miscible or that are only soluble or highly reactive in other agents.	Absorption with multiple surface contact by atomising liquid with spray nozzle or jet impaction.	Crabtree ozone analyser or midget Venturi scrubber.	Water, acid, or alkali.
Combustible gases or vapours that are not water soluble but are slowly reactive with absorbing agents.	Combustion and absorption.	Quartz or ceramic furnace with absorbers. 1 or 2 above.	Water, acid, or alkali.
Gases or vapours that are not water soluble but are slowly reactive, with absorbing agent.	Condensation.	Freezing traps or low temperature condensers. Glass or metal.	Dry

Sampling rate (l/min)	Collection efficiency (%)	Remarks
1–5	90–100	Two units in series can be used for highly reactive gases in other reagents.
1–15 (depends upon flask and sintered surface dimensions).	95–100	May plug if large particulates are present or precipitates form from reactions. Fumes require very slow rates.
1–5	90–100	May incorporate device for continually wetting column. High loadings of particulates may plug unit with packing.
5–25	60–100	Venturi scrubber satisfactory if dust is present. Atomiser absorber will plug.
1–5	90–100	Can also be used for organic halogenated fumes.
1–5	90–100 (depends on vapour pressures at reduced temperature).	Can be packed with glass beads or other extended surfaces.

Gas detection

A wide range of sensors are used in commercially-available, monitoring equipment. These instruments are best classified according to detection systems and divided into gases/vapours and particulates. They range from devices for chemical analysis in laboratory evaluation of samples collected in the field, to direct reading instruments which often rely on some chemical or physical property of the substance under analysis. Detection systems have been based on optical properties, ionisation, thermal measurements, electrode systems, semiconductors, separation characteristics, etc.

Colorimetric procedures

This technique involves the interaction of test gas with specific reagents to bring about a colour change which can be measured spectroscopically or by comparison with standards. It is a sensitive method used for detection of gas in low concentrations. Though it has been used for such gases as sulphur dioxide, chlorine, and nitrogen dioxide, it has been superseded by superior methods. Nowadays the technique is reserved for those materials for which simpler alternatives are not available or are less well established, e.g. ethylene oxide, hydrogen fluoride and formaldehyde. Portable versions are available such as the TGM 555 from Quantitec. The principle of colorimetric analysis is adopted for colorimetric tubes and paper tape systems as described below.

Colour-indicator tubes

The most commonly used portable detection system for toxic gases is based on the indicator tube. These can also be used for estimating concentrations of certain flammable gases. In the main these are used for grab samples and tubes are available for about two hundred substances. Specificity and accuracy of the tubes varies across the range and suppliers' literature must be consulted.

The system relies on a manually operated bellows or piston pump to aspirate a fixed volume of atmosphere through a glass tube (see Figs. 7.5(a) and 7.5(b)). The latter is filled with crystals (e.g. silica gel or alumina) impregnated with a reagent which undergoes a chemical reaction with a specific pollutant gas or class of gas, and undergoes a colour change as exemplified by the system used in the hydrogen sulphide detector tube:

$$H_2S + Pb^{2+} \rightarrow PbS + 2H^+$$
$$\text{(white) (brown)}$$

The indicator tube is generally designed so that the length of stain which develops in the tube is indicative of the concentration of chemical in the atmosphere and the value is read from a scale printed on the tube wall.

Older versions rely on the development of a stain throughout the entire length of the tube, the colour intensity (measured by comparison with a set of standards) is proportional to the concentration of material to which the reagent is sensitive. Some tubes contain a chemical trap layer to remove contaminants which would otherwise interfere with the determinations.

Sample lines of up to 5 m in length can be used between the pump and indicator tube to enable sampling in otherwise inaccessible locations. Attractions of this system of monitoring include low cost, simplicity, and instant read-out of concentration with no necessity for further analysis. As a result these tubes are often used by the novice who may be unaware of problems which can give rise to misinterpretation of the results so obtained. Guidance should be sought from suppliers or the literature.[17–19] Sources of inaccuracy may include:

- Both ends of the sealed tube not broken before insertion of the tube into the pump housing.
- Insertion of the tube incorrectly into the pumphousing. (The correct direction is indicated on the tube.)
- Re-use of previously-used tubes. It is advisable not to re-use tubes even if their previous use indicated zero concentration.
- Leaks in sample lines, or insufficient time allowed to lapse between pump strokes when these extensions are used.
- Use of tubes beyond expiry of the shelf-life. It is advisable to store tubes under refrigerated conditions but allow them to warm to room temperature prior to use.
- Ill-defined stain format because it is irregular, diffuse or has failed [i.e. not at right angles to tube wall]. (This can be caused by poor quality of granular support medium used by manufacturer.) It is advisable to read the maximum value indicated.
- Use of tubes under conditions of temperature, pressure or humidity outside the range for which they were calibrated.
- Blockages or faulty pumps. Pumps should be checked periodically as instructed by the manufacturer. They can be calibrated using rotameters or bubble flowmeters. Unless pumps possess a limiting orifice they should be calibrated with air indicator tube in position.
- Misuse of the pump, e.g. incomplete stroke or wrong number of strokes.
- Use of tubes with a different type of pump than that for which it was designed.

Fig. 7.6 Battery-operated pump used for prolonged sampling periods with detector tube in the instrument (Courtesy National Draeger, Inc.)

- Interference due to the presence of other contaminants capable of reacting with the tube reagent. This can result in over- or under-estimation of concentrations, but the former is the more likely and hence errs on the side of safety.
- Airborne dusts can cause tube blockage and affect the flow rate.

A smaller number of special tubes are available for long-term monitoring (Fig. 7.6); these are used in conjunction with battery-operated personal sampler pumps operating at a flow rate of 10–20 ml/second. The average concentration after a set period (e.g. 8 hours) can be read directly.

Colorimetric paper-tape detectors

The need for detection of toluene di-isocyanate was the impetus for the development of these instruments. They can be portable monitors (see

Fig. 7.7(b) or fixed, multipoint background monitors (Fig. 7.7(a)) and are based on formation of a stain due to chemical reaction between airborne contaminant and a reagent impregnated on paper tape. A cassette of the latter is electrically driven at a constant rate over a sampling orifice. Air is sampled using a self-contained pump and flow controller. Stain intensity is measured with an internal reflectometer. Use of double track design, whereby reflectance of the exposed layer is compared with that of a reference layer, allows for compensation in tape variations. Test slips are often provided for field calibrations.

A range of particularly toxic gases can be detected by choice of tape cassette, e.g. ammonia, chlorine, di-isocyanates, hydrazines, hydrides

Fig. 7.7(a) Automatic multipoint PSM analyser using colour paper-tape detection (Courtesy MDA Scientific Inc)

Fig. 7.7(b) Series 7100 portable monitor utilising colour paper-tape detection (Courtesy MDA Scientific Inc)

(arsine, diborane, germane, hydrogen selenide, phosphine, silane), hydrogen-chloride, -cyanide, -fluoride, or -sulphide, nitrogen dioxide, *p*-phenylene diamine, phosgene and sulphur dioxide. Detection limits are very low for many substances. Audible alarms can be fitted to these devices, e.g. in polyurethane production areas for isocyanate monitoring.

Infrared spectroscopy

Infrared detectors are based on the fact that molecules of the gaseous substance will absorb IR radiation over a wide spectrum of wavelengths such that characteristic graphs can be obtained akin to 'fingerprints'. This provides both qualitative and quantitative information which enables the presence of the specific material to be detected and its concentration to be determined. These analysers are common laboratory instruments; they are also used as the detection system in some fixed background monitors.

Portable IR spectrometers are available as non-dispersive models for measuring specific substances, or as a dispersive version for measuring most materials by tuning to suitable wavelengths. Thus a beam of IR radiation is passed through a gas cell and its decrease in intensity, which is proportional to the concentration of the substance in the cell, is measured. Usually these instruments are not intrinsically safe. Though detection limits down to 0.1 ppm can be achieved for certain materials, the presence of co-contaminants in some situations can cause interference. Table 7.3 lists those materials claimed to be detectable by these IR instruments and information on instrument settings and minimum detectable

Table 7.3 Compounds detectable by portable infrared analyses (after ref. 20)

Compound	MIRAN 1A			
	Analytical Wavelength-μm	Path length-m	Absorbance	Minimum Detectable Concentration-ppm (20 metre Cell)
Acetaldehyde	9.0	20.25	0.23	0.8
Acetic acid	8.5	20.25	0.044	0.1
Acetic anhydride	8.9	20.25	0.18	0.02
Acetone	8.2	2.25	0.49	0.09
Acetonitrile	9.6	20.25	0.005	5.0
Acetophenone	7.9	20.25	0.5	0.7
Acetylene	3.03	6.75	0.19	0.4
Acetylene dichloride, see 1,2-Dichloroethylene				
Acetylene tetrabromide	8.85	20.25	0.0047	0.1
Acrolein	8.6	20.25	0.0002	0.3
Acrylonitrile	10.5	20.25	0.005	0.2
Allyl Alcohol	9.8	20.25	0.0055	0.3
Allyl Chloride	10.8	20.25	0.0032	0.2
Allylglycidylether	9.1	20.25	0.08	0.07
2-Aminoethanol, see Ethanolamine				
Ammonia	10.4	20.25	0.14	0.2
n-Amyl acetate	8.0	0.75	0.12	0.02
Aniline	9.3	20.25	0.011	0.3
Arsine	4.66	20.25	0.0003	0.1
Arylam, see Carbaryl				
Benzene	14.87	20.25	0.08	0.3
p-Benzoquinone, see Quinone				
Benzyl Chloride	7.9	20.25	0.0012	0.4
Bisphenol A, see Diglycidyl ether				
Bis(chloromethyl)ether	8.99	20.25	0.0016	0.2
Boron trifluoride	6.8	20.25	0.006	0.5
Bromoform	8.7	20.25	0.005	0.05
Butadiene, (1,3-butadiene)	11.0	2.25	0.46	0.2
Butane	10.36	20.25	0.03	3.0
Butanethiol, see Butyl mercaptan				
2-Butanone (MEK)	8.5	20.25	0.55	0.15
2-Butoxyethanol (Butyl cellosolve)	8.9	9.75	0.2	0.08
Butyl acetate (n-Butyl acetate)	8.1	3.75	0.53	0.02
sec-Butyl acetate	9.9	9.75	0.48	0.15
tert-Butyl acetate	9.9	9.75	0.36	0.2
n-Butyl alcohol	9.6	8.25	0.15	0.17
sec-Butyl alcohol	10.1	9.75	0.15	0.35
tert-Butyl alcohol	8.2	20.25	0.47	0.05
Butylamine	13.0	20.25	0.025	0.3
Butyl ether	8.85	20.25	0.5	0.04
Butyl carbitol	8.9	20.25	0.32	0.08
n-Butyl glycidyl ether	8.9	20.25	0.55	0.05
Butyl mercaptan	3.37	20.25	0.04	0.08
p-tert-Butyltoluene	12.3	20.25	0.015	0.8
Carbon disulphide	4.54	20.25	0.015	0.5
Carbon dioxide	4.25	0.75	1.2	0.05

Table 7.3 (continued)

Compound	MIRAN 1A			
	Analytical Wavelength-μm	Path length-m	Absorbance	Minimum Detectable Concentration-ppm (20 metre Cell)
Carbon monoxide	4.61	20.25	0.045	0.2
Carbon tetrachloride	12.6	20.25	0.31	0.06
Carbonyl sulphide	4.85	20.25	0.5	0.02
Carbary[A]				
Chlorinated camphene[A]				
Chlorobenzene (monochlorobenzene)	9.2	20.25	0.26	0.2
Chlorobromomethane	8.1	9.75	0.29	0.2
2-Chloro-1, 3-butadiene, see Chloroprene				
Chlorodifluoromethane (Freon 22)	9.0	5.25	0.17	0.02
1-Chloro, 2,3-Epoxypropane, see Epichlorohydrin				
2-Chloroethanol, see Ethylene chlorohydrin				
Chloroethylene, see Vinyl chloride				
Chloroform (trichloromethane)	13.0	5.25	0.45	0.06
1-Chloro-1-nitropropane	12.4	20.25	0.04	1.0
Chloropentafluoroethane (Genetron 115)	8.1	5.25	0.7	0.02
Chloropicrin (Trichloronitromethane)	11.5	20.25	0.002	0.05
Chloroprene (2-chloro-1,3-butadiene)	11.4	20.25	0.08	0.4
Chlorotrifluoroethylene	9.24	20.25	0.04	0.06
Cresol (all isomers)	8.6	20.25	0.0066	0.3
Crotonaldehyde (trans-2-butenal)	8.7	20.25	0.01	0.1
Cumene (isopropyl benzene)	9.75	20.25	0.031	0.7
Cyanogen	4.7	20.25	0.016	3.0
Cyclohexane	3.4	2.25	0.56	0.03
Cyclohexanol	9.3	20.25	0.28	0.10
Cyclohexanone	8.3	20.25	0.054	0.5
Cyclohexene	8.8	20.25	0.12	1.3
Demeton V				
Deuterium oxide	3.68	20.25	0.067	1.0
DDVP, see Dichlorvos				
Diacetone alcohol (4-hydroxy-4-methyl-2-pentanone)	8.5	20.25	0.335	0.08
1,2-Diaminoethane, see Ethylenediamine				
Diborane	3.9	20.25	0.0005	0.1
Dibromochloropropane	8.27	20.25	0.19	0.2
1,2-Dibromotetrafluoroethane	8.45	20.25	0.7	0.02
o-Dichlorobenzene	13.4	20.25	0.213	0.4
p-Dichlorobenzene	9.1	9.75	0.47	0.06
Dichlorodifluoromethane (Freon 12)	9.1	0.75	1.2	0.02
1,1-Dichloroethane	9.4	9.75	0.15	0.3
1,2-Dichloroethylene	12.1	0.75	0.1	0.07
Dichloroethyl ether	8.85	20.25	0.13	0.06
Dichloromethane, see Methylene chloride				
Dichloromonofluoromethane (Freon 21)	9.3	0.75	0.42	0.08
1,1-Dichloro-1-nitroethane	9.1	20.25	0.1	0.07
1,2-Dichloropropane, see Propylene dichloride				

Table 7.3 (continued)

Compound	MIRAN 1A			Minimum Detectable Concentration-ppm (20 metre Cell)
	Analytical Wavelength-μm	Path length-m	Absorbance	
Dichlorotetrafluoroethane (Freon 114)	8.4	0.75	1.43	0.02
Dichlorvos DDVP	9.4	20.25	0.0014	0.02
Diethylamine	8.8	20.25	0.1	0.2
Diethylamino ethanol	9.4	20.25	0.08	0.1
Diethylether, see Ethyl ether				
Diethyl ketone	9.0	20.25	0.2	0.1
Diethyl malonate	9.5	20.25	0.11	0.05
Difluorodibromomethane	9.2	2.25	0.36	0.02
Diglycidyl ether[B]				
Dihydroxybenzene, see Hydroquinone				
Diisobutyl ketone	8.6	20.25	0.12	0.2
Diisopropylamine	8.5	20.25	0.026	0.1
1,2-Dimethoxyethane	8.8	5.25	0.3	0.1
Dimethoxymethane, see Methylal				
N,N-Dimethylacetamide	9.9	20.25	0.02	0.3
Dimethylamine	8.7	20.25	0.02	0.5
Dimethylaminobenzene, see Xylidene				
Dimethylaniline (N, N dimethyl aniline)	8.6	20.25	0.015	0.2
Dimethylbenzene, see Xylene				
Dimethylformamide	9.2	20.25	0.063	0.1
2,6-Dimethylheptanone, see Diisobutyl ketone				
Dimethylsulphate	9.9	20.25	0.04	0.02
Dimethyl Sulphoxide	9.0	20.25	0.5	0.2
Dioxane (diethylene dioxide)	8.9	5.25	0.38	0.05
Diphenylmethane diisocyanate, see Methylene bisphenyl isocyanate, MDI				
Enflurane	8.7	20.25	0.10	0.01
Epichlorohydrin	11.8	20.25	0.013	0.3
1,2-Epoxypropane, see Propylene oxide				
2;3-Epoxy-l-propanol, see Glycidol				
Ethanethiol, see ethyl mercaptan				
Ethane	12.1	20.25	0.15	3.0
Ethanolamine	13.0	20.25	0.0017	1.2
2-Ethoxyethanol (cellosolve)	8.9	2.25	0.24	0.06
2-Ethoxyethyl acetate (cellosolve acetate)	8.8	8.25	0.58	0.03
Ethyl acetate	8.0	0.75	0.39	0.02
Ethyl acrylate	8.4	20.25	0.33	0.04
Ethyl alcohol (ethanol)	9.5	2.25	0.5	0.2
Ethylamine	3.4	20.25	0.015	0.2
Ethyl sec-amyl ketone (5-methyl-3-heptanone)	9.0	20.25	0.04	0.3
Ethylbenzene	9.7	20.25	0.06	1.0
Ethyl bromide	8.0	5.25	0.17	0.2
Ethyl butyl ketone (3-heptanone)	9.0	20.25	0.12	0.2
Ethyl chloride	10.4	20.25	0.62	0.8
Ethyl ether	8.8	2.25	0.58	0.03
Ethyl formate	8.5	2.25	0.1	0.06
2-Ethyl hexanol	9.7	20.25	0.02	0.2
Ethyl mercaptan	3.3	20.25	0.029	0.8
Ethyl silicate	9.1	5.25	0.93	0.02

Table 7.3 (continued)

Compound	MIRAN 1A			Minimum Detectable Concentration-ppm (20 metre Cell)
	Analytical Wavelength-μm	Path length-m	Absorbance	
Ethylene	10.6	20.25	0.12	0.5
Ethylene chlorohydrin	9.3	20.25	0.026	0.08
Ethylenediamine	13.0	20.25	0.035	0.4
Ethylene dibromide (1,2-dibromoethane)	8.45	20.25	0.07	0.1
Ethylene dichloride (1,2-dichloroethane)	8.1	20.25	0.09	0.3
Ethylene glycol monomethyl ether acetate, see Methyl cellosolve acetate				
Ethylene oxide	11.8	20.25	0.12	0.4
Ethylidine chloride, see 1,1-Dichloroethane				
Fluorobenzene	8.1	20.25	0.4	0.06
Fluorotrichloromethane (Freon 11)	11.9	0.75	1.7	0.01
Fluroxene	8.5	20.25	0.086	0.01
Formaldehyde	3.58	20.25	0.015	0.2
Formic acid	8.9	20.25	0.055	0.05
Furfural	13.3	20.25	0.05	0.1
Furfuryl alcohol	9.8	20.25	0.09	0.3
Glycidol (2,3-epoxy-1 propanol)	9.9	20.25	0.020	1.3
Glycol monoethyl ether, see 2-Ethoxyethanol				
Guthion, [R], see Azinphosmethyl				
Halothane	12.3	20.25	0.027	0.08
Heptane (n-heptane)	3.4	0.75	0.29	0.01
1-Heptanol	9.5	20.25	0.2	0.1
Hexachloroethane	12.8	20.25	0.04	0.03
Hexafluoropropene	8.3	20.25	0.16	0.06
Hexane (n-hexane)	3.4	0.75	0.26	0.02
2-Hexanone	8.6	20.25	0.42	0.1
Hexone (methyl isobutyl ketone)	8.5	20.25	0.31	0.2
sec-Hexyl acetate	9.8	9.75	0.1	0.2
Hydrazine	10.55	20.25	0.001	0.25
Hydrogen chloride	3.4	20.25	0.002	1.0
Hydrogen cyanide	3.04	20.25	0.0083	0.4
Hydroquinone[A]				
Isoamyl acetate	9.4	9.75	0.27	0.14
Isoamyl alcohol	9.4	9.75	0.22	0.13
Isobutyl acetate	8.2	2.25	0.4	0.02
Isobutyl alcohol	9.6	8.25	0.30	0.08
Isodecanol	9.6	20.25	0.25	0.3
Isoflurane	8.25	20.25	0.07	0.02
Isophorone	3.4	20.25	0.12	0.4
Isoprene	11.2	5.25	0.078	0.2
Isopropyl acetate	8.0	2.25	0.63	0.02
Isopropyl alcohol	8.7	6.75	0.45	0.12
Isopropylamine	8.5	20.25	0.025	0.1
Isopropyl ether	8.9	0.75	0.17	0.07
LPG, (liquefied petroleum gas)	3.4	0.75	0.24	0.4
Mesityl oxide	8.2	20.25	0.06	0.2

Table 7.3 (continued)

Compound	MIRAN 1A Analytical Wavelength-μm	Path length-m	Absorbance	Minimum Detectable Concentration-ppm (20 metre Cell)
Methane	7.65	20.25	0.072	0.2
Methanethiol, see Methyl mercaptan				
Methoxyflurane	12.0	20.25	0.03	0.07
2-Methoxyethanol, see Methyl cellosolve				
Methyl acetate	9.5	5.25	0.31	0.2
Methyl acetylene (propyne)	7.9	20.25	0.77	1.0
Methyl acrylate	8.4	20.25	0.15	0.03
Methylal (dimethoxymethane)	9.5	0.75	0.43	0.05
Methyl alcohol (methanol)	9.5	5.25	0.3	0.1
Methylamine	3.4	20.25	0.02	0.1
Methyl amyl alcohol, see Methyl isobutyl carbinol				
Methyl n-amyl ketone (2-heptanone)	8.6	9.75	0.14	0.2
Methyl bromide	7.6	20.25	0.021	0.4
Methyl butyl ketone, see 2-Hexanone				
Methyl cellosolve	8.8	20.25	0.19	0.05
Methyl cellosolve acetate	8.0	20.25	0.39	0.02
Methyl chloride	13.4	20.25	0.14	1.5
Methyl chloroform	9.2	2.25	0.4	0.06
Methylcyclohexane	3.4	2.25	0.84	0.04
Methylcyclohexanol	9.5	20.25	0.24	0.2
o-Methylcyclohexanone	8.9	20.25	0.18	0.3
Methylene bisphenyl isocyanate, MDI[(A)]				
Methylene chloride	13.4	0.75	0.25	0.2
Methyl ethyl ketone (MEK), see 2-Butanone				
N-methyl formamide	8.4	5.25	0.065	0.2
Methyl formate	8.5	0.75	0.13	0.02
Methyl iodide	7.9	20.25	0.009	0.4
Methyl isoamyl ketone	8.6	20.25	0.6	0.04
Methyl isobutyl carbinol	8.7	20.25	0.053	0.25
Methyl isobutyl ketone, see Hexone				
Methyl isocyanate	11.6	20.25	0.00003	0.6
Methyl isopropyl ketone	8.8	20.25	0.1	0.3
Methyl mercaptan	3.38	20.25	0.008	0.4
Methyl methacrylate	8.5	2.25	0.21	0.03
Methyl propyl ketone, see 2-Pentanone				
α-Methyl styrene	11.1	20.25	0.3	0.3
Monomethylaniline	7.9	20.25	0.007	0.2
Morpholine	9.0	20.25	0.05	0.2
Nickel carbonyl	4.86	20.25	0.0003	0.005
Nitric oxide (NO)	5.3	20.25	0.015	2.0
Nitrobenzene	11.8	20.25	0.005	0.2
Nitroethane	9.0	20.25	0.08	0.6
Nitrogen dioxide (NO_2)	6.17	20.25	0.048	0.1
Nitrogen trifluoride	11.0	20.25	0.35	0.03
Nitromethane	9.3	20.25	0.056	0.9
Nitrotoluene	11.8	20.25	0.0076	0.7
Nitrotrichloromethane, see Chloropicrin				

Table 7.3 (continued)

Compound	MIRAN 1A Analytical Wavelength-μm	Path length-m	Absorbance	Minimum Detectable Concentration-ppm (20 metre Cell)
Nitrous oxide	4.50	20.25	0.3	0.07
Octane	3.4	0.75	0.40	0.02
Pentane	3.4	0.75	0.42	0.02
2-Pentanone	8.5	5.25	0.23	0.1
Perchloroethylene	10.9	5.25	0.63	0.05
Petroleum distillates	3.4	2.25	0.65	0.02
Phenyl ether-biphenyl mixture (vapour)	8.1	20.25	0.011	0.04
Phenylethylene, see styrene				
Phenyhydrazine	8.5	20.25	0.0029	0.9
Phosgene (carbonyl chloride)	11.8	20.25	0.0027	0.03
Phosphine	10.1	20.25	0.0003	1.0
Picric acid[A]				
Propane	3.35	2.25	0.79	0.03
n-Propyl acetate	8.1	0.75	0.18	0.02
Propyl alcohol	9.4	9.75	0.39	0.2
n-Propyl chloride	7.9	20.25	0.2	0.7
n-Propyl nitrate	10.4	20.25	0.11	0.1
Propylene dichloride	9.8	20.25	0.12	0.3
Propylene oxide	12.0	20.25	0.32	0.3
Propyne, see Methyl acetylene				
Pyridine	14.2	20.25	0.05	0.2
Quinone[A]				
Stoddard solvent	3.4	0.75	0.39	0.01
Styrene	11.0	12.75	0.16	0.2
Sulphur dioxide	8.6	20.25	0.003	0.5
Sulphur hexafluoride	10.7	0.75	0.7	0.02
Sulphuryl fluoride	11.5	20.25	0.10	0.04
Systox, see Demeton[R]				
1,1,2,2-Tetrachloro-1,2-difluoroethane (Freon 112)	9.7	0.75	0.19	0.06
1,1,2,2-Tetrachloroethane	8.3	20.25	0.014	0.2
1,1,1,2-Tetrachloroethane	10.4	20.25	0.4	0.2
Tetrachloroethylene, see Perchloroethylene				
Tetrachloromethane, see Carbon tetrachloride				
Tetrahydrofuran	9.2	9.75	0.47	0.2
Tetryl[A]				
Toluene	13.7	6.75	0.38	0.5
o-Toluidine	13.5	20.25	0.029	0.5
Toxaphene, see Chlorinated camphene				
Tributyl phosphate1,1,1-Trichloroethane, see Methyl chloroform				
1,1,2-Trichloroethane	10.7	20.25	0.020	0.3
Trichloroethylene	11.8	5.25	0.27	0.1
Trichloromethane, see Chloroform				
1,2,3-Trichloropropane	12.4	20.25	0.16	0.4
1,1,2-Trichloro-1,2,2-Trifluorethane (Freon 113)	8.4	0.75	0.46	0.02
Triethylamine	9.3	20.25	0.077	0.3

Table 7.3 (continued)

Compound	Analytical Wavelength-μm	MIRAN 1A Path length-m	Absorbance	Minimum Detectable Concentration-ppm (20 metre Cell)
Trifluoromonobromoethane (Freon 13B1)	8.3	0.75	1.5	0.03
2,4,6-Trinitrophenol, see Picric acid				
2,4,6-Trinitrophenylmethyl-nitramenine, see Tetryl				
Turpentine	3.4	2.25	0.13	0.02
Vinyl benzene, see Styrene				
Vinyl acetate	8.1	20.25	0.37	0.02
Vinyl bromide	10.85	20.25	0.096	0.4
Vinyl chloride	10.9	20.25	0.0017	0.3
Vinylcyanide, see Acrylonitrile				
Vinylidene chloride	9.18	20.25	0.24	0.1
Vinyl toluene	11.1	20.25	0.2	0.4
Xylene (xylol)	12.6	20.25	0.33	0.6
Xylidene	7.2	20.25	0.056	0.2

Analytical wavelength: The analytical wavelength has usually been chosen as that of the strongest band in the spectrum which is free from interference due to atmospheric water and CO_2. If more than one infrared absorbing material is present in the air in significant concentration, the use of another analytical wavelength may be necessary.

Pathlengths: Indicated for the MIRAN-1A were chosen to optimise readings at the exposure limits.

Slit width: All measurements were made using a 1 mm slit.

Minimum detectable concentration: Is the concentration which would produce an absorbance equal to the peak-to-peak noise of the instrument. MIRAN-1A minimum detectable concentrations are lower than those for the MIRAN-101, 103 or 104 by approximately a factor of 3.

Absorbance (equivalent to the OSHA limit): Absorbance less than the tabulated values indicate concentrations under the exposure limits regardless of the presence of interfering compounds.

Key: (A) solid or vapour pressure too low for analysis.
  (B) difficult to obtain commercially.

concentrations.[20] Figure 7.8 depicts a microprocessor-controlled, single beam, infrared spectrometer equipped with audible alarms; it contains in its memory a library of calibration data. Outputs can be linked to a chart recorder.

Ultraviolet spectroscopy

The ability of UV sources to induce photo-ionisation is referred to in Chapter 2. Also when exposed to UV radiation the absorbance of energy at specific wavelengths by some materials is strong and proportional to their concentration. UV techniques have found use in meters, e.g. for mercury (see Figs 7.9(a) and 7.9(b)), ozone, chlorine and phosgene.

374 *The Safe Handling of Chemicals in Industry*

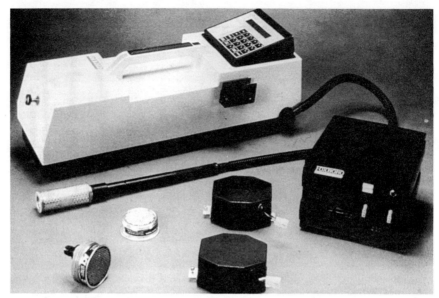

Fig. 7.8 Miran 1B portable ambient air analyser using infrared detection (Courtesy Quantitec Ltd)

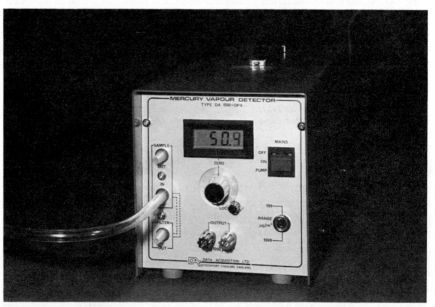

Fig. 7.9(a) Mercury vapour detector model DA1500-DP6 using UV detection (Courtesy Data Acquisition Ltd)

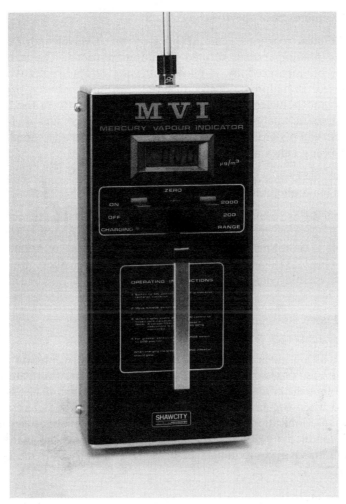

Fig. 7.9(b) MVI mercury vapour indicator (Courtesy Shaw City Ltd)

Fluorescence

The ability of some materials to 'fluoresce' is exploited in some oxygen meters. A fluorescent chemical absorbed in a porous glass disc is excited by a UV light source. The degree of fluorescence, which is influenced by the oxygen concentration, is measured by a photo-conductive cell with a reference cell to compensate for fluctuations in UV intensity.

Luminescence spectroscopy

Some chemical reactions generate light of mixed wavelength instead of heat. The phenomenon is termed chemiluminescence. The intensity of light can be measured by a photomultiplier through selective filters. Examples of chemiluminescence reactions are those between ozone and ethylene, and between ozone and nitric oxide. These form the basis for some automatic analysers for ozone (Fig. 7.10) and oxides of nitrogen. These instruments are sensitive and suitable for continuous sampling with short response times and high specificity.

Flame photometry

These systems use a hydrogen flame, as do flame ionisation detectors, but measure emissions due to sulphur ions at 394 nm. Though sensitive the indication is non-linear. They find use in detection of total sulphur content (Fig. 7.11) but can be used with selective filters for specific sulphur-containing inorganic or organic compounds.

Flame ionisation detectors (FIDs)

These detectors are useful for most organic gases/vapours with sensitivity of 1 ppm with a linear response with weight per cent over about 5 orders of magnitude. Good precision may be obtained. They operate on the principle that the concentration of ions in an air/hydrogen flame is greatly increased when organic vapours are introduced. Polarised electrodes are used to sense ionisation and convert the response to a voltage signal. Like thermal conductivity detectors, FIDs require linking to chromatographic units to improve compound specificity. Intrinsically safe designs are available.

Photo ionisation

These detectors, though not intrinsically safe, are more sensitive than FIDs with limits of detection of approximately 0.1 ppm with linear response over 6 or 7 orders of magnitude. Their operating principle is similar to that for FIDs though ionisation is achieved using a UV light source. Since most permanent gases have ionisation potentials above the excitation energy of the UV source their presence does not cause interference.

Electron capture

A small radioactive source of nickel or tritium is used to ionise air contained in a sample cell. Electronegative contaminants present in the

Ambient air analysis for hazardous substances 377

Fig. 7.10 CSI model 2000 portable ozone meter using chemiluminescence detection (Courtesy Techmation Ltd)

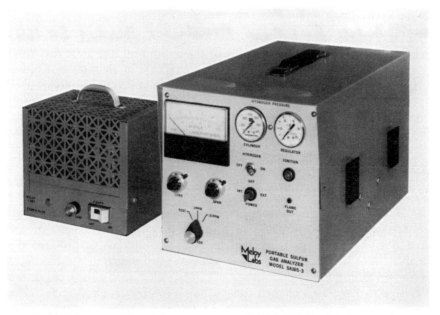

Fig. 7.11 Meloy portable sulphur analyser model SA 165-3 using flame photometry detection (Courtesy Techmation Ltd)

sample capture some of the free electrons produced during ionisation. The ionisation current is reduced in proportion to the concentration of contaminant. This sensor is selective for species such as compounds containing halogen, oxygen and unsaturated moieties. It finds greatest use in gas chromatography especially for the detection of halogenated organic solvents to which it is very sensitive (e.g. trichloroethylene can be detected at levels below 1 part per 10,000 million).

Because it is adversely affected by water vapour the sensor cannot be used directly as an atmospheric monitor.

Mass spectrometry

This technique involves bombardment of molecules with electrons to produce fragments of ions of different masses which are separated by an electric or magnetic field in a vacuum. This structural breakdown is reproducible and the spectrum produced is characteristic of the chemical. The technique is versatile, highly sensitive and highly selective, and can be coupled to gas chromatography units to aid analysis of complex mixtures. Generally, because of its complexity, mass spectrometry is normally reserved for laboratory use, although developments have led to application on industrial plants and mini-transportable units are available.[21]

Thermal conductivity detectors

The concentration of a gas or vapour in a carrier such as argon can be related to the thermal conductivity of the substance. The heat loss from a hot filament to the gas is registered as a change in electrical resistance which is measured using a Wheatstone Bridge circuit. The method is non-specific and most useful for measuring high concentrations. It is most commonly used for leak tracing or in chromatographic detection systems.

Catalytic elements

Some detectors known as 'Pellistors' comprise short coils of platinum wire encapsulated in a pellet of refractory oxide such as aluminium. These beads which operate at high temperature, are rendered inert with aqueous potassium hydroxide to produce a reference element, or are coated with platinum, palladium or thorium materials to produce the 'active' element. Flammable gas is oxidised on the heated catalytic wire element causing the temperature, and hence electrical resistance, of the Pellistor to rise. This change can be measured by comparison with the electrical resistance of the reference bead using a Wheatstone Bridge circuit.

The method is employed in an explosimeter when the instrument is usually calibrated in the range of 0–100% lower explosive limit (LEL) for a particular substance. Recalibration, or application of correction factors, is required when other gases are involved. Portable equipment should be explosion-proof to prevent ignition of surrounding atmospheres. For fixed point systems it may be necessary to rely on remote sensing heads with conventional electrics housed in an adjoining switchroom.

For zero adjustment, instruments must either be taken to 'uncontaminated' air or use made of activated carbon filters to remove flammable vapours (other than methane).

Sources of error include inadequate calibration, drift due to age, design not fail-safe (i.e. no indication of component failure), poisoning of Pellistor (common poisons are silicones, halocarbons, leaded petrol), too high a sampling rate (causing cooling of the elements), sampling lines and couplings not airtight, condensation of high-boiling components in the line between sample head and sensor, and hostility of atmosphere (e.g. presence of corrosive gases, high humidity or ambient temperature).

An example of an intrinsically safe, fail-safe explosimeter utilising Pellistors is given in Fig. 7.12.

Semiconductors resemble Pellistors. A platinum/rhodium filament is used to heat a pellet of 'doped' oxide which absorbs any flammable gas passing over it. This causes a change in electrical conductivity, and

Fig. 7.12 Digi Flam 850 intrinsically-safe flammable gas monitor using Pellistor detection (Courtesy Neotronics Ltd)

Fig. 7.13 Combined intrinsically-safe monitoring set for detection of flammable gases, hydrogen sulphide or carbon monoxide, and oxygen-deficiency (Courtesy Neotronics Ltd)

subsequently the voltage, in the electrodes attached to the pellet. Amplification of this signal registers as a deflection on a meter. The system is versatile and can be very selective and sensitive enabling low concentrations of gas to be measured. Errors can arise when the instrument is used for gases, other than that for which it was calibrated.

Catalytic detectors are also used in some oxygen monitors. Here an atmosphere of methanol vapour is maintained by means of a wick to soak up methanol from an internal reservoir. In the presence of oxygen the organic vapour is oxidised on a catalytic element which functions as a thermocouple. The rate of oxidation, and hence temperature of the element, is dictated by the concentration of oxygen in the test atmosphere. The meter is graduated in per cent concentration of oxygen. One development has been a flammable-gas monitor and oxygen-deficiency and toxic gas monitor mounted in a common carrier case (Fig. 7.13): oxygen and toxic gas detection utilises electrochemical sensors while Pellistors are used for flammable-gas detection.

Electrical conductivity

The concentration of acid or alkaline gases can be measured by monitoring their effect on the conductivity of a reagent. This non-selective technique has been used for sulphur dioxide, hydrogen chloride, ammonia, etc. Systems can also be 'dry', e.g. as used for monitoring carbon monoxide which is oxidised to carbon dioxide as the air sample is driven over catalytic electrodes.

Coulometry

This technique relies on precise measurement of the quantity of electricity flowing through a solution during an electrochemical process when gases are drawn through an electrolytic cell. The method is capable of precision down to ppb concentrations of certain reactive inorganic gases. Although basically non-specific, the technique can be made more selective for specific contaminants by adjustment of the concentration, pH and composition of the electrolyte used in the cell. The method has been used for oxides of nitrogen, sulphur dioxide, ozone, hydrogen sulphide and chlorine.

Ion-selective electrodes and polarographic cells

When two electrodes are immersed in a reagent electrolyte contained in a special cell, current is generated when gas diffuses through the porous cell walls or through a porous membrane separating the electrode liquid from the atmosphere. Variation of electrode material enables cells to be made electrode specific. Generally these sensors are used for detection of inorganic gas. Monitors based on polarographic cells are available for measuring hydrogen fluoride, hydrogen chloride, or carbon monoxide or for oxygen content. Figures 7.14(a) and (b) show a range of meters for measuring the oxygen content of air. Diffusion of oxygen to the air cathode is controlled by a diffusion barrier. The oxygen becomes reduced to hydroxyl ions at the cathode which in turn oxidises the metal anode. The resulting current is proportional to the rate of consumption of oxygen.

Figure 7.15 shows a portable toxic-gas alarm monitor with interchangeable sensor heads for carbon monoxide, hydrogen sulphide and sulphur dioxide. The reactions as the toxic gas diffuses at the sensing electrode (anode) and the corresponding reaction at the counter electrode (cathode) are shown below.

Toxic gas	Reaction of anode	Reaction of cathode
SO_2	$SO_2 + 2H_2O = H_2SO_4 + 2H^+ + 2e$	$\frac{1}{2}O_2 + 2H^+ + 2e = H_2O$
H_2S	$H_2S + 4H_2O = H_2SO_4 + 8H^+ + 8e$	$2O_2 + 8H^+ + 8e = 4H_2O$
CO	$CO + H_2O = CO_2 + 2H^+ + 2e$	$\frac{1}{2}O_2 + 2H^+ + 2e = H_2O$

Fig. 7.14(a) Otox 90 series digital oxygen analysers using ion electrode detection (Courtesy Neotronics Ltd)

Fig. 7.14(b) Oxywarn 110R: oxygen meter utilising a potentiostatic polarographic cell (Courtesy Draeger Safety)

Ambient air analysis for hazardous substances 383

Fig. 7.15 Toxiguard 1/R with interchangeable electrochemical sensing heads (Courtesy Sabre Gas Detection)

Instruments with ion-selective, pH electrodes are available for detection of carbon dioxide, ammonia, sulphur dioxide and oxides of nitrogen.

Paramagnetic susceptibility

Several types of device are available which exploit the magnetic properties of oxygen. Under 'normal' conditions oxygen is paramagnetic, a property which is inversely proportional to its absolute temperature, and when heated it becomes diamagnetic. As oxygen passes through a cell it is attracted to the magnetic field of the detector, an electrically heated resistor, where it then looses its magnetic properties. The cooling effect produced on the detector is related to the oxygen content of the test atmosphere. Another approach depends on the rotation of suspended magnetic objects in test cells, which is proportional to the oxygen content of the atmosphere. Thus a nitrogen-filled diamagnetic glass-bell is suspended in a strong magnetic field from which it is naturally deflected to a degree depending on the magnetic susceptibility of the surrounding gas. In the Munday cell a magnetic restoring force to a null is achieved by a platinum coil which also forms part of the suspension. Deflection from the null is monitored by photocells. In yet a third variation, sample and reference gas flow from opposite directions into a magnetic gap, the field of which is generated by an intermittently operated electromagnet. When the magnet is activated each of the two gases undergoes a change in pressure proportional to the oxygen content of the respective gas in the inhomogeneous part of the field. When the oxygen content of the two gases differs a pressure difference results causing a deflection on a diaphragm detected capacitively.

Gas chromatography (GC)

Though not really a detection system, GC is used to identify substances qualitatively and quantitatively. A mixture of carrier gas and test sample is injected into the head of a column of fine solid. As the sample starts to pass through the column the different components, which have varying affinities with the stationary phase, begin to separate and are eluted at different rates and pass through a detection system such as FID, or thermal conductivity, electron capture, etc.

By comparing peak heights and retention times with standards for a similar GC operating condition materials can be identified quantitatively and their concentrations accurately determined.

Figure 7.16 shows a portable GC system in use in the field. This uses a FID detection system with facility for continuous monitoring and has

Fig. 7.16 Century organic vapour analyser in use in the field. This portable GLC unit is equipped with flame ionisation detectors (Courtesy Quantitech Ltd)

interchangeable columns for known organic pollutants. Table 7.4 provides guidance on column details for the Century organic vapour analyser.[22]

Particulate monitoring

Airborne particulate matter such as dust, aerosol, mist and fume can pose fire/explosion and toxic hazards, or be of nuisance value. Justification for air quality monitoring for particulates parallels that for gases, though the equipment required differs in detail. Thus, it is often important to distinguish between total dust levels and concentrations of respirable fractions (e.g. of aerodynamic diameters in the range 2–7 μm).

Many particles of dust are too small to be seen with the naked eye under normal lighting. However, when illuminated with a strong beam of light a cloud of particles is rendered visible due to reflection. This is the 'Tyndall effect' and the technique is useful for quick assessment as to whether dust exists – and if so its flow pattern – examining the effect of ventilation, or identification of likely sources of leakage.

For quantitative data more sophisticated approaches are needed and whether personal, spot or static sampling is adopted will depend upon the nature of the information required.

The air in the general working atmosphere, or in the breathing zone of individuals, may be collected using a pump coupled to a means of isolating particulate matter for subsequent analysis or determination. Techniques for separating dust or aerosol particles of respirable dimensions from non-respirable fractions include horizontal elutriation and centrifugation. Equipment for personal monitoring comprises a lapel-mounted filter holder connected to a portable pump with a flow rate of about 3 l/min. Respirable matter can be separated by use of a small cyclone (as shown in Fig. 7.17). In order to ensure uniformity of fractionation, smooth and constant flow rates are essential. The dust collection and analytical stages are separate operations.

For background monitoring miniaturisation is unimportant and as a consequence equipment incorporates pumps of higher flow rates, typically up to 100 l/min (such as that shown in Fig. 7.18(a)). This enables sampling times to be short and larger samples to be obtained (e.g. for laboratory analysis). Both direct reading and absolute methods are available.

The main principles of instrument design (summarised in Table 7.5 and based on ref. 23) are impaction/impingement, thermal or electrostatic precipitation, filtration, beta-attenuation photometry and piezoelectric mass detection.

386 The Safe Handling of Chemicals in Industry

Table 7.4 Chromatographic column guide for Century organic vapour analyser

A blank in the table indicates that no data are available for the analysis.
Compound names: Wherever possible compounds are listed under the chemical name and used in the Merck Index.
Relative response: Century organic vapour analysers are factory calibrated to measure 'total organic' vapours according to a standard (methane). Since different organic vapours interact with the flame ionisation detector (FID) to varying extents, it is vital that the instrument user be aware of the magnitude of the variation in order to obtain the most accurate data. Each user must determine relative responses for his individual instrument.
Chromatographic retention time: For chromatographic work, the OVA can be used with a variety of column lengths and packing materials. For highest accuracy, temperature control for the column is mandatory. This is accomplished using the portable isothermal pact (PIP) kit which is supplied with three 8 inch columns packed with B, G, and T materials respectively. (See following chart for packing material description.) Isothermal control is accomplished non-electrically using an ice/water mixture for 0 °C and a seeded eutectic mixture for 40 °C. The data listed are for comparison purposes only since retention time for a compound can vary due to the condition of the column packing material, packing procedure, and chemical interaction among the components of a vapour mixture.

Column packing material:
B – 3% Diisodecyl phthalate on Chromosorb W, AW, 60/80 mesh
E – 20% Carboxax 400 on Chromosorb P, AW, 60/80 mesh
G – 10% SP – 2100 on Supelcoport, 60/80 mesh
T – 10%, 1,2,3-Tris(2-cyanoethoxy) propane on Chromosorb P, AW, 60/80 mesh
PT – Porapak T, 60/80 mesh

Compound	Relative response (%)	Column temperature (°C)	Column packing retention time (min)		
			B-8	G-8	T-8
Acetone	60	0	0:37	0:27	1:50
		40	0:20	0:15	0:30
Acetonitrile	70	0	0:40	0:30	5:24
		40	0:20	0:18	1:15

Compound	Relative response (%)	Column temperature (°C)	Column packing retention time (min)		
			B-8	G-8	T-8
Chlorobenzene	200	0	5:45	8:00	11:20
		40	1:08	1:24	1:35
Chlorodifluoromethane (Freon 22)	40	0	0:11	0:11	0:15
		40	—	—	—

Ambient air analysis for hazardous substances

Compound					
Acrylonitrile	70	0	0:35	0:30	3:45
		40	0:19	0:17	0:51
Allyl alcohol	30	0	1:15	0:44	11:45
		40	0:37	0:26	1:35
Allyl chloride	50	0	0:16	0:28	0:31
		40	0:08	0:16	0:15
Benzene	150	0	0:50	1:19	1:43
		40	0:22	0:25	0:32
2-Bromo-2-chloro-1,1,1-trifluoroethane (Halothane)	45	0	0:37	0:35	0:58
		40	0:17	0:14	0:19
Bromoethane	75	0	0:15	0:26	0:23
		40	0:05	0:15	0:14
1-Bromopropane	75	0	0:21	0:58	0:40
		40	0:08	0:22	0:18
2-Butane	60	0	0:15	0:15	0:10
		40	—	—	—
n-Butanol	50	0	4:10	2:20	15:07
		40	0:24	0:53	1:44
2-Butanol	65	0	1:45	0:55	6:07
		40	0:24	0:24	0:53
n-Butyl acetate	80	0	4:15	7:30	11:40
		40	0:50	1:14	1:31
n-Butyl acrylate	60	0	—	—	—
		40	2:30	2:15	2:40
2-Butyl acrylate	70	0	1:22	1:45	1:46
		40	2:15	2:57	4:40
n-Butyl formate	50	0	0:28	0:34	0:49
		40	1:22	2:00	3:31
2-Butyl formate	60	0	0:20	0:26	0:37
		40	—	—	—
n-Butyl methacrylate	60	0	4:03	5:46	5:10
		40	—	—	—
2-Butyl methacrylate	80	0	2:46	3:40	3:34
		40	0:20	1:24	0:37
Carbon tetrachloride	10	0	0:10	0:25	0:17
		40	—	—	—
Chloroform	65	0	0:55	0:57	2:00
		40	0:20	0:20	0:31
1-Chloropropane	75	0	0:16	0:31	0:23
		40	0:05	0:16	0:14
2-Chloropropane	90	0	0:15	0:23	0:18
		40	0:05	0:05	0:05
2-Chloro-1,1,2-trifluoroethyl difluoromethyl ether (Ethrane)	150	0	0:36	0:26	1:22
		40	0:13	0:12	0:19
Cumene	100	0	11:00	20:00	12:45
		40	2:20	3:03	2:19
Cyclohexane	85	0	0:36	1:25	0:19
		40	0:18	0:26	0:14
Cyclohexanone	100	0	18:00	12:45	—
		40	3:00	2:00	—
n-Decane	75	0	2:57	6:20	1:35
		40	—	—	—
o-Dichlorobenzene	50	0	8:06	10:00	11:20
		40	0:10	0:11	0:12
Dichlorodifluoromethane (Freon 12)	15	0	—	—	—
		40	—	—	—
1,1 Dichloroethane	80	0	0:17	0:37	0:45
		40	0:08	0:17	0:18
1,2 Dichloroethane	80	0	1:14	1:08	3:50
		40	0:23	0:22	0:43
trans 1,2-Dichloroethylene	50	0	0:16	0:35	0:31
		40	0:05	0:18	0:16
Dichlorofluoromethane (Freon 21)	70	0	0:16	0:15	0:23
		40	—	—	—
Dichloromethane	100	0	0:27	0:29	1:08
		40	0:10	0:10	0:22
1,2-Dichloropropane	90	0	0:41	1:49	2:56
		40	0:18	0:29	0:36
1,3-Dichloropropane	80	0	1:32	4:12	4:24
		40	0:26	0:47	1:20

Table 7.4 (continued)

Compound	Relative Response (%)	Column temperature (°C)	Column packing retention time (min) B-8	G-8	T-8
1,2-Dichloro 1,1,2,2-tetrafluoroethane (Freon 114)	110	0	0:12	0:12	0:14
	50	40	—	—	—
Diethyl ether	50	0	0:20	0:26	0:19
		40	0:05	0:05	0:05
Diethyl ketone	80	0	2:00	2:01	3:16
		40	0:29	0:30	0:49
p-Dioxane	30	0	3:15	2:09	6:40
		40	0:44	0:34	1:19
Ethane	80	0	0:15	0:15	0:15
		40	—	—	—
Ethanethiol	30	0	0:18	0:24	0:26
		40	0:13	0:12	0:14
Ethanol	25	0	0:59	0:31	0:26
		40	0:26	0:22	0:43
Ethyl acetate	65	0	0:48	1:00	2:20
		40	0:20	0:20	0:31
Ethyl acrylate	40	0	1:40	2:10	4:08
		40	0:30	0:30	0:50
Ethyl benzene	100	0	6:10	9:31	7:44
		40	1:15	1:35	1:35
Ethyl butyrate	70	0	4:34	6:22	8:00
		40	0:48	1:07	1:24
Ethyl formate	40	0	0:25	0:29	1:05
		40	0:16	0:17	0:20

Compound	Relative response (%)	Column temperature (°C)	Column packing retention time (min) B-8	G-8	T-8
Methyl propyl ketone	70	0	2:20	2:05	6:14
		40	0:33	0:32	0:52
Nitromethane	35	0	0:51	0:40	3:00
		40	0:25	0:19	1:31
1-Nitropropane	60	0	4:50	2:50	25:00
		40	0:46	0:41	4:05
2-Nitropropane	70	0	2:53	1:51	10:00
		40	0:35	0:31	1:52
Nonane	90	0	7:26	8:00	2:32
		40	1:08	2:40	0:42
Octane	80	0	2:47	7:39	1:07
		40	0:39	1:09	0:27
Pentane	65	0	0:18	0:25	0:14
		40	0:12	0:12	0:12
Pentanol	40	0	12:00	6:00	20:00
		40	2:44	1:17	3:36
Propane	80	0	0:05	0:11	0:05
		40	—	—	—
n-Propanol	40	0	2:50	1:00	6:30
		40	0:35	0:30	1:05
2-Propanol	65	0	1:13	0:30	3:43
		40	0:25	0:20	0:38
n-Propyl acetate	75	0	2:04	2:52	5:52
		40	0:30	0:38	0:51
n-Propyl ether	65	0	1:37	2:06	1:15

Ambient air analysis for hazardous substances

Compound					
Ethyl methacrylate	70	0	4:13	6:13	6:20
		40	0:47	1:01	1:01
Ethyl propionate	65	0	1:40	2:48	4:10
		40	0:25	0:35	0:50
Ethylene dibromide	50	0	5:20	4:51	15:00
		40	1:00	0:55	2:43
Ethylene dichloride	60	0	1:07	1:08	3:45
		40	0:20	0:21	0:50
Ethylene oxide	70	0	0:10	0:13	0:20
		40	—	—	—
Fluorotrichloromethane (Freon 11)	10	0	0:15	0:18	0:17
		40	—	—	—
Heptane	75	0	0:50	2:20	0:27
		40	0:20	0:30	0:16
Hexane	70	0	0:23	0:50	0:20
		40	0:13	0:20	0:13
Isoprene	50	0	0:10	0:23	0:20
		40	0:05	0:05	0:05
Methane	100	0	0:05	0:05	0:05
		40	—	—	—
Methyl alcohol	12	0	0:37	0:21	2:23
		40	0:22	0:14	0:45
Methyl acetate	41	0	0:30	0:30	1:30
		40	0:17	0:15	0:24
Methyl acrylate	40	0	0:50	0:58	2:30
		40	0:20	0:21	0:37
Methyl cyclohexane	100	0	0:54	2:37	0:25
		40	0:18	0:33	0:17
Methyl cyclopentane	80	0	0:22	1:02	0:17
		40	0:05	0:19	0:05
Methyl ethyl ketone	80	0	1:00	0:50	3:30
		40	0:22	0:20	0:43
Methyl isobutyl ketone	80	0	4:20	3:15	7:30
		40	0:42	0:40	1:25
Methyl methacrylate	50	0	1:41	2:22	4:08
		40	0:27	0:31	0:55
n-Propyl formate	50	0	0:43	0:29	0:18
		40	0:46	1:07	2:10
Pyridine	128	0	0:17	0:18	0:27
		40	8:00	4:15	—
Styrene	85	0	—	1:26	—
		40	20:00	25:00	28:00
1,1,1,2-Tetracholoroethane	100	0	2:06	2:26	2:39
		40	8:25	7:48	10:00
1,1,2,2-Tetrachloroethane	100	0	1:16	1:27	1:37
		40	32:00	14:00	50:00
Tetrachloroethylene	70	0	4:09	2:37	7:51
		40	3:00	5:45	2:10
Tetrahydrofuran	40	0	0:41	1:06	0:33
		40	1:05	1:05	1:45
Toluene	110	0	0:23	0:23	0:30
		40	2:30	4:05	4:30
1,1,1 Trichloroethane	105	0	0:38	0:47	0:53
		40	0:31	1:10	0:47
1,1,2 Trichloroethane	85	0	0:15	0:23	0:20
		40	5:13	3:43	15:00
Trichloroethylene	70	0	0:40	0:45	2:30
		40	1:17	2:02	1:25
Trichlorotrifluoroethane (Freon 113)	80	0	0:23	0:28	0:28
		40	0:15	0:30	0:16
Triethylamine	70	0	0:13	0:14	0:13
		40	—	2:49	—
Vinyl acetate	50	0	0:34	0:43	1:55
		40	0:15	0:17	0:30
Vinylidene chloride	40	0	0:20	0:25	0:22
		40	0:13	0:14	0:14
m-Xylene	111	0	2:39	12:03	8:31
		40	0:39	1:43	1:17
o-Xylene	116	0	3:29	15:07	8:40
		40	0:48	1:58	1:45
p-Xylene	116	0	2:46	12:25	8:23
		40	0:39	1:42	1:19

Fig. 7.17 Personal dust collector fitted with cyclone to separate respirable and non-respirable dust (Courtesy Rotheroe & Mitchell Ltd)

Impaction and impingement

This entails trapping particles via impaction in a cascade system, on a gel-coated disc, or by bubbling the air through a liquid.

Thermal precipitators

These rely on the property of hot bodies to create dust-repellent zones. Dust-laden air is sampled through a vertical slot between two microscope

Ambient air analysis for hazardous substances 391

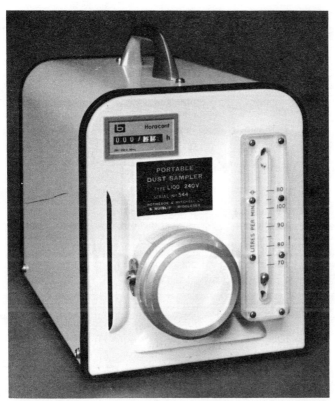

Fig. 7.18(a) High-volume sampling pump (Courtesy Rotheroe & Mitchell Ltd)

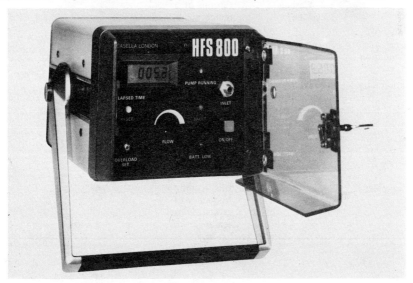

Fig. 7.18(b) High-volume sampling pump (flow rate adjustable over 4 to 10 litres min^{-1}) (Courtesy Casella London Ltd)

Table 7.5 Particulates monitoring – principles of apparatus

Principle	Examples	Collection	Sampling rate (l/min)	Collection efficiency (%)
Impinger	Midget impinger	By bubbling through liquid phase		
Impactor	1. Konimeter	Impaction on gel-coated disc	1–37 (depending on type)	60–100
	2. Cascade impactor	Impaction in 4 stages on glass disc		
	3. Andersen Sampler	Impaction in 8 stages on to glass or metal discs		
Electrostatic or thermal precipitation	Casella thermal precipitator	Deposition on glass slides or discs	1–85	90–100
Filtration	1. Fibrous filter		1–50	85–100 Depends on particle size values stated for usually encountered
	2. Membrane filter			
Respirable dust separation	1. Hexhlet (horizontal elutriator)	Fibrous filter	1–50	60–100
	2. Casella cyclone	Fibrous or membrane filter		
Beta attenuation		Impaction on disc or filtration		
Photometry	1. Number concentraton, e.g. Royco			
	2. Mass concentration, e.g. Simslin			
Piezoelectric		Electrostatic frequency of crystal. Direct reading.		

Analysis	Advantages/disadvantages
Microscopy	Aggregates broken up; only particles 1 μm collected.
Built in microscope Microscopy	Underestimates small particles, overestimates large particles. Particles between 0.5–5.0 μ collected.
Gravimetric or chemical	Instrument inflexible.
Microscopy	Poor for large particles. Collection efficiency increases as particle size decreases.
Gravimetric or chemical	Fibrous filter good for gravimetric analysis (fast and relatively easy).
Microscopy gravimetric or chemical	Membrane filter good for microscopy identification of particles and counting where required.
Gravimetric or chemical	Instrument must be kept horizontal for sampling; relatively large quantities of dust collected in short period.
Gravimetric, microscopic or chemical	The only instrument for carrying out personal respirable dust sampling.
Attenuation of Beta radiation Direct reading	Provides short- or long-term TWA (e.g. up to 8 hours, depending on model) of dust or fume mass concentration.
Light scattered on to a photomultiplier Direct reading.	Gives automatic particle sizing but accuracy only guanranteed if calibrated for particulate of interest.
Light scattered onto a photomultiplier Direct reading	Very versatile – only accurate continuous long-term mass monitoring instrument; sample may also be collected on a filter.
Change in resonant frequency	

cover glasses. Between these is stretched horizontally a heated fine wire. Samples are deposited on the cover glass for examination microscopically. The dimensions and geometric arrangement of the sample head are critical, and flow rate is limited to 7 ml/min for standard meters and 2 ml/min for long-running instruments.

Electrostatic precipitators

A central electrode is maintained at a potential of 10 to 20 kV and surrounded by a trembler-collecting electrode. Between the electrodes a corona discharge is produced in which particles become highly charged and attracted to the collecting electrode, removal of which enables the dust to be weighed directly or washed off for chemical analysis.

Filtration

This is the most common method of collecting particulates. Two main versions are available, based on fibre or membrane design. The latter comprise films with uniform pore size and thickness made from polymeric materials such as cellulose derivatives, PVC, or polycarbonate. This provides efficient filtration, and analysis of material collected at the surface is easy. Fibrous filters consist of a compressed matrix of fibres which act as a depth filter. Though these collect a greater weight of dust before flow resistance increases, they are less efficient for small particles and analysis can be more difficult.

When selecting filter material consideration must be given to filter strength, collection efficiency, compatibility with pump, electrical resistance – which must be low to avoid sampling losses due to electrostatic charges, etc.

Glass-fibre filters of nominal pore size 8 μm are used for gravimetric studies whereas membrane (e.g. cellulose nitrate), or silver filters, are more suitable if analysis of the dust is required or if the weight of dust to be collected is very small.

Residues are analysed gravimetrically,[24] chemically or microscopically. Changes in weight of filter due to water loss or uptake from the atmosphere must be considered in gravimetric analysis. Since uptake of water in the workplace atmosphere is rapid, desiccation prior to weighing is of little value and a humidity-controlled balance-room is preferred. Use of control filters is more common, when control and sample filters should be allowed to equilibrate for at least 2 hours in the balance room before weighing. Since the total weight of dust collected by personal sampling may be as little as 100 μg a microbalance capable of weighing down to at least 10 μg is required for adequate accuracy. Careful handling techniques are essential.

Though optical microscopy is declining in popularity as a means of monitoring dust, some instruments are equipped with built-in microscope detectors. Furthermore, optical microscopy is especially useful for determining particle size and shape. It is particularly relevant for differentiating between the different forms of asbestos (e.g. chrysotile fibres tend to have a long, curly morphology whereas the others are mainly straight) and other particles (as illustrated by Fig. 7.19)[25] and in counting asbestos fibres. Indeed it is the method currently adopted in the UK by the HSE, as described more fully in other publications.[26-28] However, scanning electron microscopy and energy dispersive methods may be required when more sensitive techniques are required because the hygiene standard is very low or airborne asbestos concentrations are low (e.g. assessing risk from asbestos dust in public buildings).[29]

Beta attenuation

β-radiation is emitted from a carbon-14 source within the monitor and detected by a Geiger tube. As air is drawn into the device and particles collected by impaction or filtration, the radiation intensity reaching the detector is reduced according to the mass of dust collected, and a direct reading in mg/m^3 provided.

Use of impactors restricts sampling periods to a few minutes whereas filter versions are useful for sampling periods of hours.

Photometers

Particles in an airstream scatter light from an intense source relative to the incident beam at an angle and intensity determined by the particle size and surface characteristics. Data on the number and particle-size distribution can be obtained with instruments based on this technique.

Two examples of direct-reading dust monitors based on light-scattering techniques are given in Fig. 7.20. For accuracy monitors require calibration with the dust to be measured in a dust cloud of known concentration. In the SIMSLIN, the sample is collected on a membrane filter so that calibration is achieved by comparison of the average recorded dust concentration with that calculated from the weight gain on the filter.

Sampling strategies and data interpretation

Sampling strategies must reflect the hazard under investigation. Most flammable gases are, as discussed in Chapter 4, heavier than air and pockets can accumulate in pipe bends, gulleys, manholes, etc. Samples must be collected appropriately. Similarly, analysis of atmospheres containing high flashpoint materials at room temperature will give low

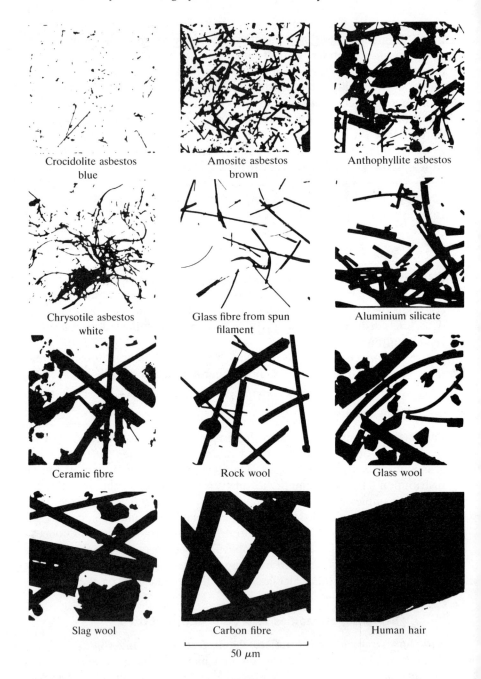

Fig. 7.19 Microphotographs of common industrial fibres. Unless present in very high concentrations these cannot be seen with the naked eye in normal lighting conditions (Medical Research Council, Pneumoconiosis Unit)

Ambient air analysis for hazardous substances 397

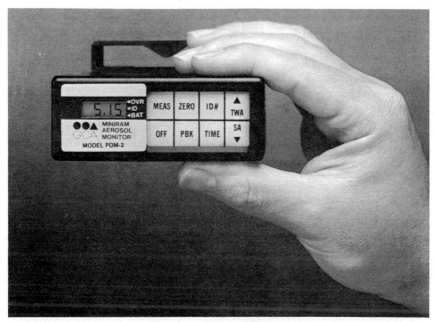

Fig. 7.20(a) Miniram PDM-3 direct reading respirable aerosol sampler, based on light-scattering detector (Courtesy Analysis Automation Ltd).

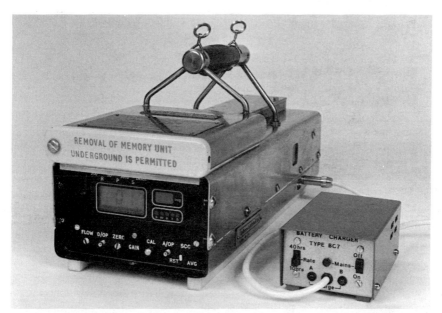

Fig. 7.20(b) SIMSLIN II direct reading dust sampler based on light-scattering (Courtesy Rotheroe Mitchell Ltd)

readings but if the environment is to be subjected to hot work hazardous conditions could be created.

Some instruments give zero readings in atmospheres of flammable gas at concentrations above the upper explosive limit. Careful observation of meter reaction may be revealing. Thus rapid fluctuation in output from zero, to positive, to zero as the probe penetrates into the test atmosphere may be indicative of a high concentration of flammable gas or vapour.

Besides spasmodic monitoring requirements for trouble-shooting purposes, sampling should be programmed to include an initial survey followed by regular checks on background levels, leaks and personal exposures.

When assessing risk of exposure to toxic chemicals initial surveys comprising visual inspection and administrative checks may indicate whether environmental analysis is required (see Fig. 7.21).[30] Detection by odour may serve as an indication but is unreliable because of physiological phenomena, masking, etc., as discussed in Chapter 5. Cognisance should be taken of what substances are in use and the form in which they are encountered (gas, mist, dust, etc.). Visible mist or dust clouds, or evidence of significant dust deposits, e.g. on roof beams or ledges, may require investigation. The hazardous properties of the chemicals and the likelihood of exposure to them must be assessed. When exposure is possible the offending processes should be identified together with the groups of individuals at greatest risk.

When environmental monitoring is required the importance of adopting a good sampling strategy cannot be over-emphasised.

> An environmental health officer visiting a beauty salon had detected the familiar smell of methyl methacrylate. There were four workbenches in the shop each of which was served by an extractor fan. Using a detector tube the inspector took a reading behind the manicurist who was using the chemical and this showed a level of about 110 ppm. Another reading at the other side of the shop showed a concentration of 75 ppm. He then issued an Improvement Notice relating to all four workbenches, although he had taken only one sample from each of two places. The Notice stated that the ventilation was ineffective in that a dangerous level of methyl methacrylate vapour was recorded in the breathing zone of the manicurist and that readings showed levels in excess of 100 ppm, the 8-hour TLV.
>
> The owner of the salon appealed and the tribunal considered that the issue was whether the inspector had established grounds for the opinion that there was both danger in the premises from use of the chemical and an exposure of employees to dangerous concentrations. It deduced that the use of the chemical was momentary and intermittent, and that while exposure might be for more than 10 minutes in each day, there was no evidence to indicate that the levels in the atmosphere would exceed 100 ppm for that length of time. It concluded that 'it is quite inadequate to make a finding

Ambient air analysis for hazardous substances 399

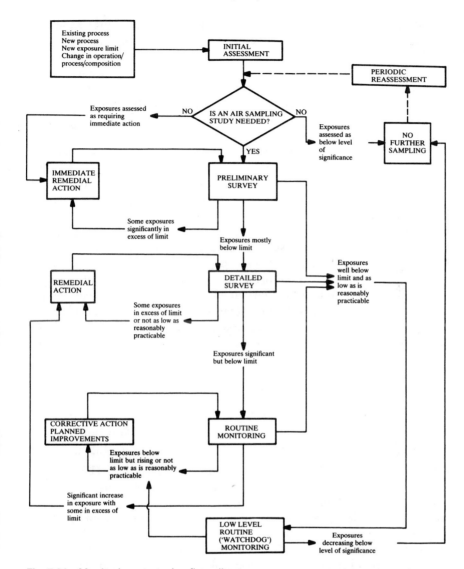

Fig. 7.21 Monitoring strategies flow diagram

which could lead to prosecution for a criminal offence on the basis of one sample only without any indication of a time period during which the relevant maximum was exceeded'. It cancelled the Notice, and added that had it affirmed the Notice it would have done so only in relation to the one workbench where the reading was in excess of 100 ppm and not the other three.[31]

The pattern and duration of group and individual exposures should be studied. Thus, atmospheric concentrations of substances are influenced by the number of sources, the rate of release, the type and position of source and dispersion patterns. Exposure levels are further influenced by the position of the operator relative to the source, the duration of exposure and whether personal protection is worn. Fluctuations in exposure can occur, e.g. within shifts, between shifts, between processes and between individuals.

Where doubt exists about the level of exposure to airborne chemicals a crude assessment can be made by determining levels under 'worst case' conditions, paying due attention to the foregoing variations and to measurement errors associated with the sampling and analytical techniques. Ideally this study should be by personal dosimetry but reliance is often placed on data from background monitors or grab-samples. Both peak and time-weighted average exposure should be considered. Depending upon the outcome more detailed studies and even routine monitoring may be required when personal monitoring becomes even more essential.

Sampling times should be long enough to overcome fluctuations but short enough for results to be meaningfully associated with specific activites and for corrective action to be taken. Usually, sampling times of one shift are most appropriate but occasionally longer, or shorter, durations are desirable.

For monitoring of particulates the minimum sampling time should be calculated from the following equation[23]:

$$\text{Minimum vol (m}^3\text{)} = \frac{10 \times \text{sensitivity of analytical method (mg)}}{\text{suitable hygiene standard (mg/m}^3\text{)}}$$

Choice of 'suitable hygiene standard' will depend upon circumstances but where there is doubt it has been suggested that $0.1 \times$ TLV is used.

For asbestos monitoring it is suggested[26] that the sampling time ideally should be such that the fibre density on the filter is between 100 and 400 fibres/mm^2. Table 7.6(a) lists the minimum and maximum sampling times for different fibre concentrations, using a flowrate of 1 l/min.

To achieve the optimum fibre density in atmospheres containing less than 0.2 fibre/ml over any 4-hour sampling period, or when monitoring is required over much shorter periods than 4 hours a different flow rate will be required. Table 7.6(b) gives the minimum sampling times using a sampling rate of 2 l/min.

When compliance with hygiene standards is assessed using short-term sampling (e.g. 15 min) then the maximum number of samples obtainable within an 8-hour shift is 32 per instrument. The average of these will indicate whether or not exposures have exceeded the time-weighted average hygiene standard.

Table 7.6a Minimum and maximum sampling times at a flowrate of 1 l/min for different fibre concentrations

Personal exposure	Optimum sampling time at 1 l/min	
(fibre/ml)	Minimum time (100 fibre/mm^2)	Maximum time (400 fibre/mm^2)
0.1	6.6 hours	—
0.2	3.3 hours	—
0.5	80 minutes	5.4 hours
1.0	40 minutes	2.7 hours
2.0	20 minutes	80 minutes
5.0	10 minutes	32 minutes

Table 7.6b Minimum sampling times using a 2 l/min rate of sampling

Personal exposure	Minimum sampling times at 2 l/min at given fibre density	
(fibre/ml)	(50 fibre/mm^2)	(100 fibre/mm^2)
0.05	3.2 hours	—
0.1	1.6 hours	3.3 hours
0.2	50 minutes	1.6 hours
0.5	20 minutes	40 minutes
1.0	10 minutes	20 minutes
2.0	—	10 minutes

When fewer than 32 results are available it is necessary to lower the acceptable upper limit for their average below the hygiene standard to compensate for lack of data. The following equation defines the upper acceptable limit to the average, x (max), thus

$$\bar{x}\ (\text{max}) = \text{TWA hygiene standard} - 1.6 \left(\frac{1}{\sqrt{n}} - \frac{1}{32} \right) \sigma$$

Where: n = number of results; σ = standard deviation.

Where σ is unknown from previous results an estimate can be made from a few samples, and the maximum acceptance limit is TLV - ($k \times$ range) where k is obtained from Table 7.7.[32]

When the calculated limit is below the observed range, compliance can be assumed for that day. Since variations in concentrations also occur between days and weeks the statistics of compliance testing for longer periods need consideration. Thus the average exposure for a week may be below the time-weighted average hygiene standard but occasional shift

Table 7.7 Values for the range

Number of results	k
32	0
10–31	0.1
6–9	0.2
5	0.3
4	0.4
3	0.8
2	0.9

Table 7.8 Test of tolerability of group exposure

No. of persons	Critical average body burden of group exceeded at
1	2 × TLV-TWA
10	1.5 × TLV-TWA
25	1 × TLV-TWA
50	0.5 × TLV-TWA
100	0.25 × TLV-TWA

excursions may have occurred. Provided that the average exposure for the week is below the standard, and that no more than one shift exposure exceeds the standard, the area can be considered to be in compliance. In deciding whether the grand average is acceptable the k values in Table 7.7 may be applied.

It may be that not every individual exposure need be below the Occupational Exposure Limit or Threshold Limit Value for a group of individuals to be in compliance. A factor such as that shown in Table 7.8 can be applied to account for the size of the group.[32]

For routine monitoring the frequency will be influenced by the level of exposure. One guide[32] suggests that personal monitoring should be carried out at least once per month if time-weighted personal exposures are 1–2 × TLV, once per quarter if 0.5–1 or 2–4 × TLV, and once per annum if between 0.1–0.5 or 4–20 × TLV. Routine personal monitoring becomes superfluous when exposures are below × 0.1 or above × 20 the TLV-TWA, assuming there has been no change in the process, the materials handled, the ventilation, or other means of control since the previous survey.[32] These frequencies refer to monitoring of a whole shift for every 10 employees undertaking a particular job. Where less than 10 are employed 10 is assumed. Changes in the general milieu are likely to require more frequent checks to classify the new situation.

Data analysis is best presented on control charts with clearly defined 'warnings' and 'action' levels to indicate deterioration of, and loss of,

Table 7.9 Recommended guidance for routine monitoring

Survey	Results	Action
Preliminary	All < 0.1 × OEL*	None if exposures are low as reasonably practicable.
	0.1–1.5 × OEL	Investigate and carry out a more detailed survey.
	Some > 1.5 × OEL	Investigate, take remedial action and repeat survey.
Detailed	All < 0.25 × OEL	None if exposure is as low as reasonably practicable.
	≤ 1.25 × OEL	Investigate, take remedial action and repeat survey.
	Arithmetic mean < 0.5 × OEL	Consider routine monitoring and the appropriate frequency.
	Mean > 0.5 × OEL (or with individual results scattered above the limit)	Investigate, assess control measures, improve where possible, repeat survey and consider routine monitoring.
Routine	All	Check values, mean, range, etc. Is there compliance with OEL and if so what is confidence?
	Differ significantly from previous survey	Investigate, consider remedial action detailed survey or change protocols.

* Higher values can be used if standard is based on nuisance or odour and there are no known effects of the maximum exposure concentration measured.

control respectively. In the case of 8-hour sampling periods the positions of 'warning' lines are such that on average excursions will occur at the rate of one shift per month, while 'action' levels would be set such that breaches would occur once every six months. More detailed considerations of the statistical aspects of environmental monitoring are given in references 6, 30 and 33.

Data from the surveys will dictate the level of action required including the need for routine monitoring. As a rule of thumb, for substances assigned a control limit (CL)[5] routine monitoring should be introduced when exposures exceed 0.1 CL. Where the occupational exposure limit (OEL) is a recommended value[5] the guidance shown in Table 7.9 has been proposed.[34]

Calculation of exposure

When single substances are involved, calculating the exposure from analytical data and assessing the risk is relatively straightforward. For mixtures of chemicals the situation is more complex and the hazard depends upon the individual components and the relative proportion of each in the mixture. The composition of the atmospheric mixture of dusts or vapours may differ markedly from that of the bulk powder or liquid.

Also the toxic effects of the individual constituents in the mixture may act additively, independently, synergistically or antagonistically. In the absence of evidence to the contrary it should never be assumed that components of the mixture behave independently. Some limited guidance on dealing with mixtures has been published.[4,35-37]

Single substances

Example 1

An operator is exposed to 80 ppm of ethyl acetate for 6 hr 45 min. During the rest of the shift his exposure is known to be zero. The 8-hr time-weighted average exposure is calculated thus:
6.75 hr @ 80 ppm
1.25 hr @ 0 ppm

$$8 \text{ hr time-weighted average} = \frac{(6.75 \times 80) + (0 \times 1.25)}{8} = 67.5 \text{ ppm}$$

This is below the current OEL, 8-hour TWA value of 400 ppm.)

Example 2

An operator is exposed to ethyl acetate vapour during his working day. Samples of air were collected periodically and analysed with the following results:

Work period	Analysis (ppm)	Duration of sampling (hr)
8.00–10.30	20	2.5
10.30–12.30	30	2.0
21.30–14.30	45	2.0
14.30–16.00	100	1.5

The 8-hr time-weighted average is computed thus

$$= \frac{(2.5 \times 20) + (2 \times 30) + (2 \times 45) + (1.5 \times 100)}{8} = 44 \text{ ppm}$$

Furthermore, it is noted that the exposure is relatively high for a short period between 14.30 and 16.00. This should be further investigated with a view to identifying the cause and reducing the exposure.

Mixtures

Where toxic effects are known to be additive and in the absence of synergism the ratio of the concentration (C) to the exposure limit (EL) for each compound may be added as shown by Example 3. If the sum does not exceed unity the exposure limit is considered not to have been breached.

Example 3
For an airborne mixture containing
150 ppm acetone (exposure limit 1000 ppm)
75 ppm sec-butyl acetate (exposure limit 200 ppm)
25 ppm of methyl ethyl ketone (exposure limit 200 ppm); then:
$$\frac{150}{1000} + \frac{75}{200} + \frac{25}{200} = 0.650, \text{ i.e.} < 1$$

Alternatively, a composite exposure limit (CEL) can be calculated for the mixture if all components are known, using the fractional airborne composition (*f*) of each compound coupled with the respective individual exposure limit (EL) i.e.

$$\text{CEL} = \frac{1}{\frac{f_1}{EL_1} + \frac{f_2}{EL_2} \cdots \frac{f_n}{EL_n}}$$

Example 4
For the mixture quoted in Example 3 containing

150 ppm of acetone, the fraction in the mixture $= \frac{150}{250} = 0.6$

75 ppm of sec-butyl acetate, the fraction in the mixture $= \frac{75}{250} = 0.3$

25 ppm of methyl ethyl ketone, the fraction in the mixture
$$= \frac{25}{250} = 0.1$$
$$\underline{.250} \text{ Total}$$

$$\text{CEL} = \frac{1}{\frac{0.6}{1000} + \frac{0.3}{200} + \frac{0.1}{200}} = 385 \text{ ppm}$$

The additive approach need not be used when it is known that such residual risk would not be increased significantly by treating the components independently and ensuring that no individual exposure limit is exceeded as illustrated by Example 5.

Example 5
For an airborne mixture containing
0.035 mg/m³ of cadmium oxide (exposure limit 0.05 mg/m³)
0.65 mg/m³ of sulphuric acid (exposure limit 1 mg/m³)
Then by inspection of the data exposures are within individual exposure limits.

When a mixture contains some constituents considered to have additive toxic effects and others which need to be treated independently a CEL is calculated just for the additive components. Each independently acting component is considered in turn to determine compliance with its individual exposure limit. If the ratio of the concentration (C) of the independently acting substance in the mixture to the exposure limit (EL) of that substance is less than 1 then the limit will be exceed by a factor $\frac{C}{EL}$ and the CEL must be reduced accordingly.

Example 6

For a mixture containing the following:
150 ppm acetone
75 ppm sec-butyl acetate
25 ppm methyl ethyl ketone
25 ppm benzene
275 ppm Total

The first three substances are treated additively as shown in Example 4 to give a value of 385 ppm so that exposure to the mixture in respect of these compounds would be acceptable if all reasonably practicable steps had been taken to minimise exposure. Benzene exposure is checked independently thus

$$\frac{C}{EL} = \frac{25}{10} = 2.5$$

So that the limit for benzene would be exceeded by 2.5 times. The airborne concentration of the mixture therefore must be reduced to below

$$\frac{275}{2.5} = 110 \text{ ppm}$$

For materials which act with synergism, potentiation, or antagonism, professional assessment is required in each case.

References

(*All places of publication are London unless otherwise stated*)
1. The Royal Society for the Prevention of Accidents, *RoSPA Bull.*, 1982 (Jan.), 2.
2. *Sampling and Analytical Guide for Airborne Health Hazards*. Applied Technology Division, Du Pont Co.
3. National Institute for Occupational Safety and Health, *Manual of Analytical Methods*. NIOSH: Cincinnati, USA. 1985
4. American Conference of Governmental Industrial Hygienists, *Threshold Limit Values for 1984*. ACGIH: Cincinnati, Ohio, USA.
5. Health and Safety Executive, *Guidance Note EH 40: Occupational Exposure Limits*, 1986.

6. Thain, W., *Monitoring Toxic Gases in the Atmosphere for Hygiene and Pollution Control*. Pergamon 1980.
7. Cullis, C. F., Firth, J. G. (eds), *Detection and Measurement of Hazardous Gases*. Heinenson.
8. American Conference of Governmental Industrial Hygienists, *Air Sampling Instructions for Evaluation of Atmospheric Contaminants* (6th edn). ACGIH: Ohio, 1983.
9. Nelson, G. O., *Controlled Test Atmospheres*. Ann Arbor Science: Ann Arbor, Michigan, 1971.
10. *Tentative Methods of Sampling Atmospheres for Analysis of Gases and Vapors*. American Society for Testing and Materials: Philadelphia, July 1956.
11. *Chemical Tube User Guide* (2nd edn). MDA Scientific.
12. Coker, D. T., *Intern. Environ. and Safety*, 1981 (June) 13.
13. (a) Benson, G. B., Boyce, G. E. & Haile, D. M., *Annals Occup. Hyg*. 1981, **24** (4), 367.
    (b) Kavanagh, S., Muller, S. L., Seal, J., Stevens, A. J., Swale, J. & Reaveley, *Annals Occup. Hyg*. 1980, **23** (2), 133.
14. Coker, D. T. & Jones, A. L., *Annals Occup. Hyg*. 1981, **24** (4), 399.
15. Benson, G. B. & Boyce, G. E., *Annals Occup. Hyg*. 1981, **24** (4), 401.
16. Olishifski, J. B. (ed.), *Fundamentals of Industrial Hygiene*. National Safety Council: Chicago, 1971.
17. American Industrial Hygiene Association, *Direct Reading Colorimetric Indicator Tubes Manual*. AIHA: Akron, Ohio, 1976.
18. Walton, J. A., 'Detector tubes', in *Handbook of Occupational Hygiene*. Kluwer Publishing 1980.
19. British Standards Institution, *BS 5343: 1976: Gas Detector Tubes*.
20. *OSHA Concentration Limits for Gases: Incorporating Infrared Analytical Data for Compliance Testing and other Applications*. Foxborough Analytical: South Newalk, USA, 1981.
21. Ottely, T. W., *Int. Environ. and Safety*, 1980 (Dec.), 13; *Safety Surveyor*, 1974 (Sept.), 5.
22. *Chromatographic Column Selection Guide for Century Organic Vapour Analyser*. Foxborough Analytical: South Newalk, USA, 1981.
23. Chemical Industry Safety and Health Council, *A Guide to the Evaluation and Control of Toxic Substances in the Working Environment*. Chemical Industries Assoc. Ltd 1980.
24. Health and Safety Executive, *General Methods for the Gravimetric Determination of Respirable and Total Dust* MDHS 14, 1983.
25. *Asbestos-Killer Dust*. BSSRS Publications Ltd 1979, 16.
26. Health and Safety Executive, *Guidance Note EH10: Asbestos-Control Limits, measurement of airborne dust concentration and the assessment of control measures*. HMSO.
27. Health and Safety Executive, *Asbestos Fibres in Air. Determination of Personal Exposures by the European Reference Version of the Membrane Filter Method*, Methods for the Determination of Hazardous Substances, MDHS 39.
28. Vaughan, N. P., Rooker, S. J. & Le Guen, J. M., *Annals Occup. Hyg*. 1981, **24** (3), 28.

29. Le Guen, J. M. & Burdett, G., *Annals Occup. Hyg.*, 1981, **24**(2), 185.
30. Health and Safety Executive, *Guidance Note EH. 42: Assessment and Sampling Strategies in Occupational Hygiene.*
31. Industrial Relations Services, *Health and Safety Information Bulletin*, 1984 (107), 13.
32. Roach, S. A., 'Sampling for airborne toxic hazards, in *Handbook of Occupational Hygiene.* Kluwer Publishing, Middlesex, 1980.
33. Bush, K. A. & Leidd, N. A., in W. Cralley (ed.), *Pattys Industrial Hygiene and Toxicology*, Vol. III. Wiley 1979.
34. National Institute of Occupational Safety and Health, *Occupational Exposure Sampling Strategy Manual.* US Dept of Health Education and Welfare: Cincinnati, Ohio, 1977.
35. Chemical Industries Association, *Occupational Exposure Limits For Mixtures of Airborne Substances Hazardous To Health.* CIA 1985.
36. World Health Organisation, *Health Effects of Combined Exposures in the Working Environment*, Report of a WHO Expert Committee, 1981.
37. Ballantyne, B., *J. Occup. Med.* 1985, **27** (2), 85.

Appendix 7.1
Methods for sampling and analysis for a range of airborne toxic chemicals

(An abreviated version of *Sampling and Analytical Guide For Airborne Health Hazards*, published by the Applied Technology Division of Du Pont Company.)

Key to Table

Sampling methods

AT	– Alumina tube	F+I	– Impinger preceded by a filter
F	– Filter	M	– Miscellaneous sampler (dosimeter, etc)
GF	– Glass fibre filter		
I	– Impinger or fritted bubbler	OT	– Other tubes (long duration detector tubes, porous polymer tubes etc)
MCEF	– Mixed Cellulose ester membranes filter		
P	– PRO-TEK™ Colorimetric Air Monitoring Badge System for SO_2 NO_2 and NH_3	PVC	– Polyvinyl-chloride filter
		SM	– Silver membrane filter
		ST	– Silica gel tube
CT	– Charcoal tube or PRO-TEK™ organic vapour		

Min sample size (l)

The minimum air sample (litres) that will provide enough of the substance for the most accurate analysis at the TLV concentrations using the analytical procedures listed.

Sugg. Max. sample flow rate (ml/min)

The maximum flow rate recommended for the given collection method. Sampling could be done at lower rates as long as min. sample size is met.

Analytical technique

AAS	– Atomic Adsorption Spectroscopy	C	– Colorimetric
		GLC	– Gas liquid chromatography
G	– Gravimetric	LDDT	– Long-duration detector tubes
ISE	– Ion specific electrode	UV	– Ultraviolet
PC	– PRO TEK™ PT_3 Colorimetric Readout		

Appendix 7.1

Substance	Sampling method	Min sample size (*l*)	Sugg. max. sample rate (ml/min)	Analytical technique	Ref.
Abate	F (5 μm PVC)	250	2,000	G	1,3
Acetaldehyde	I	10	2,000	C	1
Acetic acid	I	100	2,500	Titration	
Acetic anhydride	I	100	1,000	C	1, 2, 5
Acetone	CT	2	200	GLC	1, 2
Acetonitrile	CT	10	200	GLC	1, 2
Acetylene dichloride (see 1,2-dichloroethylene)					
Acetylene tetrabromide	ST	100	1,000	GLC	1, 2
Acrolein	I	10	1,500	C	1
	AT	6	200	GLC	
Acrylamide	I		1,000		10, 12
Acrylonitrile	CT	10	200	GLC	1, 2
Aldrin	CT	20	200	GLC	1, 2
Allyl alcohol	F+I (GF)	180	1,000	GLC	1, 2
Allyl chloride	CT	10	200	GLC	1, 2
Allyl glycidyl ether (AGE)	CT	100	1,000	GLC	1, 2, 10
Allyl propyl disulphide	CT	10	50	GLC	1
Alundum® (Al_2O_3)	CT	10	1,000	GLC	1, 12
4-Aminodiphenyl	F (5 μm PVC)	250	2,000	G	
	F (0.8 μm MCEF)	540	4,000	C	3
2-Aminoethanol (see ethanolamine)					
2-Aminopyridine	I	5	1,000	C	1
Ammonia	OT	1	100	LDDT	1, 10, 11
	P			PC	39
Ammonium chloride-fume	F (0.8 μm MCEF)	100	2,000	C	7
Ammonium sulphamate (ammate)	F (5 μm PVC)	250	2,000	G	1, 7

Substance	Sampling		Method	Notes
n-Amyl acetate	CT	10	GLC	1, 2
sec-Amyl acetate	CT	10	GLC	1, 2
Aniline	ST	20	GLC	1, 2
Anisidine (o-, p-isomers)	ST	15	GLC	1, 2
Antimony and compounds (as Sb)	F (0.8 μm MCEF)	360	AAS	1, 2, 8
ANTU (α-naphthylthiourea)	F (0.8 μm MCEF)	300	GLC	1
Arsenic and compounds (as As)	F (0.8 μm MCEF)	90	AAS	1, 2
Arsenic trioxide production (as As)	CT	30	C	1
Arsine	F (0.8 μm MCEF)	10	AAS	1, 2, 10
Asbestos		100	(microscopic fibre count)	1, 2, 10
Asphalt (petroleum) fumes	F (2 μm PVC)	250	G	
Azinophos methyl	I	100	GLC	13
Barium (soluble compounds)	F (0.8 μm MCEF)	180	AAS	1, 2
Benzene	CT	12	GLC	1, 2, 10, 11
	OT	2	LDDT	
Benzidine production	F (0.8 μm MCEF)	480	C	1, 9
p-Benzoquinone (see quinone)				
Benzoyl peroxide	I	30	C	1
Benzo(a)pyrene				1
Benzyl chloride	CT	10	GLC	1, 2
Beryllium	F (0.8 μm MCEF)	270	AAS (30 min. allowed at 0.025 mg/m^3)	1, 2, 10
Biphenyl	CT	30	GLC	1, 12
Bismuth telluride	F (0.8 μm MCEF)	100	AAS	1
Boron oxide	F (2 μm PVC)	60	G	1, 2
Boron trifluoride	F+I	30	C	10

Appendix 7.1 (continued)

Substance	Sampling method	Min sample size (l)	Sugg. max. sample rate (ml/min)	Analytical technique	Ref.
Bromine	I	45	1,000	C	7
Bromine pentafluoride	I	15	2,500	ISE	
Bromoform	CT	10	200	GLC	1, 2
Butadiene (1, 3-Butadiene)	CT	1	50	GLC	1,2
Butanethiol (see butyl mercaptan)					
2-Butanone	CT	10	200	GLC	1, 2
2-Butoxyethanol (butyl cellosolve)	CT	10	200	GLC	1, 2
n-Butyl acetate	CT	10	200	GLC	2
sec-Butyl acetate	CT	10	200	GLC	1, 2
tert-Butyl acetate	CT	10	200	GLC	1, 2
n-Butyl alcohol	CT	10	200	GLC	2
sec-Butyl alcohol	CT	10	200	GLC	1, 2
tert-Butyl alcohol	CT	10	200	GLC	1, 2
Butylamine	I	15	1,000	C	
Butyl cellosolve (see 2-butoxyethanol)					
tert-Butyl chromate (as CrO$_3$)	F (0.8 μm MCEF)	23	1,500	C	
n-Butyl glycidyl ether	CT	10	200	GLC	1, 2
n-Butyl lactate	F (0.8 μm MCEF)	30	1,000	GLC	
Butyl mercaptan	CT	10	1,000	GLC	1, 12
p-tert-Butyltoluene	CT	10	200	GLC	1, 2
Cadmium, dust and salts (as Cd)	F (0.8 μm MCEF)	15	1,000	AAS	1, 2
Cadmium, fume (as Cd)					1
Cadmium oxide fume (as Cd)	F (0.8 μm MCEF)	25	1,500	AAS	1, 2
Calcium arsenate (as As)	F (0.8 μm MCEF)	500	1,500	AAS	1
Calcium carbonate/marble	F			C	
				G (nuisance dust)	

Substance	Sampling	Volume	Method	Refs
Calcium oxide	F (0.8 μm MCEF)	85	1,500	AAS	1, 2
Camphor, synthetic	CT	10	200	GLC	1, 2
Caprolactam					
Dust	F (0.8 μm MCEF)			GLC	
Vapour	I			GLC	
Carbaryl (Sevin®)	F (GF)	90	1,500	C (Do not use Tenite holder.)	1, 2, 10
Carbon black	F (2 μm PVC)	200	1,700	G (stainless steel support screen)	1, 2
Carbon dioxide	(air bag)	4	50	GLC	1, 2, 10
Carbon disulphide	CT	6	200	GLC (used with drying tube)	1, 2
Carbon monoxide	M		100	(dosimeter)	1, 11
	OT	1	20	LDDT	
Carbon tetrachloride	CT	5	1,000	GLC (10 min allowed at 200 ppm)	1, 2, 10
Carbonyl chloride (phosgene)					
Cellulose (paper fibre)	F			G (nuisance dust)	1
Chlordane	I	100	1,000	GLC	13
Chlorinated camphene	F (0.8 μm MCEF)	15	1,000	GLC	1, 2
Chlorinated diphenyl oxide	F (0.8 μm MCEF)	90	1,000	GLC	1, 2
Chlorine	I	30	2,000	C	1, 10
Chlorine dioxide	I	40	1,000	C	16
Chlorine trifluoride	I	15	1,000	ISE	17
Chloroacetaldehyde	CT	10	1,000	GLC	12
Chloroacetophenone (phenacyl chloride)	CT	100	1,000	GLC	12
Chlorbenzene (monochlorobenzene)	CT	10	200	GLC	1, 2
o-Chlorobenzylidene Malononitrile (OCBM)	CT	10	1,000	GLC	1, 12
Chlorobromomethane/bromochloromethane	CT	5	200	GLC	1

Appendix 7.1 (continued)

Substance	Sampling method	Min sample size (l)	Sugg. max. sample rate (ml/min)	Analytical technique	Ref.
2-Chloro-1,3-butadiene (see β-chloroprene)					
Chlorodiphenyl (42% chlorine)	F(0.8 μm MCEF)	100	1,500	GLC	1
	OT	50	200	GLC	1
Chlorodiphenyl (54% chlorine)	F(0.8 μm MCEF)	100	1,500	GLC	1, 2
	OT	50	200	GLC	1
1-Chloro, 2,3-epoxy-propane (see epichlorhydrin)					
2-Chloroethanol (see ethylene chlorhydrin)					
Chloroethylene (see vinyl chloride)					
Chloroform (trichloromethane)	CT	15	1,000	GLC	1, 2
bis-Chloromethyl ether	CT	30	200	GLC	12
1-Chloro-1-nitropropane	CT	10	200	GLC	18
Chloropicrin	I	100	1,000	C	2
β-Chloroprene	CT	3	50	GLC	12
2-Chloro 6-trichloromethyl pyridine (N-Serve®)	CT	10	200	GLC	
Chromates, certain insoluble forms	F (0.8 μm MCEF)	90	1,500	AAS	2
Chromic acid and chromates (as Cr)	F (5 μm PVC)	22.5	1,000	C	1, 2
Chromium, soluble chromic and chromous salts (as Cr)	F (0.8 μm MCEF)	90	1,500	AAS	1, 2
Coal tar pitch volatiles (see particulate polycyclic aromatic hydrocarbons (PPAH), as benzene solubles)					
Cobalt metal, dust and fume (As Co)	F (0.8 μm MCEF)	720	1,500	AAS	1, 2

Substance	Sampling	Volume	Method	Refs
Copper				
Fume	F (0.8 μm MCEF)	720	AAS	1
Dusts and mists (as Cu)	F (0.8 μm MCEF)	90	AAS	1, 2
Corundum (Al$_2$O$_3$)	F		G (nuisance dust)	
Cotton dust (raw)	F (5 μm PVC)	540	G	2
Crag® herbicide	F	100	G	1
Cresol, all isomers	ST	20	GLC	1, 2
Crotonaldehyde	CT	20	GLC	1, 3, 12
	I	100	C	
Cumene	CT	10	GLC	1, 2
Cyanides (as CN)	F+I (0.8 μm MCEF)	90	ISE	1, 2, 11
		1	LDDT	
Cyanogen	I	15	C	19
Cyclohexane	CT	2.5	GLC	1, 2
Cyclohexanol	CT	10	GLC	1, 2
Cyclohexanone	CT	40	GLC	1, 2
Cyclohexene	CT	5	GLC	1, 2
Cyclohexylamine	I	10	C	1, 15
Cyclopentadiene	CT	10	GLC	1, 12
2, 4-D (2,4-Diphenoxy-acetic acid)	I	100	GLC	1, 13
DDT (Dichlorodiphenyltri-chloroethane)	F(GF)	90	GLC	1, 2
DDVP (see dichlorvos)				
Diacetone alcohol (4-hydroxy-4-methyl-2-pentanone)	CT	10	GLC	1, 2
1, 2-Diaminoethane (see ethylenediamine)				
Diazomethane	OT	10	GLC (XAD-2 resin, tube)	1, 2
Dibrom®	I	100	GLC	13
1,2-Dibromoethane	CT	1	GLC (5 min allowed at 50 ppm)	1, 2

Appendix 7.1 (continued)

Substance	Sampling method	Min sample size (l)	Sugg. max. sample rate (ml/min)	Analytical technique	Ref.
2-n-Dibutylaminoethanol					1
Dibutyl phosphate					1
Dibutyl phthalate	F (0.8 μm MCEF)	30	1,000	GLC	1, 2
Dichloracetylene					
o-Dichlorobenzene	CT	10	200	GLC	1, 2
p-Dichlorobenzene	CT	3	50	GLC	1, 2
Dichlorobenzidine					1
Dichlorodifluoromethane	CT	3	50	GLC	1, 2
1,3-Dichloro-5,5-dimethyl hydantoin	L	100	1,000	C	
1,1-Dichloroethane	CT	10	200	GLC	1, 2
1,2-Dichloroethane	CT	3	200	GLC (12 min allowed at 200 ppm)	1, 2, 10
1,2-Dichloroethylene	CT	3	200	GLC	1, 2
Dichloroethyl ether	CT	15	1,000	GLC	1, 2
Dichloromethane (see methylene chloride)					
Dichloromonofluoromethane	CT	3	50	GLC (Use 2 large char. tubes back to back.)	1, 2
1,1-Dichloro-1-nitroethane	CT	15	1,000	GLC (15 min. sample)	1, 2
1,2-Dichloropropane (see propylene dichloride)					
Dichlorotetrafluoroethane	CT	3	50	GLC	1, 2
Dichlorvos (DDVP)	I	100	1,000	GLC	1
Dieldrin	F (GF)	180	1,500	GLC	1, 2
Diethylamine	ST	50	1,000	GLC	1, 2
Diethylaminoethanol	I	30	1,000	C	1, 20
Diethylene triamine	I	100	1,500	C	1, 8

Ambient air analysis for hazardous substances 417

Diethyl ether (see ethyl ether)				
Diethyl phthalate	F (0.8 µm MCEF)	30	1,000	GLC	2
Difluorodibromomethane	CT	10	200	GLC	1, 2
Diglycidyl ether	CT	15	1,000	GLC	12
Dimethoxyethane (see methylal)					
Diisobutyl ketone	CT	12	200	GLC	1, 2
Diisopropylamine	I	15	200	C	1, 21
Dimethyl acetamide	ST	50	1,000	GLC	1, 2
Dimethylamine	ST	50	200	GLC	1, 2
Dimethylaminobenzene (see xylidene)					
Dimethylaniline (N,N-dimethylaniline)	CT	20	200	GLC	1, 2
Dimethylbenzene (see xylene)	ST	5	200		
Dimethyl-1,2-dibromo-2-dichloroethyl phosphate (see Dibrom®)					
Dimethylformamide	ST	50	1,000	GLC	1, 2
2,6-Dimethyl-4-heptanone (see diisobutyl ketone)					
1,1-Dimethylhydrazine	I	100	1,000	C	1, 2
Dimethylphthalate	CT	20	200	GLC	12
Dimethyl sulphate					1
Dinitrobenzene (all isomers)	I	100	1,000	C	1, 22
Dinitro-o-cresol	CT	60	200	GLC	1, 12
Dinitrotoluene	I	100	1,000	C	1, 38
Dioxane (tech. grade)	CT	10	200	GLC	1, 2
Diphenyl (see biphenyl)					
Diphenylmethane diisocyanate (see methylene bisphenyl Isocyanate [MDI])					
Dipropylene glycol methyl ether	CT	10	200	GLC	1, 2
Di-sec, Octyl phthalate (di-2-ethylhexylphthalate)	CT	30	200	GLC	12
Dust, inert or nuisance					
Respirable	F (5 µ PVC)	102	1,700	G (Use cyclone but no Tenite holders.)	2
Total					

Appendix 7.1 (continued)

Substance	Sampling method	Min sample size (l)	Sugg. max. sample rate (ml/min)	Analytical technique	Ref.
Emery	F			G (nuisance dust)	13
Endrin	I	100	1,000	GLC	1, 2, 10
Epichlorhydrin	CT	20	200	GLC	1, 2
EPN	F (GF)	120	1,500	GLC	
1,2-Epoxypropane (see propylene oxide)					
2,3-Epoxy-1-Propanol (see glycidol)					
Ethanethiol (see ethyl mercaptan)					
Ethanolamine	CT	20	200	GLC	1, 12
2-Ethoxyethanol	CT	10	200	GLC	12
2-ethoxyethyl acetate (cellosolve acetate)	CT	10	200	GLC	1, 2
Ethyl acetate	CT	6	200	GLC	1, 2
Ethyl acrylate	CT	10	200	GLC	1, 2
Ethyl alcohol (ethanol)	CT	1	50	GLC	1, 2
Ethylamine	ST	30	200	GLC	1, 2
Ethyl sec-amyl ketone (5-methyl-3-heptanone)	CT	10	200	GLC	1, 2
Ethyl benzene	CT	10	200	GLC	1, 2
Ethyl bromide	CT	4	200	GLC	1, 2
Ethylbutyl ketone (3-heptanone)	CT	10	200	GLC	1, 2
Ethyl chloride					1
Ethyl ether	CT	3	200	GLC	1, 2
Ethyl formate	CT	10	200	GLC	1, 2
Ethyl silicate	OT	9	50	GLC	1, 2
Ethylene chlorohydrin	CT	20	200	GLC	1, 2
Ethylenediamine	I	30	1,000	C	1, 15
Ethylene dibromide (see 1,2-dibromomethane)					
Ethylene dichloride (see 1,2-dichloroethane)					

Substance	Sampling	Volume	Method	Ref
Ethylene glycol dinitrate and/or nitroglycerine	OT	15	GLC (Tenax tube, 5 min. sample.)	1, 2
Ethylene glycol monomethyl Ether acetate (methyl cellosolve acetate)	CT	20	GLC	2
Ethylene oxide	CT	5	GLC (Use 2 back-to-back large tubes.)	1, 2
Ethylidene chloride (see 1,1-dichloroethane)				
N-Ethylmorpholine	CT	10	GLC	1, 12
Ferbam	F (0.8 μm MCEF)	150	C	23
Ferrovandium dust	F (0.8 μm MCEF)	100	AAS	24
Fluoride (as F)	I	10	ISE	1, 10
	F			
Fluorotrichloromethane	CT	4	GLC	1, 2
Formaldehyde	I	25	C (10 min. allowed at 10 ppm.)	1, 10
Formic acid	AT	6	GLC	1, 25
	I	100	GLC	
			C	
Furfural	CT	20	GLC	1, 12
	AT	6	GLC	
	CT	10	GLC	1, 12
Furfuryl alcohol				
Glass, fibrous or dust	F		G (nuisance dust)	
Glycerin mist	F		G (nuisance dust)	
Glycidol (2,3-epoxy-1-propanol)	CT	50	GLC	1, 2
Glycol monoethyl ether (see 2 ethoxyethanol)				
Graphite (synthetic)	F		G (nuisance dust)	
Guthion® (see azinphos-methyl)				
Gypsum	F		G (nuisance dust)	
Hafnium	F	720	AAS	1, 24
Heptachlor	F	100	GLC	1, 13
Heptane (n-heptane)	I		GLC	1, 2
Hexachlorocyclopentadiene	CT	4	GLC	1

Appendix 7.1 (continued)

Substance	Sampling method	Min sample size (l)	Sugg. max. sample rate (ml/min)	Analytical technique	Ref.
Hexachloroethane	CT	10	200	GLC	1, 2
Hexachloronaphthalene	F (0.8 μm MCEF)	30	1,000	GLC (1,000 cc/min. only)	1, 2
Hexane (n-hexane)	CT	4	200	GLC	1, 2
2-Hexanone (methyl butyl ketone)	CT	10	200	GLC	1, 2
Hexone (methyl isobutyl ketone)	CT	10	200	GLC	1, 2
sec-Hexyl acetate	CT	10	200	GLC	12
Hexylene glycol					1
Hydrazine	I	100	1,000	C	1, 2
Hydrogen bromide	I	100	1,000	ISE (1,000 cc/min. only)	1, 2
Hydrogen chloride	I	15	1,000	ISE (1,000 cc/min. only)	1, 2
Hydrogen cyanide	F+I (0.5 μm MCEF)	10	2,000	ISE	1, 10, 11
	OT	1	20	LDDT	
Hydrogen fluoride	I	45	1,500	ISE (1,500 cc/min. only)	1, 2, 10
Hydrogen selenide	I	100	1,000	AAS	24
Hydrogen sulphide	I	30	2,000	C	1, 2, 11
	OT	1	20	LDDT	
Hydroquinone	I	75	1,000	C	1, 26
Indium and compounds (as In)	F (0.8 μm MCEF)	300	1,500	C	1, 27
Iodine	I	15	1,000	C	28
Iron oxide fume (NOC)	F (0.8 μm MCEF)	30	1,500	AAS	1, 2
Iron pentacarbonyl	F (0.8 μm MCEF)	175	1,500	AAS	29
Iron salts, soluble (as Fe)	F (0.8 μm MCEF)	4	1,500	AAS	

Ambient air analysis for hazardous substances 421

Substance			Method	
Isoamyl acetate	CT	10	200	GLC	1, 2
Isoamyl alcohol	CT	10	200	GLC	1, 2
Isobutyl acetate	CT	10	200	GLC	1, 2
Isobutyl alcohol	CT	10	200	GLC	1
Isophorone	CT	12	200	GLC	1, 2
Isopropyl acetate	CT	9	200	GLC	1, 2
Isopropyl alcohol	CT	3	200	GLC	1, 2
Isopropylamine	I	100	1,000	GLC	1, 2
Isopropyl ether	CT	3	50	GLC	1, 2
Isopropyl glycidyl ether	CT	10	200	GLC	1, 2
Kaolin	F			G (nuisance dust)	
Ketene	I	50	1,000	C	1, 2
Lead, inorg., fumes and dusts (as Pb)	F (0.8 μm MCEF)	100	4,000	AAS	1, 2, 10
Limestone	F			G (nuisance dust)	
Lindane	F+I	90	1,500	GLC	1, 2
Lithium hydride	F (0.8 μm MCEF)	720	1,500	AAS	24
LPG (liquefied petroleum gas)		direct reading combustible gas meter			
Magnesite	M				
Magnesium oxide fume (as Mg)	F			G (nuisance dust)	
	F (0.8 μm MCEF)	150	1,500	AAS	1, 2
Malathion	F (GF)	120	1,000	GLC	1, 2, 10
Manganese and compounds (as Mn)	F (0.8 μm MCEF)	22.5	1,500	AAS (1,500 cc/min. only)	1, 2
Manganese cyclopentadienyl tricarbonyl (as Mn)	F (0.8 μm MCEF)	25	1,500	AAS	2
Marble/calcium carbonate	F			G (nuisance dust)	
Mercury (alkyl compounds) (as Hg)	OT	3	50	AAS (chromosorb)	1, 2
Mesityl oxide	CT	10	200	GLC	1, 2
Methanethiol (see methyl mercaptan)					
Methoxychlor	I	100	1,000	GLC	1, 13
2-Methoxyethanol (methyl cellosolve)	CT	50	1,000	GLC	1, 2

Appendix 7.1 (continued)

Substance	Sampling method	Min sample size (l)	Sugg. max. sample rate (ml/min)	Analytical technique	Ref.
Methyl acetate	CT	7	200	GLC	1, 2
Methyl acetylene (propyne)	CT	2	50	GLC	12
Methyl acetylene propadiene mixture	CT	2	50	GLC	12
Methyl acrylate	CT	5	200	GLC	1, 2
Methylal (dimethoxymethane)	CT	2	200	GLC	1, 2
Methyl alcohol (methanol)	ST	3	50	GLC	1, 2, 10
Methylamine	I	50	2,000	C	1, 15
Methyl amyl alcohol (see methyl isobutyl carbinol)					
Methyl n-amyl ketone (2-heptanone)	CT	10	200	GLC	1, 2
Methyl bromide	CT	11	1,000	GLC (2 large char. tubes in series.)	1, 2
Methyl butyl ketone (see 2-hexanone)					
Methyl cellosolve (see 2-methoxyethanol)					
Methyl cellosolve acetate (see ethylene glycol monomethyl ether acetate)					
Methyl chloride				(10 min. allowed at 300 ppm.)	1
Methyl chloroform (1,1,1-Trichloroethane)	CT	6	200	GLC	1, 2
Methylcyclohexane	CT	4	200	GLC (for Methylcyclohexane.)	1, 2
Methylcyclohexanol	CT	10	200	GLC	1, 12
o-Methylcyclohexanone	CT	10	200	GLC	1, 12
Methylene bisphenyl isocyanate (MDI)	I	20	1,000	C	1
Methylene chloride (dichloromethane)	CT	5	1,000	GLC	1, 2, 10
4,4' Methylene bis (2-chloraniline) (MOCA)	ST	3	500	GLC (Use prefilter)	1, 2

Substance	Sampler	Volume	Method	Refs
Methyl ethyl ketone (MEK) (see 2-butanone)				
Methyl formate	CT	10	GLC	1, 12
Methyl iodide	CT	50	GLC	1, 2
Methyl isoamyl ketone	CT	5	GLC	
Methyl isobutyl carbinol	CT	10	GLC	1, 2
Methyl isobutyl ketone (see hexone)				
Methyl mercaptan	CT	15	GLC	12
	I	30	C	30
Methyl methacrylate	CT	5	GLC	2
Methyl parathion	I	50	GLC	10
Methyl propyl ketone (see 2 pentanone)				
α-Methyl styrene	CT	3	GLC	2
Molybdenum (as Mo) Soluble compounds	F (0.8 μm MCEF)	90	AAS	1, 2
Insoluble compounds	F (0.8 μm MCEF)	90	AAS	1, 2
Monomethyl aniline	I	100	GLC	1, 2
Monomethyl hydrazine	I	22.5	C	1, 2
Morpholine	ST	20	GLC	1, 2
Naphtha (coal tar)	CT	10	GLC	1, 2
Naphthalene	CT	200	GLC	1, 2
β-Naphthylamine	I	200	AAS	1
Nickel carbonyl				24
Nickel metal				1
Nickel, soluble compounds (as Ni)	F (0.8 μm MCEF)	90	AAS	2
Nicotine	OT	100	GLC (XAD-2 resin tube)	1, 2
Nitric acid	I	200	C (2,800 cc/min. only)	1, 2, 10
Nitric oxide	OT	1	C (LDDT) (MOL sieve tube)	1, 2, 11
		1		
		20		
		50		

Appendix 7.1 (continued)

Substance	Sampling method	Min sample size (l)	Sugg. max. sample rate (ml/min)	Analytical technique	Ref.
p-Nitroaniline	ST	15	1,000	GLC	1
Nitrobenzene	ST	50	1,000	GLC	1, 2
p-Nitrochlorobenzene	ST	50	1,000	GLC	1, 2
Nitroethane	CT	10	200	GLC	1, 12
Nitrogen dioxide	OT	1	20	(LDDT)	1, 2, 11
	OT	1	50	(MOL sieve tube)	
	I	10	4,000	C	
Nitrogen trifluoride	P			PC	39
Nitroglycerine	OT	15	1,000	GLC (Tenax tube)	1, 2
Nitromethane	CT	10	200	GLC	12
1-Nitropropane	CT	10	200	GLC	1, 12
2-Nitropropane	CT	10	200	GLC	1, 12
Nitrotoluene	ST	20	200	GLC	1, 2
Nitrotrichloromethane (see chloropicrin)					
Octachloronaphthalene	F (0.8 μm MCEF)	30	1,000	GLC	1, 2
Octane	CT	4	200	GLC	1, 2
Oil mist, mineral	F (0.8 μm MCEF)	100	1,500	C (fluor. spect.)	1, 2
Ozone	I	45	1,000	C	1, 2
Paraffin wax fume	F (SM)	200	2,000	G	
Paraquat, respirable sizes	F (0.8 μm MCEF)	150	1,500	C	1, 13
Parathion	I	25	1,500	GLC	1, 2
	F (GF)	120	1,500	GLC	
Particulate polycyclic aromatic hydrocarbons (PPAH), as benzene solubles	F (0.8 μm SM)	720	1,500	G (Use GF prefilter)	2

Substance	Collection	Volume (L)	Analysis	References
Pentachloronaphthalene	F+I (GF)	250	GLC	1, 2
Pentachlorophenol	F (0.8 μm MCEF)	150	GLC	1, 13
Pentaerythritol	F		G (nuisance dust)	
Pentane	CT	2	GLC	1, 2
2-Pentanone	CT	10	GLC	1, 2
Perchloroethylene	CT	1	GLC (10 min. allowed at 300 ppm.)	1, 2
Perchloromethyl mercaptan	CT	30	GLC	12
Perchloryl fluoride	I	30	ISE	17
Petroleum distillates (naphtha)				1
Phenol	I	100	GLC	1, 2, 10
p-Phenylene diamine	I	101	C	31
Phenyl ether (vapour)	CT	10	GLC	1, 2
Phenyl ether-diphenyl mixture (Vapor)	ST	10	GLC	2
Phenylethylene (see styrene, monomer)				
Phenyl glycidyl ether	CT	50	GLC	1, 2
Phenylhydrazine	I	100	C	1, 2
Phosdrin (Mevinphos®)	I	100	GLC	13
Phosgene (carbonyl chloride)	I	25	C	1, 10
Phosphine	F (0.8 μm MCEF)	5	C (GF prefilter)	2
Phosphoric acid	F (0.8 μm MCEF)	15	C (four 2-hr. samples)	1, 2
Phosphorus (yellow)			C	1
Phosphorus pentachloride	I	100	ISE	1, 16, 17
Phosphorus trichloride	I	100	C	1, 16, 17
			ISE	
Phthalic anhydride	F (0.8 μm MCEF)	100	GLC	1, 2

Appendix 7.1 (continued)

Substance	Sampling method	Min sample size (l)	Sugg. max. sample rate (ml/min)	Analytical technique	Ref.
Picric acid	I	100	1,000	GLC	1
Pival® (2-pivalyl-1,3-indandione)	F			G (nuisance dust)	13
Plaster of Paris					
Platinum, soluble salts (as Pt)	F (0.8 μm MCEF)	720	1,500	AAS	1, 2
Polychlorobiphenyls (see chlorodiphenyls)					
Polytetrafluoroethylene decomposition products				(See Fluoride procedure.)	
Propane					1
Propargyl alcohol	CT	30	200	GLC	12
n-Propyl acetate	CT	10	200	GLC	1, 2
Propyl alcohol	CT	10	200	GLC	1, 2
n-Propyl nitrate	CT	70	1,000	GLC	1, 2
Propylene dichloride (1,2-dichloropropane)	CT	10	200	GLC	1, 12
Propylene glycol monomethyl ether	CT	10	200	GLC	
Propylene oxide	CT	5	200	GLC	1, 2
Propyne (see methyl acetylene)					
Pyrethrum	I	100	1,000	GLC	23
Pyridine	CT	100	1,000	GLC	1, 2
Quinone	I	100	1,000	C	1, 32
RDX	F (0.8 μm MCEF)	150	1,500	UV	33
Rhodium (as Rh)					
Metal fume and dusts	F (0.8 μm MCEF)	720	1,500	AAS	1, 2
Soluble salts	F (0.8 μm MCEF)	370	1,500	AAS	1, 2
Ronnel	I	100	1,000	GLC	13

Substance	Sampler	Volume (L)	Method	Ref.
Rosin core solder pyrolysis products (as formaldehyde)	I	25	1,000	C	10
Rotenone (commercial)	I	60	1,000	UV	1, 34
Rouge	F			G (nuisance dust)	1, 2
Selenium compounds (as Se)	F (0.8 μm MCEF)	360	1,500	AAS	1, 2
Selenium hexafluoride (as Se)	I	60	1,000	AAS	24
Sevin® (see carbaryl)					
Silica					
Respirable	F (5 μm PVC)	816	1,700	(cyclone, X-ray diffraction)	1, 10
Total	F (SM, 1,2, 25)			G (nuisance dust)	1
Silicon	F			G (nuisance dust)	1, 2
Silicon carbide	F (0.8 μm MCEF)	45	1,500	G (nuisance dust)	1, 2
Silver, metal and soluble compounds (as Ag)					
Sodium fluoroacetate	I	350	1,000	C (for titration)	1
Sodium hydroxide	F			G (nuisance dust)	1, 2, 10
Starch	CT	100	200	AAS	1, 24
Stibine				C	
Stoddard solvent	CT	3	200	GLC	1, 2
Strychnine	F (0.8 μm MCEF)	360	1,500	UV	34
Styrene, monomer (phenylethylene)	CT	1	200	GLC	1, 2
Succinaldehyde (see glutaraldehyde)					
Sucrose	F			G (nuisance dust)	
Sulphur dioxide	OT	1	20	LDDT	1, 11
	F+I (0.8 μm MCEF)	100	2,000	C (titration)	2, 10
Sulfur hexafluoride	P			PL	39
Sulphuric acid	I	15	1,000	UV	1
	F (0.8 μm MCEF)	10	1,500	C (titration)	1, 2, 10
Sulphur monochloride	I	15	1,000	ISE	17

Appendix 7.1 (continued)

Substance	Sampling method	Min sample size (l)	Sugg. max. sample rate (ml/min)	Analytical technique	Ref.
Sulphuryl fluoride	I	15	1,000	GLC	13
2, 4, 5-T	I	100	1,000	GLC	13
Tantalum	F (0.8 μm MCEF)	720	1,500	AAS	1, 24
TEDP	I	100	1,000	GLC	13
Teflon® decomposition products				(See Fluoride procedure.)	
Tellurium and compounds (as Te)	F (0.8 μm MCEF)	670	1,500	AAS	1, 2
Tellurium hexafluoride (as Te)	CT	360	1,000	AAS (MCEF prefilter)	1,2
TEPP	I	100	1,000	GLC	13
Terphenyls	F(0.8 μm MCEF)	15	1,000	GLC	1, 2
1,1,1,2-Tetrachloro 2,2-Difluoroethane	CT	2	35	GLC	1, 2
1,1,2,2-Tetrachloro 1,2-Difluoroethane	CT	2	50	GLC	1, 2
1,1,2,2-Tetrachloroethane	CT	10	200	GLC	1, 2, 10
Tetrachloroethylene (see perchloroethylene)					
Tetrachloromethane (see carbon tetrachloride)					
Tetrachloronaphthalene	F+I (GF)	100	1,300	GLC	1
Tetraethyl lead (as Pb)	CT	60	1,000	AAS	1, 35
	F+I (0.8 μm MCEF)	60	2,000	C	
Tetrahydrofuran.	CT	9	200	GLC	1, 2
Tetramethyl lead (as Pb)	CT	60	1,000	AAS	1, 35
	F+I (0.8 μm MCEF)	60	2,000	C	
Tetramethyl succinonitrile	CT	50	1,000	GLC	1, 2
Tetranitromethane	I	250	1,000	GLC	1, 2

Tetryl (2,4,6-trinitrophenyl-methylnitramine)	F (0.8 μm MCEF)	100	1,500	C	1, 2
Thallium, soluble compounds (as Tl)	F (0.8 μm MCEF)	540	1,500	AAS	
Thiram	F (0.8 μm MCEF)	180	1,500	AAS	1, 24
Tin, inorganic compounds, except SnH₄ and SnO₂ (as Sn)	F (0.8 μm MCEF)	180	1,500	C	1, 23
Tin, organic compounds (as Sn)	F (0.8 μm MCEF)	240	1,000	AAS	1, 2
Tin oxide (as Sn)	F			G (nuisance dust)	1
Titanium dioxide (as Ti)	F (0.8 μm MCEF)	100	1,500	AAS (nuisance dust)	1, 2
Toluene (toluol)	CT	2	1,000	GLC (10 min. allowed at 500 ppm.)	1, 2, 10
Toluene-2, 4- diisocyanate (TDI)	I	20	2,000	C	1, 2, 10
o-Toluidine	ST	50	1,000	GLC	1, 2
Toxaphene (see chlorinated camphene)					
Tributyl phosphate	F (0.8 μm MCEF)	100	1,500	GLC	1, 2
1,1,1-Trichloroethane (see methyl chloroform)					
1,1,2-Trichloroethane	CT	10	200	GLC	1, 2
Trichloroethylene	CT	10	1,000	GLC (10 min. allowed at 300 ppm.)	1, 2, 10
Trichloromethane (see chloroform)					
Trichloronaphthalene	F+I (GF)	100	1,300	GLC	1, 2
1,2,3-Trichloropropane	CT	10	200	GLC	1, 2
1,1,2-Trichloro 1,2,2-Trifluoroethane	CT	1.5	50	GLC	1, 2
Triethylamine	I	100	1,000	GLC	1, 2
Trifluoromonobromomethane	CT	1	50	GLC (2 char., tubes)	1, 2

Appendix 7.1 (continued)

Substance	Sampling method	Min sample size (l)	Sugg. max. sample rate (ml/min)	Analytical technique	Ref.
Trimethyl benzene	CT	5	200	GLC	
2,4,6-Trinitrophenol (see picric acid)					
2,4,6-Trinitrophenyl-methylnitramine (see tetryl)					
2,4,6-Trinitrotoluene (TNT)	F (0.8 μm MCEF)	360	1,500	C	
Triorthocresyl phosphate	F (0.8 μm MCEF)	100	1,500	GLC	1, 2
Triphenyl phosphate	F (0.8 μm MCEF)	100	1,500	GLC	1, 2
Tungsten and compounds (as W)					
Soluble					1
Insoluble					1
Turpentine	CT	10	200	GLC	1, 2
Uranium, natural (as U)					
Soluble	F (0.8 μm MCEF)	45	1,500	(fluorometric)	36
Insoluble compounds	F (0.8 μm MCEF)	45	1,500	(fluorometric)	36
Vanadium (V_2O_5) (as V)					
Dust	F (0.8 μm MCEF)	25	1,700	AAS	1, 2
Fume					1
Vinyl acetate	CT	10	200	GLC	1
Vinyl benzene (see styrene)					
Vinyl chloride	CT	5	50	GLC (Use 2 tubes.)	1, 2, 11
	OT	1	20	LDDT	11
Vinyl cyanide (see acrylonitrile)					
Vinylidene chloride					1

Vinyl toluene	CT	10	200	GLC	1, 2
VM and P naphtha					
Warfarin					
Welding fumes (total particulate) (NOC)	F (0.8 μm MCEF)	720	1,500	G	
Xylene (o-, m-, p-isomers)	CT	12	1,000	GLC	1, 2, 10
Xylidene	ST	20	200	GLC	1, 2
Yttrium	F (0.8 μm MCEF)	500	1,500	AAS	1, 2
Zinc chloride fume	F (0.8 μm MCEF)	25	1,500	AAS	2
Zinc oxide fume	F (0.8 μm MCEF)	360	1,500	AAS (also X-ray diffraction)	1, 2, 10
Zinc stearate	F (PVC)			G (nuisance dust)	
Zirconium compounds (as Zr)	F (0.8 μm MCEF)	720	1,500	AAS	1, 2

Appendix 7.1 References for Table

(*All places of publication are London unless otherwise stated*)
1. US Department of Health, Education and Welfare, *NIOSH Manual of Analytical Methods*. Public Health Service: Cincinnati. Vols 1, 2 & 3, Apr. 1977; Vol. 4, Aug. 1978; Vol. 5, Aug. 1979.
2. US Department of Health, Education, and Welfare, *NIOSH Manual of Sampling Data Sheets* (1977 edn). Public Health Service: Cincinnati 1977.
3. Altshuller, A. P. & Cohen, I. R., 'Spectrophotometric determination of crotonaldehyde with 4-Hexylresorcinol', *Analytical Chemistry*, 1961, **33**, 1180.
4. Miller, Franklin, *et al.*, 'Determination of acetic acid in air', *American Industrial Hygiene Association Journal*, 1956, **17**, 221.
5. Diggle, W. M. & Gage, J. C., 'Determination of keten and acetic anhydride in the atmosphere', *The Analyst*, 1953, **78**, 473.
6. American Conference of Governmental Industrial Hygienists, *TLVs Threshold Limit Values for Chemical Substances in Workroom Air Adopted by ACGIH for 1978*. ACGIH: Cincinnati, 1978.
7. Elkins, H. B., *The Chemistry of Industrial Toxicology* (2nd edn) Wiley: New York 1959.
8. McChesney, E. W., 'Colorimetric micromethod for determination of antimony in biological materials', *Industrial and Engineering Chemistry Analytical Edition*, 1946, **18**,146.
9. Leithe, W., *The Analysis of Air Pollutants*. Ann Arbor-Humphrey Science: Ann Arbor, 1970.
10. US Department of Health, Education, and Welfare, *NIOSH Criteria for a Recommended Standard. ... Occupational Exposure to...* Public Health Service: Cincinnati 1974.
11. Leichnitz, K. R., 'Detector tubes and prolonged air sampling', *National Safety News*, 1977, **115**, 59.
12. White, L. D., *et al.*, 'A convenient optimized method for the analysis of selected solvent vapors in the industrial atmosphere', *American Industrial Hygiene Association Journal*, 1970, **31**, 225.
13. *Analysis of Pesticide Residues*. Primate and Pesticides Effects Laboratory: Perrine.
14. Siggia, S, *Quantitative Organic Analysis via Functional Groups* (3rd edn). Wiley Press: New York, 1963.
15. Scherberger, R. F., *et al.*, 'The determination of n-butylamine in air', *American Industrial Hygiene Association Journal*, 1960, **21**, 471.
16. *Standard Methods for the Examination of Water and Wastewater* (13th edn) American Public Health Association: Washington, D.C., 1971.
17. *Orion Research: Analytical Methods Guide*. (6th edn). Orion Research: Cambridge, 1973.
18. Feinsilver, L. & Oberst, F. W., 'Microdetermination of chloropicrin vapor in air', *Analytical Chemistry*, 1953, **25**, 820.
19. Rich, W. E., *Quantitative Analysis of Gaseous Pollutants*. Ann Arbor-Humphrey Science: Ann Arbor, 1970.

20. Miller, F. A., 'Determination of microgram quantities of diethanolamine, 2-methylaminoethanol, and 2-diethylaminoethanol in air', *American Industrial Hygiene Association Journal*, 1967, **28**, 330.
21. Dahlgren, G., 'Spectrophotometric determination of ethyl-, diethyl- and triethylamine in aqueous solution', *Analytical Chemistry*, 1964, **36**, 596.
22. Schrenk, H. H., *et al.*, *A Microcolorimetric Method for the Determination of Benzine*. US Bureau of Mines, R.I. 3287 (1935).
23. *Pesticide Analytical Manual*, Vol. J. Food and Drug Administration: Washington, DC, n.d.
24. *Analytical Methods for Atomic Absorption Spectrophotometry*. Perkin-Elmer: Norwalk, n.d.
25. Warner, B. R. & Raptis, L. Z., 'Determination of formic acid in presence of acetic acid', *Analytical Chemistry*, 1955, **27**, 1783.
26. Houghton, J. A. & Lee, G., 'Ultraviolet spectrophotometric and fluorescence data', *American Industrial Hygiene Association Journal* 1961, **22**, 296.
27. Sandell, E. B., *Colorimetric Determination of Traces of Metals* (3rd edn). Interscience: New York, 1959.
28. Johannesson, J. K., 'Determination of microgram quantities of free iodine using *o*-tolidine reagent', *Analytical Chemistry*, 1956, **28**, 1475.
29. Brief, R. S., *et al.*, 'Iron pentacarbonyl: its toxicity detection, and potential for formation', *American Industrial Hygiene Association Journal*, 1967, **28**, 21.
30. Moore, H., *et al.*, 'Spectrophotometric method for the determination of mercaptans in air', *American Industrial Hygiene Association Journal*, 1960, **21**, 466.
31. Stewart, J. T., *et al.*, 'Spectrophotometric determination of primary aromatic amines with 9-chloroacridine', *Analytical Chemistry*, 1969, **41**, 360.
32. Oglesby, F. L., 'Quinone vapors and their harmful effects: plant exposure associated with eye injuries', *Journal of Industrial Hygiene and Toxicology*, 1947, **29**, 74.
33. Schroeder, W. A., *et al.*, 'Ultraviolet and visible absorption spectra in ethyl alcohol', *Analytical Chemistry*, 1951, **23**, 1740.
34. Horwitz, W. (ed.), *Official Methods of Analysis of the Association of Official Analytical Chemists* (11th edn). AOAC: Washington, DC, 1970.
35. Moss, R. & Browett, E. V., 'Determination of tetra-alkyl lead vapour and inorganic lead dust in air', *The Analyst*, 1966, **91**, 428.
36. *Hygienic Guide Series* . . . American Industrial Hygiene Association: Detroit, 1958–onwards.
37. *Code of Federal Regulations*, Title 29, Section XVII, Part 1910. GPO: Washington, DC, 1976.
38. Jacobs, M. B., *Analytical Chemistry of Industrial Poison, Hazards, and Solvents*. Interscience: New York, 1949
39. *PRO-TEK*™ Colorimetric Monitoring Badge System Operating Instructions. E.I. du Pont de Nemours 1979.

Appendix 7.2 Compounds for which there are analytical methods recommended by NIOSH

Acetaldehyde
Acetic acid
Acetic anhydride
Acetone
Acetone cyanohydrin
Acetonitrile (Methyl cyanide)
Acetylene dichloride (1.2-Dichloroethylene)
Acetylene tetrabormide (Tetrabromoethane)
Acid mists
Acrolein
Acrylonitrile
ALAD (δ-Aminolevulinic acid dehydratase)
Aldrin
Allyl alcohol
Allyl chloride
Allyl glycidyl ether
Aluminum
Amines, aliphatic
Amines, aromatic
4-aminobiphenyl
Aminoethanol compounds
bis (2-Aminoethyl) amine (Diethylenetriamine)
δ-Aminolevulinic acid dehydratase
p-Aminophenylarsonic acid
2-Aminopyridine
Ammonia
Ammonium sulfamate
Amorphous silica
n-Amyl acetate
sec-Amyl acetate (α-Methylbutyl acetate)
Aniline
Anisidine
Anthanthrene
Anthracene
Antimony
ANTU (δ-Naphthyl Thiourea)
Arsenic
Arsine
Asbestos
Azelaic acid

Aziridine
Azo dyes

Barium
Benz(c)acridine
Benz(a)anthracene
Benz(a,h)anthracene
Benz(a)anthrone
Benzene
Benzene, chlorinated
Benzene-solubles
Benzidine
Benzidine-based dyes
Benzo Azurine G
Benzo(b)fluoranthene
Benzo(j)fluoranthene
Benzo(k)fluoranthene
Benzo(g,h,i)perylene
Benzopurpurine 4B
Benzo(a)pyrene
Benzo(c)pyrene
Benzoyl peroxide
Benzyl chloride
Beryllium
Bibenzyl
Biphenyl (Diphenyl)
Biphenyl-phenyl ether mixture (Phenyl ether-biphenyl vapor mixture)
Bis(chloromethyl)ether
2,2-Bis[4-(2,3-epoxypropoxy)phenyl] propane
Bismuth
Bisphenol A
2,2-Bis(p-chlorophenyl) 1,1,1-trichloroethane (DDT)
Boron carbide
Boron oxide
Bromoform
Butadiene (1,3-Butadiene)
1-Butanethiol (n-Butyl mercaptan)
2-Butanone (Methyl ethyl ketone or MEK)
2-Butoxy ethanol (butyl cellosolve)
sec-Butyl acetate

tert-Butyl acetate
Butyl acetate (n-Butyl acetate)
sec-Butyl alcohol
tert-Butyl alcohol
Butyl alcohol (n-Butyl alcohol)
Butyl cellosolve (2-Butoxy ethanol)
n-Butyl glycidyl ether
n-Butyl mercaptan
n-Butylamine
p-tert-Butyltoluene

Cadmium
Calcium
Calcium arsenate
Calcium oxide
Camphor
Carbaryl® (Sevin)
Carbon black
Carbon dioxide
Carbon disulfide
Carbon monoxide
Carbon tetrachloride
Carbonyl chloride (Phosgene)
Chlordane
Chloride
Chlorinated camphene (Toxaphene)
Chlorinated diphenyl oxide
Chlorine
2-Chloro-1,3-butadiene (Chloroprene)
1-Chloro-2,3-epoxypropane (Epichlorohydrin)
1-Chloro-1-nitropropane
Chloroacetaldehyde
α-Chloroacetophenone
Chlorobenzene (Monochlorobenzene)
o-Chlorobenzylidine malononitrile
Chlorobromomethane
Chlorodiphenyl (42% chlorine)
Chlorodiphenyl (54% chlorine)
2-Chloroethanol (Ethylene chlorohydrin)
Chloroform
Chloromethyl methyl ether
p-Chlorophenol
Chloroprene (2-Chloro-1,3-butadiene)
Chromic acid

Chromium
Chromium fume
Chrysene
Chrysotile
Coal-tar naphtha (Naphtha, coal tar)
Coal-tar pitch volatiles
Cobalt
Cobalt, metal, dust, and fume
Congo Red
Copper
Copper dust and mists
Copper fume
Crag herbicide I
Cresol, all isomers
Cristobalite
Crotonaldehyde
Cumeme
Cyanide
Cyclohexane
Cyclohexanol
Cyclohexanone
Cyclohexene
Cyclohexylamine
Cyclopentadiene

DBPC
2,4 D
DDT (2,2-Bis(p-chlorophenyl)-1,1,1-trichloroethane)
DDVP
Demeton
Diacetone alcohol (4-Hydroxy-4-methyl-2-pentanone)
1,2-Diaminoethane (Ethylene diamine)
o-Dianisidene-based dyes
Diazomethane
Diazonium salts
Diborane
1,2-Dibromoethane (Ethylene dibromide)
2-Dibutylaminoethanol (Aminoethanol compounds)
Dibutyl phosphate
Dibutylphthalate
1,1-Dichloro-1-nitroethane
o-Dichlorobenzene

p-Dichlorobenzene
3,3'-Dichlorobenzidine
Dichlorodifluoromethane
  (Refrigerant 12)
1,1-Dichloroethane (Ethylidene
  chloride)
1,2-Dichloroethane (Ethylene
  dichloride)
Dichloroethyl ether
1,2-Dichloroethylene (Acetylene
  dichloride)
Dichloromethane (Methylene
  chloride)
Dichloromonofluoromethane
  (Refrigerant 21)
2,4-Dichlorophenoxyacetic acid and
  salts
1,2-Dichloropropane
Dichlorotetrafluoroethane
  (Refrigerant 114)
Dichloro-5-triazine-2,4,6-trione,
  sodium salt
Dichlorovos
Dieldrin
2-Diethylaminoethanol
Diethylamine
Diethylcarbamoyl chloride
Diethylene dioxide (Dioxane)
Diethylenetriamine
Difluorodibromomethane
Difluorodichloromethane
Diglycidyl ether of Bisphenol A
Diisobutyl ketone (2,6-Dimethyl-4-
  heptanone)
Diisopropylamine
Dimethoxymethane (Methylal)
Dimethyl acetamide
Dimethyl benzene (Xylene)
1,3-Dimethyl butyl acetate (sec-
  Hexyl acetate)
Dimethyl formamide
2,6-Dimethyl-4-heptanone
  (Diisobutyl ketone)
Dimethyl sulfate
N,N-Dimethyl-p-toluidine
Dimethylamine
4-Dimethylaminoazobenzene

2,4-Dimethylaminobenzene
  (Xylidine)
N,N-Dimethylaniline
Dimethylarsenic acid
1,1-Dimethylhydrazine
Dimethylnitrosamine
bis (Dimethylthiocarbamoyl) disulfide
Dinitrobenzene (all isomers)
Dinitro o-cresol
Dinitrotoluene
Dioxane
Diphenyl
4,4'-Methylenebis (phenyl isocyanate)
  (MDI)
Dipropylene glycol methyl ether
Direct Black 38
Direct Blue 6
Direct Blue 8
Direct Brown 95
Direct Red 2
Direct Red 28
2,6-Di-tert-butyl-p-cresol
Dowtherm A (Phenyl
  etherbiphenyl vapour mixture)

Endrin
Epichlorohydrin (1-Chloro-2,3-
  epoxypropane)
EPN (O-Ethyl-O-p-nitrophenyl
  phenyl-phosphonothiolate)
2,3-Epoxy-1-propanol (Glycidol)
1,2-Epoxypropane (Propylene
  oxide)
2,2-bis[4-(2,3-Epoxypropoxy) phenyl]
  propane
Ethanol (Ethyl alcohol)
Ethanolamine (Aminoethanol
  compounds)
Ether
2-Ethoxyethanol
2-Ethoxyethylacetate
Ethyl acrylate
Ethyl acrylate
Ethyl alcohol (Ethanol)
Ethyl benzene
Ethyl bromide
Ethy butyl ketone (3-Heptanone)

Ethyl chloride
Ethyl ether
Ethyl formate
Ethyl sec-amyl ketone (5-Methyl-3-heptanone)
Ethyl silicate
O-Ethyl-O-*p*-nitrophenyl phenyl-phosphonothiolate phosphonate (EPN)
Ethylamine
Ethylene chloride (Ethylene dichloride)
Ethylene chlorohydrin (2-Chloroethanol)
Ethylenediamine
Ethylene dibromide (1,2-Dibromoethane)
Ethylene dichloride (1,2-Dichloroethane)
Ethylene glycol
Ethylene glycol dinitrate
Ethylene oxide
Ethylene thiourea
Ethylenimine
di-2-Ethylhexylphthalate (di-sec-Octyl phthalate)
Ethylidene chloride (1,1-Dichloroethane)
N-Ethylmorpholine

Fluoranthene
Fluoride
Fluoroacetate, sodium
Fluorotrichloromethane (Refrigerant 11)
Formaldehyde
Formic acid
Furfural
Furfuryl alcohol

Galena
Gallium
Glycidol (2,3-Epoxy-1-propanol)
Hafnium
Heptachlor
Heptane
3-Heptanone (Ethyl butyl ketone)

2-Heptanone (Methyl (*n*-amyl) ketone)
Hexachlorobutadiene
Hexachlorocycloapentadiene
Hexachloroethane
Hexachloronaphthalene
Hexamethylenetetramine
Hexane
Hexavalent chromium
2-Hexanone (Methyl butyl ketone or MBK)
Hexone (Methyl isobutyl ketone or MIBK)
sec-Hexyl acetate (1,3-Dimethyl butyl acetate)
Hippuric acid
Hydrazine
Hydrogen bromide
Hydrogen chloride
Hydrogen cyanide
Hydrogen fluoride
Hydrogen sulfide
Hydroquinone
4-Hydroxy-4-methyl-2-pentanone (Diacetone alcohol)

2-Imidazolidinethione (Ethylene thiourea)
Indium
Iron
Iron oxide fume
Isoamyl acetate
Isoamyl alcohol
Isobutyl acetate
Isobutyl alcohol
Isophorone
Isopropanol (Isopropyl alcohol)
Isopropyl acetate
Isopropyl alcohol (Isopropanol)
Isopropyl benzene (Cumeme)
Isopropyl glycidyl ether
Isopropylamine
Isopropylether
4,4'-Isopropylidenediphenol

Kepone
Ketene

Lead
Lead sulfide
Lindane
Liquified petroleum gas (LPG)
Lithium
LPG (Liquified petroleum gas)
MAPP (Methyl acetylene/propadiene)
Magnesium
Magnesium oxide fume
Malathion
Maleic anhydride
Manganese
Manganese fume
MBK (2-Hexanone)
MDI (4,4'-Methylenebis-(phenyl isocyanate))
MEK (2-Butanone)
Mercury
Mesityl oxide
Methanol (Methyl alcohol)
2-Methoxyethanol (Methyl cellosolve)
Methozychlor
Methyl(n-amyl)ketone (2-Heptanone)
5-Methyl-3-heptanone
Methyl acetate
Methyl acetylene
Methyl acetylene propadiene mixture
Methyl acrylate
Methyl alcohol (Methanol)
Methyl bromide
α-Methyl butyl acetate (sec-Amyl acetate)
Methyl butyl ketone (2-Hexanone)
Methyl cellosolve (2-Methoxyethanol)
Methyl cellosolve acetate
Methyl chloride
Methyl chloroform (1,1,1-Trichloroethane)
Methyl cyanide (Acetonitrile)
Methyl ethyl ketone (2-Butanone)
Methyl ethyl ketone peroxide
Methyl formate
Methyl iodide
Methyl isoamyl acetate
Methyl isobutyl carbinol

Methyl isobutyl ketone (Hexone)
Methyl methacrylate
α-Methyl styrene
Methylal (Dimethoxymethane)
Methylamine
Methylcyclohexane
Methylcyclohexanol
Methylcyclohexanone
4,4'-Methylenebis (2-chloroaniline)
4,4'-Methylenebis (phenyl isocyanate) (MDI)
Methylene chloride (Dichloromethane)
Methylhydrazine
Mevinphos®
MIBK (Hexone)
MOCA
Molybdenum
Molybdenum insoluble compounds
Molybdenum soluble compounds
Monochloroacetic acid
Monochlorobenzene (Chlorobenzene)
Monomethyl aniline
Monomethyl hydrazine
Monomethylarsonic acid
Morpholine

Naphtha, coal tar (Coal-tar naphtha)
Naphthalene
Naphthylamines
α-Naphthyl thiourea
Nickel
Nickel carbonyl
Nickel fume
Nicotine
Nitric acid
Nitric oxide
p-Nitroaniline
Nitrobenzene
4-Nitrobiphenyl
p-Nitrochlorobenzene
Nitroethane
2-Nitropropane
Nitrogen dioxide
Nitroglycerin
Nitroglycol
Nitromethane

N-Nitrosodimethylamine
Nitrotoluene

Octachloronaphthalene
Octane
di-sec-Octyl phthalate (di-2-
  Ethylhexylphthalate)
Oil mist
Organic solvents
Organo(alkyl)mercury
Organoarsenicals
Ozone

Palladium
Paraquat
Parathion
PCBs
Pentachlorobenzene
Pentachloroethane
Pentachloronaphthalene
Pentachlorophenol
Pentane
2-Pentanone
Perchloroethylene (Tetra-
  chloroethylene)
Perylene
Petroleum distillates (Petroleum
  naphtha)
Petroleum naphtha (Petroleum
  distillates)
Phenacylchloride
Phenanthene
Phenol
Phenyl ether
Phenyl ether-biphenyl vapor mixture
  (Dowtherm A)
Phenyl ethylene (Styrene)
Phenyl glycidyl ether
Phenylhydrazine
Phenyloxirane
Phosdrin
Phosgene (Carbonyl chloride)
Phosphate
Phosphine
Phosphoric acid
Phosphorus (white, yellow)
Phosphorus pentachloride
Phosphorous trichloride

Phthalic anhydride
Picric acid
Platinum, soluble salts
PNAs
Polychlorinated biphenyls (PCBs)
Polymethylsiloxane
Polynuclear aromatic hydrocarbons
  (PNAs)
Potassium
Propane
Propanol (Propyl alcohol)
n-Propyl acetate
Propyl alcohol (Propanol)
n-Propyl nitrate
Propylene dichloride
Propylene oxide (1,2-Epoxypropane)
Propyne
Pyrene
Pyrethrum
Pyridine

Quartz (Silica, crystallaine)
Quinone

Refrigerant 11
  (Fluorotrichloromethane)
Refrigerant 113 (1,1,2- trichloro-
  1,2,2-fluoroethane)
Refrigerant 114
  (Dichlorotetrafluoromethane)
Refrigerant 12
  (Dichlorodifluoromethane)
Refrigerant 21
  (Dichloromonofluromethane)
Rhodium
Rhodium, metal fume and dust
Rhodium, soluble salts
Ronnel
Rotenone
Rubidium

Selenium
Sevin (Carbaryl)
Silica, amorphous
Silica, crystallaine
Silicon
Silver
Silver, metal and soluble compounds
Sodium

Sodium dichloroisocyanate dihydrate
Sodium fluoroacetate
Sodium-2,4-dichlorophenoxyethyl sulfate
Sodium hydroxide
Stibine
Stoddard Solvent
Strontium
Strychnine
Styrene (Vinyl benzene)
Styrene oxide
Sulfate
Sulfite
Sulfur dioxide
Sulfur hexafluoride
Sulfuric acid
Sulfuryl fluoride
Systox

2,4,5-T
TDI (Toluene 2,4-diisocyanate)
Tantalum
Tellurium
Tellurium hexafluoride
TEPP
Terphenyl
Tetrabromoethane (Acetylene tetrabromide)
1,1,1,2-Tetrachloro-2,2-difluoroethane
1,1,2,2-Tetrachloro-1,2-difluoroethane (Refrigerant 112)
1,2,4,5-Tetrachlorobenzene
1,1,2,2-Tetrachloroethane
Tetrachloroethylene (Perchloroethylene)
Tetrachloromethane (Carbon tetrachloride)
Tetrachloronaphthalene
Tetraethyl lead
Tetraethyl pyrophosphate
Tetrahydrofuran
Tetramethyl lead (as Pb)
Tetramethyl succinonitrile
Tetramethyl thiourea
Tetramethyl thiuram disulfide (Thiram)
Tetranitromethane

Tetryl (2,4,6-Trinitrophenylmethylnitramine)
Thallium
Thiophene
Thiram(Tetramethyl thiuram disulfide)
Tin
Tin, organic compounds
Tissue preparation
Titanium
Titanium diboride
Titanium dioxide
o-Tolidine based dyes
Toluene
Toluene-2,4-diisocyanate (TDI)
o-Toluidine
Toxaphene (Chlorinated camphene)
Tributyl phosphate
1,1,2-Trichloro-1,2,2-trifluoroethane (Refrigerant 113)
1,2,4-Trichlorobenzene
1,1,1-Trichloroethane (Methyl chloroform)
1,1,2-Trichloroethane
Trichloroethylene
Trichloroisocyanuric acid
Trichloromonofluoromethane (Fluorotrichloromethane)
Trichloronaphthalene
2,4,5-Trichlorophenoxyacetic acid and salts
1,2,3-Trichloropropane
1,3,5-Trichloro-s-triazine-2,4,6-trione
Tridymite
Triethylamine
Trifluoromonobromomethane
Trimellitic anhydride
2,4,7-Trinitro-9-fluorenone
2,4,6-Trinitrophenylmethylnitramine (Tetryl)
Triorthocresyl phosphate
Triphenyl phosphate
Tungsten
Turpentine

Vanadium
Vanadium, V$_2$O$_5$ fume

Vinyl acetate
Vinyl benzene (Styrene)
Vinyl bromide
Vinyl chloride
Vinyl toluene
Vinylidene chloride
Warfarin
Xylene (Xylol or Dimethyl Benzene)
Xylidine (2,4-Dimethylamino-benzene)
Xylol (Xylene)
Yttrium
Zinc
Zinc fume
Zinc oxide
Zirconium compounds
Zirconium oxide

Appendix 7.3 HSE Methods for the Determination of Hazardous Substances

MDHS 1 Acrylonitrile in air
Laboratory method using charcoal adsorption tubes and gas chromatography

MDHS 2 Acrylonitrile in air
Laboratory method using porous polymer adsorption tubes, and thermal desorption with gas chromatographic analysis

MDHS 3 Generation of test atmospheres of organic vapours by the syinge injection technique
Portable apparatus for laboratory and field use

MDHS 4 Generation of test atmospheres of organic vapours by the permeation tube method
Apparatus for laboratory use

MDHS 5 On-site validation of sampling methods

MDHS 6 Lead and inorganic compounds of lead in air
Laboratory method using atomic absorption spectrometry

MDHS 7 Lead and inorganic compounds of lead in air
Laboratory method using X-ray fluorescence spectrometry

MDHS 8 Lead and inorganic compounds of lead in air
Colorimetric field method using sym-diphenylthio-carbazone (dithizone)

MDHS 9 Tetra alkyl lead compounds in air
Personal monitoring method

MDHS 10 Cadmium and inorganic compounds of cadmium in air
Laboratory method using atomic absorption spectrometry

MDHS 11 Cadium and inorganic compounds of cadmium in air
Laboratory method using X-ray fluorescence spectrometry

MDHS 12 Chromium and inorganic compounds of chromium in air
Laboratory method using atomic absorption spectrometry

MDHS 13 Chromium and inorganic compounds of chromium in air
Laboratory method using X-ray fluorescence spectrometry

MDHS 14 General method for the gravimetric determination of respirable and total dust

MDHS 15 Carbon disulphide

MDHS 16 Mercury vapour in air
Laboratory method using hopcalite adsorbent tubes, and acid dissolution with cold vapour atomic absorption spectrometric analysis

MDHS 17 Benzene in air
Laboratory method using charcoal adsorbent tubes, solvent desorption and gas chromatography

MDHS 18 Tetra alkyl lead compounds in air
Continous on-site monitoring method using PAC
Check *atomic* absorption spectrometry

MDHS 19 Formaldehyde in air
Colorimetric field method using 4,5-dihyroxy-2,7-naphthalenedisulphonic acid

MDHS 20 Styrene in air
Laboratory method using charcoal adsorbent tubes, solvent desorption and gas chromatography

MDHS 21 Glycol ether and glycol ether vapours in air
Laboratory method using charcoal adsorbent tubes, solvent desorption and gas chromatography

MDHS 22 Benzene in air
Laboratory method using porous polymer adsorbent tubes, thermal desorption and gas chromatography

MDHS 23 Glycol ether and glycol ether vapours in air
Laboratory method using Tenax adsorbent tubes, thermal desorption and gas chromatography

MDHS 24 Vinyl chloride in air
Laboratory method using charcoal adsorbent tubes, solvent desorption and gas chromatography

MDHS 25 Organic isocyanates in air
Laboratory method using 1-(2-methoxyphenyl) piperazine solution and high-performance liquid chromatography

MDHS 26 Ethylene oxide in air
Laboratory method using charcoal adsorbent tubes, solvent desorption and gas chromatography

MDHS 27 Protocol for assessing the performance of a diffusive sampler

MDHS 28 Chlorinated hydrocarbon solvent vapours in air
Laboratory methods using charcoal adsorbent tubes, solvent desorption and gas chromatography

MDHS 29 Beryllium and inorganic compounds of beryllium in air
Laboratory method using atomic absorption spectrometry

MDHS 30 Cobalt and inorganic compounds of cobalt in air
Laboratory method using atomic absorption spectrometry

MDHS 31 Styrene in air
Laboratory method using porous polymer adsorbent tubes, thermal desorption and gas chromatography

MDHS 32 Dioctyl phthalates in air
Laboratory method using Tenax adsorbent tubes, solvent desorption and gas chromatography

MDHS 33 Adsorbent tube standards
Preparation by the syringe loading technique

MDHS 34 Arsine in air
Colorimetric field method using silver diethyl-dithiocarbamate in the presence of excess silver nitrate

MDHS 35 Hydrogen fluoride and inorganic fluorides in air
Laboratory method using an ion selective electrode

Appendix 7.4 A selection of relevant British Standards

BS 893   Methods for the measurement of the concentration of particulate material in ducts carrying gases.

BS 1747   Methods for the measurement of air pollution.
- Pt 1 – Deposit gauges.
- Pt 2 – Determination of concentration of suspended matter.
- Pt 3 – Determination of sulphur dioxide.
- Pt 4 – The lead dioxide method.
- Pt 5 – Directional dust gauges.
- Pt 6 – Sampling equipment used for the determination gaseous sulphur compounds in ambient air.
- Pt 7 – Determination of mass concentration of sulphur dioxide in ambient air. Thorin spectrophotometric method.

BS 1756   Methods for the sampling and analysis of flue gases.
- Pt 1 – Methods of sampling.
- Pt 2 – Analysis by the Orsat apparatus.
- Pt 3 – Analysis by the Haldane apparatus.
- Pt 4 – Miscellaneous analysis.
- Pt 5 – Semi-routine analysis.

BS 3405   Method for the measurement of particulate emission including grit and dust (simplified method).

BS 4947   Specification for test gases for gas appliances.

BS 5243   General principles for sampling airborne radioactive materials.

BS 5343   Specification for gas-detection tubes.

BS 6020   Instruments for the detection of combustible gases.
- Pt 1 – Specification for general requirements and test methods.
- Pt 2 – Specification for safety and performance requirements for Group I instruments reading up to 5% methane in air.
- Pt 3 – Specification for safety and performance requirements for Group I instruments reading up to 100% methane.
- Pt 4 – Specification for performance requirements for Group II instruments reading up to 100% lower explosive limit.
- Pt 5 – Specification for performance requirements for Group II instruments reading up to 100% gas.

Drafts for Development

(DD 54)   – Methods for the sampling and analysis of fume from welding and allied processes.
- Pt 1 – Particulate matter.
- Pt 2 – Gases.

CHAPTER 8

Radioactive chemicals

This chapter provides an introduction to the safe handling of sources of ionising radiation in industry. An understanding of the nature of radiation and its biological effects is an essential prerequisite to devising appropriate control measures and, therefore, these are summarised first.

Only the elements of the subject are covered and the list of references at the end of the chapter should be consulted for detailed information.

Ionising radiation

Energy transmitted, emitted or absorbed in the form of waves or particles is termed 'radiation'. The electromagnetic spectrum describes the different forms of radiation according to the wavelength as illustrated by Fig. 2.4. Radiation can be ionising or non-ionising. The latter includes noise, heat and light; the ionising forms represent that radiation capable of producing ions directly or indirectly by interaction with matter, including human tissue. It is this property which renders exposure to such radiation potentially hazardous.

Exposures to ionising radiation arise from both natural and man-made sources. Natural ionising radiation stems from cosmic rays and radioactive nucleides such as potassium-40, carbon-14 and isotopes of thorium and uranium. The latter appear in rocks, earth and building materials such as bricks. Industrial or medical uses of potential sources of ionising radiation include nuclear reactors, high-energy particle accelerators, X-ray tubes, X-ray diffractometers, X-ray radiographic and fluorescent equipment, and some level gauges, guides for moving tools, electron microscopes, high-voltage rectifiers, static eliminators, welding, polymer curing (cross-linking), chemical analysis, chemical and biological tracers, sterilisation of food or medical supplies, cardiac pacemakers, etc. Significant exposures may also result from mining radioactive ores, mainly from short-lived solid decay products and airborne radon gas.

Atomic structure

All matter is composed of chemical elements of which 92 are naturally occurring and about 12 man-made. Elements can be linked in definite proportions to produce chemical compounds. Each element is composed of atoms which comprise a central nucleus containing protons (which are positively charged) and neutrons (which are electrically neutral) and around which orbit negatively charged electrons.

The radius of a nucleus is of the order of 10^{-14} to 10^{-15} m and its density is about 10^{17} kg/m^3. In fact almost the whole of the mass of the atom is concentrated in this minute nucleus which only occupies about 10^{-4} or 10^{-5} times the volume of the atom (one of the principal reasons for 'rarity' of nuclear collision processes). The remainder of the volume of the atom is thinly populated by the relatively light orbital electrons.

Chlorine atom (Cl)   *Chloride ion (Cl$^-$)*

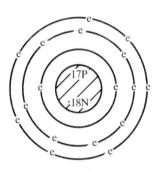 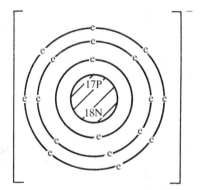

Sodium atom (Na)   *Sodium ion (Na$^+$)*

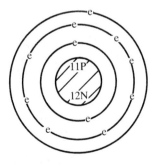 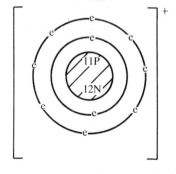

P = protons
N = neutrons } nucleus
e = electrons

Fig. 8.1 Examples of atoms and ions

446   *The Safe Handling of Chemicals in Industry*

It is the latter which are primarily of interest to the chemist since they determine chemical properties and hence the different forms of elemental matter.

In their normal state atoms contain an equal number of protons and electrons so that the atom as a whole is electrically neutral. When atoms lose electrons they become positively charged and when electrons are gained the atom becomes negatively charged as illustrated by Fig. 8.1. These charged particles are termed ions.

The number of protons in the nucleus is called the atomic number (Z). Since each proton carries one unit of positive charge, Z equals the total positive charge of the nucleus and also (in an electrically neutral, i.e. non-ionised atom) the number of orbital electrons. If N is the number of neutrons in the nucleus, then the mass number A of the nucleus is defined as $Z + N$. Since the masses of the proton and the neutron are each approximately one, A is taken as the whole number nearest to the atomic weight of the element. Thus $A = Z+N$. Atoms which have the same atomic number, Z, have the same chemical properties and belong to the same element. The mass of an atom is determined by $Z + N$ and if N is changed the result will be to change the mass (or atomic weight) of the element but *not* its chemical properties. Thus atoms of the same element having different mass numbers may occur. These are called isotopes or nucleides. Examples are given in Table 8.1 (by convention isotopes are

Table 8.1 Nuclear composition of isotopes

Element	Symbol	Atomic number	Protons	Neutrons	Total number of protons and neutrons	Atomic weight
Hydrogen	$^{1}_{1}H$	1	1		1	1.0080
(Deuterium)	$^{2}_{1}H$	1	1	1	2	
Carbon	$^{12}_{6}C$	6	6	6	12	12.010
	$^{13}_{6}C$	6	6	7	13	
	$^{14}_{6}C$	6	6	8	14	
Nitrogen	$^{13}_{7}N$	7	7	6	13	14.008
	$^{14}_{7}N$	7	7	7	14	
	$^{15}_{7}N$	7	7	8	15	
Chlorine	$^{35}_{17}Cl$	17	17	18	35	35.457
	$^{37}_{17}Cl$	17	17	20	37	
Lead	$^{206}_{82}Pb$	82	82	124	206	207.21
	$^{207}_{82}Pb$	82	82	125	207	
	$^{208}_{82}Pb$	82	82	126	208	
Uranium	$^{234}_{92}U$	92	92	142	234	238.07
	$^{235}_{92}U$	92	92	143	235	
	$^{238}_{92}U$	92	92	146	238	

identified by the atomic number as subscript to the symbol and the atomic weight as superscript).

Many isotopes are completely stable but others are radioactive. Radioactivity is characterised by the emission of particles or radiation from the atomic nucleus, thus changing the nuclear structure of the atom which often results in the production of a different element.

Types of radiation

Five categories of ionising radiation occur industrially, namely alpha, beta and gamma radiation (as emitted by radioactive chemicals), X-radiation (as generated by electrical devices) and neutron radiation (as emitted in fission). Each form of radiation possesses its own characteristic and penetrating power, thus affecting the human body to a differing extent and necessitating a different type of shielding material.

Alpha particles or rays

These are emitted mainly by radio-isotopes of the heavier elements and comprise a stream of α-particles which consist of helium nuclei (i.e. 2 neutrons plus 2 protons) and so carry a double positive charge. On emission of α-radiation the original isotope, is transformed into an isotope of an element two atomic numbers less. For example, uranium-238 is transformed to thorium-234 on emission of α-particles.

Similarly, plutonium-239 gives rise to uranium-235 and polonium-218 decays to lead-214.

These are expressed thus:

$$^{238}_{92}U \xrightarrow{\alpha} {}^{234}_{92}Th$$

$$^{239}_{94}Pu \xrightarrow{\alpha} {}^{235}_{92}U$$

$$^{218}_{84}Po \xrightarrow{\alpha} {}^{214}_{82}Pb$$

Transformations involving α-radiation are usually accompanied by either γ-radiation or X-radiation.

The velocity of α-particles is about one-tenth that of light and they have a range in air of from 3 to 9 cm depending on the emitter. They do not penetrate matter very readily because of their relatively large particle size and their double positive charge: radiation is stopped completely by paper, cellophane, aluminium foil and even the skin. However, α-emitters may be inhaled, ingested, or absorbed through the skin and absorption of the α-particles within tissues then causes intense local ionisation. Use is made of the strong ionising power in the detection and counting of α-particles.

Beta particles or rays

These consist of negative electrons. They are very much lighter than α-particles and move with a velocity approaching that of light; their maximum range depends on the energy characteristics of the emission but can be of the order of several metres in air.

The energy of β-particles ranges from 0 to 4 MeV (see page 450). β-Particles of energy less than 0.07 MeV do not penetrate the epidermis whereas those of energy 2.5 MeV penetrate about 12.5 mm of soft tissue. Thus β-particles have considerably more penetrating power than α-particles and about 1 mm of aluminium is required to stop these more energetic particles. The rapid change of momentum associated with their passage through dense materials can give rise to Bremsstrahlung (or braking)-radiation which is in effect very penetrating electromagnetic radiation similar to γ-radiation (see later). Most transformations emitting β-radiation are accompanied by γ-radiation or X-radiation. On emission of β-radiation the original isotope is transformed into an isotope of the element of one atomic number either higher or lower, but with the same atomic mass. For example:

$$^{14}_{6}C \longrightarrow ^{14}_{7}N + e$$

$$^{30}_{15}P \longrightarrow ^{30}_{14}Si + e$$

Gamma (γ) rays

Many radio elements emit γ-radiation. This is an electromagnetic radiation similar to, but shorter in wavelength than, X-rays. (It is customary to use the term γ-rays when the radiation originates within the nucleus and X-rays when it is formed outside.)

No 'gamma only' emitters exist and γ-radiation is associated with many α- or β-radiations, each radioisotope emitting γ-radiation of specific energy. For example the energy of γ-radiation from potassium-42 is four times that possessed by γ-radiation from gold-198. Gamma radiation does not transform the nucleus into that of another isotope or element. For all practical purposes γ-rays are the same as X-rays; they are very penetrating and require the same kind of heavy shields. Thus, the γ-rays from ^{60}Co will readily penetrate 15 cm of steel. On the other hand its ionising power is considerably less than that of β-radiation. Unlike X-radiation, however, γ-radiation emits radiation continuously and cannot be switched off.

Absorbing material does not completely stop γ-rays but only reduces their intensity. From a radium source, γ-radiation is reduced to 50% by 1.3 cm of lead, to 25% by 2.5 cm, and to 12.5% by 3.8 cm. Thus efficiency of absorbers for γ-radiation, is expressed in half-thicknesses.

X-radiation

Like γ-radiation, X-radiation is electromagnetic in nature. It can be emitted when β-particles react with atoms of other matter. More often, however, X-rays are electrically generated by accelerating electrons in a vacuum tube and are therefore capable of being 'switched-off'. Though similar to γ-rays the energy of the X-rays can be much higher, as dictated by the generating voltage. As with γ-radiation, X-rays are extremely penetrating, the radiation merely being attenuated by distance or shielding.

Neutron radiation

This consists of a stream of neutrons. Radionucleides do not emit neutrons spontaneously, although in a small number of very heavy radionucleides such as plutonium the nucleus undergoes spontaneous fission (i.e. splits into two fragments) with emission of neutrons. Usually neutrons are emitted by bombarding beryllium atoms with an α-emitter such as radium-226, or polonium-210, or with a high-energy γ-emitter like antimony-124. Neutron radiation is also obtained using commercial generators. Neutrons decay into a proton and an electron with a half-life of 11.7 minutes. Because neutron radiation is particulate it eventually dissipates all its energy but, because it is uncharged, it is extremely penetrating.

Data on all isotopes are published in summary version in the *Radio-chemical Manual* which includes the symbol, mass number, half-life, modes of decay, energy of emitted radiation, etc.

Rates of radiodecay

Of the 1,300 or so different nucleides about 1,000 are unstable (radio-active), some of which are naturally occurring. Some decay products are themselves radioactive (e.g. thorium-234 from uranium-238) while others are not (e.g. argon-40 from potassium-40). The rate of isotope decay is exponential according to the equation

$$N = N_o e^{-\lambda t}$$

where N_o is the number of nuclei present initially, N represents the number of nuclei present at time t, and λ is the radioactive decay constant. This relationship is illustrated by Fig. 8.2.

The time taken for half of the nuclei of a radioactive species in a sample to decay is known as the half-life, ($T_{\frac{1}{2}}$). In two half-lives the activity drops to a quarter, etc. The half-life relationship is $T_{\frac{1}{2}} =$

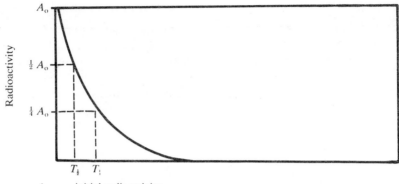

A_o = initial radioactivity
$T_\frac{1}{2}$ = time for activity to drop to half initial level

Fig. 8.2 Rate of radioactive decay

$0.693/\lambda$. Depending upon the isotope, half-lives range from billions of years to billionths of a second.

The energy of radiation, which is an indication of its quality or penetrating power, is measured in electron volts, eV. This is so small that it is normally expressed in kilo electron volts (KeV) or mega electron volts (MeV).

Units of radiodecay and radiological protection

The rate of disintegration for a given nucleide is constant and unaffected by temperature. Traditionally, radioactivity is measured in curies (Ci) or more commonly millicuries (mCi) or microcuries (μCi).

$$1 \text{ Ci} = 3.7 \times 10^{10} \text{ transformations per second or } 3.7 \times 10^{10} \text{ Bq}$$

The SI unit of activity is the becquerel (Bq)

$$1 \text{ Bq} = 2.703 \times 10^{-11} \text{ Ci}$$

Each transformation may emit α- and γ- or β- and γ-radiation, all of different energies, and therefore the quantity is no guide to radiation dose. The units of radiation dose are the roentgen, the rad and the rem. For X-rays and γ-rays (but not α-, β- or neutron radiation) the unit is the roentgen (R). This corresponds to the production of ions carrying a charge of 2.58×10^{-4} coulombs per kilogram, equivalent to 87.7 ergs per gram, i.e. this is essentially a measure of output. However, the amount of energy absorbed in an exposure of 1 R is influenced by the nature of the absorbing material and the energy of radiation. For this reason the 'rad' is used to describe the energy (from all types of ionising radiation) actually absorbed in any medium.

$$1 \text{ rad} = 100 \text{ ergs per gram}$$

Since the deposition of energy in air and on human tissue is very similar, the roentgen and rad are considered numerically equivalent when assessing radiological protection.

Different types of radiation inflict different degrees of biological damage, even at the same absorbed dose: heavier particles produce more biological damage for a given dose than do X-, γ- or β-rays. In assessing the total risk from exposure to a mixture of radiations this quality difference must be accounted for by application of a quality factor (QF) the unit of such dose equivalents being the rem.

It is often necessary to know the rate at which dose is experienced and this is expressed as rem per hour, R per hour or rad per hour.

SI units of radiation dose are the gray (Gy) and the sievert (Sv) where 1 Gy = 100 rad and 1 Sv = 100 rem.

An operator in a year receives a dose of 1.5 rad of γ-radiation plus 0.2 rad of α-radiation. The QF for γ-rays is 1 and the QF for α-rays is 10. This total dose equivalent for the year is computed thus

γ dose equivalent is 1.5 × 1	= 1.5 rem
α dose equivalent is 0.2 × 10	= 2.0 rem
Total dose equivalent	= 3.5 rem
	= i.e. 0.035 Sv

Biological effects of ionising radiation

The effects of radiation on the body are influenced by the dose, duration of exposure, exposed organ and route of entry. Sealed sources are those which are so contained or bonded wholly with material, so as to remain intact. These sources ordinarily can only irradiate the body externally, and their biological effects depend upon their ability to penetrate the skin. Unsealed sources such as radioactive dusts, aerosols, liquids, gases and vapours, on the other hand, could enter the body by ingestion, inhalation or skin absorption. These pose both an external and internal radiation hazard. Once in the body the radio isotopes will irradiate tissues and organs continuously and will be metabolised in the same way, and possess the same target organ specificity as their 'cold' counterparts as described in Chapter 5. Thus some materials would distribute themselves uniformly throughout various parts of the body while the majority concentrate in specific organs (e.g. iodine in the thyroid, and plutonium in the bone). Thus α-particles are stopped completely by the skin and therefore present no external radiation hazard. Once in the body the protection barrier no longer exists and intense ionisation can result. β-emitters can be both internal and external hazards since the most energetic can penetrate at least the outer layers of skin. Skin burns and malignancies can result. Once inside the body they are extremely hazardous though less ionising than α-rays. High-energy X- and γ-radiation can penetrate the whole body.

In the main the biological effects of ionising radiation stem from damage to individual cells following ionisation of the water content. Oxidising agents such as hydrogen peroxide form together with ions and reactive species known as free-radicals. These all attack key organic molecules within the cell such as nucleic acids. The outcome may be early cell death, prevention or delay of cell division, or permanent modification capable of being passed on to daughter cells. Thus effects can be somatic or genetic.

Somatic effects

These effects relate to damage to ordinary cells and affect the individual alone. The most rapidly dividing cells such as lymphocytes and granulocytes (see Ch. 5) are most susceptible. Acute effects are fortunately rare. The symptoms arising from acute exposures to 400–600 rem are summarised in Table 8.2.

Table 8.2 Effects of acute exposure to ionising radiation

Time after exposure to 400–600 rems	Symptoms
0–48 hr	Loss of appetite, nausea, vomiting, fatigue and prostration.
2 days to 2–3 weeks	The above disappear and the patient appears well.
2–3 weeks to 6–8 weeks	Purpura and haemorrhage, diarrhoea, loss of hair, fever and severe lethargy. It is during this period that fatalities occur.
6–8 weeks	The recovery stage in which the patient shows general recovery and severe symptoms begin to subside.

When an adult male is acutely and uniformly irradiated with a few hundred rads of X- or γ-rays the effects noticed within a few weeks relate to damaged stem cells in the bone marrow. Those cells recovering will divide until the cell population returns to a normal level. The intervening period, however, is critical for the individual's survival.

> A male technologist walked past an unshielded cobalt-60 source of 7,700 curies. One hour after the accidental exposure to γ-radiation he was sent to hospital. His exposure was ascertained from his dosimeter to be 260 roentgens. It was believed that he had been exposed for about 40 seconds with an estimated mid-line dose of 127 rad and that dose for the marrow was 118 rad. His hands had received between 500 and 1,200 rad. At $2\frac{1}{2}$ hr after exposure he began vomiting. This recurred ten times during the next 24 hr. After this time he complained of itching and burning of his eyes and reddening was evident. These signs disappeared within a further 24 hr. After 7 days he returned home with daily visits to the hospital for blood counts. On the 25th day post-exposure he was readmitted to hospital to reduce bacteriological contamination. Haematological depression was

observed between the 25th and 34th day. By day 48 his blood condition had fully recovered. Three days after exposure he developed a dull aching sensation in the fingers and palm of his hand. Sometimes the pain woke him at night. This pain, became intermittent and eventually disappeared. For about 4 months after the exposure he was easily fatigued but eventually made a full recovery.

The acute effects of different doses of X- and γ-radiation are summarised in Table 8.3.

Chronic effects include loss of hair, fibrosis of the lungs and connective tissue, skin damage and increased pigmentation of the skin, and cataract formation with opacity occurring on the posterior surface of the lens. For radiological protection purposes an annual dose of 15 rem per year over a 50-year working lifetime (i.e. 750 rem) is believed unlikely to produce lens opacification that would impair vision. Other tissues susceptible to chronic effects include the embryo and foetus, the gonads, blood and thyroid.

A lecturer gave a series of evening talks on X-rays and on each occasion exposed his hands to the rays for several hours. Two or three weeks later the skin around the roots of the nail was red and painful. This effect disappeared only to be replaced by transverse lines on the nails. The latter became brittle, and broke and degenerated, and eventually fell off. Warty growths developed. Two or three years later he began to experience pain. Sores developed over a finger which became extremely painful and sensitive to touch. The warts continued to grow and cracks formed which were painful and resistant to healing. Pain developed in his arm and he noted a loss of power.

Genetic effects

The possible consequence of irradiating germ cells is cell death, chromosome damage and gene mutation which could manifest itself by hereditary disease.

Table 8.3 Effects of acute exposures to ionising radiation

Dose (Gy)	Effects
Below 1	No clinical effects but small depletions in normal white cells count and in platelets likely within 2 days.
1	About 15% of those exposed show symptoms indicated in Table 8.2. 1 Gy delivered to whole body or 5 Gy delivered to bone marrow produces leukaemia.
2	Some fatalities occur.
3.5–4	LD_{50} (see Ch. 5) death occurring within 30 days. Erythema (reddening of skin) within 3 weeks.
7–10	LD_{100} death occurring within 10 days.

Maximum permitted dose

All radiations are potentially harmful and exposures must be kept as low as possible. Furthermore, maximum permitted doses (MPDs) have been defined for different body organs. Current standards are based on specific recommendations of the International Commission on Radiological Protection (ICRP). For occupationally exposed persons these are:

Organ	Rem in a year
Gonads and red bone marrow (and in the case of uniform exposure, the whole body)	5 rem
Skin, thyroid, bone	30 rem
Hands, forearm, feet and ankles	75 rem
Other single organs	15 rem

Up to one-half of these maxima may be allowed in a quarter-year and in exceptional cases half the whole body, gonad, red bone marrow dose may be allowed in consecutive quarters as long as $5(N-18)$ rem, where N is the age of the person exposed, is not exceeded.

In order to allow adequate control of radiation hazards derived working limits have been suggested for, e.g. adequate shielding. This is defined as shielding or demarcation barriers outside which the dose rate received by designated or classified workers does not exceed 2.5 mR/hr. For other workers the dose rate must not exceed 0.75 mR/hr.

National legislation must be consulted for interpretation.

In the UK, MPDs are at present included in various schedules to the Ionising Radiations (Unsealed Radioactive Substances) Regulations of 1968 and the Ionising Radiations (Sealed Sources) Regulations of 1969, though such values must be read in conjunction with the appropriate sections of the Regulations. Furthermore, changes are likely as a consequence of an EEC Directive. The HSE also proposes introduction of new Regulations, an approved Code of Practice and Guidance Notes in the near future. For unsealed sources which constitute internal hazards maximum permissible annual intake values have been derived for specific nucleides and different routes of entry.

Radiation monitors and maximum levels of contamination

Radiation monitoring may be required to check the following:
1. *Surface contamination* The aims here include preventing spread of contaminant, detecting failures in containment or departures from good practice, to restrict levels as low as reasonably achievable and

avoid exposures approaching specified limits, and to provide data needed for planning further monitoring programmes.
2. *Air contamination* Routine monitoring of ambient air will be required:
   (a) when volatiles are handled in quantity (e.g. tritium and its compounds in large-scale production);
   (b) for the handling of any radioactive isotope in conditions of frequent and substantial workplace contamination;
   (c) for the processing of plutonium and other transuranic elements;
   (d) for the handling of unsealed radionucleides in hospitals in therapeutic amounts, the use of hot cells, and reactors, and critical facilities.

Individual monitoring

Monitoring requirements for external radiation will be dictated by classification of the work environment (as defined by the ICRP) and on circumstances including accidental exposure. Monitoring of skin routinely may also be required, notably the hands. If simple precautionary measures are followed and if the radioactive source is never held in bare hands, then routine monitoring of individuals for external radiation is unlikely to be required for sources below specified levels.

Also, if it can be established by a survey that operating procedures are of high standard, routine individual monitoring is unlikely to be necessary for:
(a) non-radiological work in departments dealing with X-rays and γ-rays;
(b) dental radiography;
(c) industrial processes involving radiological control or measurement.

At the other extreme, regular monitoring of workers for external and internal contamination may be needed in certain circumstances. For example, from experience, routine individual monitoring for internal contamination will be required:
(a) when handling large quantities of volatiles (e.g. tritium and its compounds in large-scale production, in luminising, and in heavy-water reactors);
(b) in natural and enriched uranium processes;
(c) in the processing of plutonium and other transuranic elements;
(d) in uranium milling and refining;
(e) in production of large quantities of radionucleides.

The nature of biological monitoring depends upon the type of isotope inside the body, e.g. whole body monitoring for γ-emitters, and faeces, urine or breath monitoring for α- or β-emitters. Many devices are available for monitoring ionising radiation ranging from portable to fixed installations. These are used to determine background levels of radiation or to assess employee exposure.

Table 8.4 Maximum permissible levels of contamination and methods of assessment

Category	Surface	Max permissible level ($\mu Ci/cm^2$)		
		From alpha emitters		Other emitters
		In Class I*	In Class II–IV*	
A	Surfaces of the interiors and contents of total enclosures and fume-cupboards	The minimum that is reasonably practicable		
B	Surfaces (other than those above) of active areas and plant apparatus, equipment (including personal protection), materials and articles within active areas.	10^{-4}	10^{-3}	10^{-3}
C	Surfaces of the body	10^{-5}	10^{-5}	10^{-4}
D	All other surfaces	10^{-5}	10^{-4}	10^{-4}

* As defined is Schedule 3 of The Ionising Radiations (Unsealed) Radioactive Regulations 1968.

Excessive background radiation may result from spillages, leaks and general contamination. In the UK maximum permissible levels of contamination of surfaces (other than contamination which cannot be removed by normal methods) are shown in Table 8.4. Employee exposure limits were discussed on page 454.

The main types of instrument used for measuring background intensity of ionising radiation are the scintillation counter and the Geiger–Muller counter. The former consists of a screen which, when bombarded with α-, β-, γ-, and, or neutron radiation, emits flashes of light which in turn are converted (and amplified) into electric current. The Geiger–Muller counter consists of a tube filled with gas and containing a positively charged wire; the tube wall comprises the negative electrode. Electric current is generated (and amplified) when α-, β-, or γ-rays entering the tube cause ionisation of the gas. In both cases the magnitude of the electric current is proportional to the radioactive intensity.

For monitoring airborne radioactive particulate matter the dust (or aerosol) is collected by drawing air through a filter paper using a pump (see Ch. 7).

The filter paper is then counted in a low background area and the level of airborne activity (A) is calculated thus:

$$A = C_c \times \frac{100}{E_c} \times \frac{1}{F} \times \frac{1}{v}$$

where C_c = counts per minute
E_c = overall efficiency of counting system (as percentage)
F = conversion factor from dis/min to μCi, i.e. 2.22×10^6
v = volume of sampled air in cm^3.

The two main methods used for monitoring employee dose are film badges or thermoluminescent dosimeters (TLDs). Film badges consist of a sensitive photographic film housed in a plastic case equipped with windows of various materials to shield selectively certain radiations yet permit others to penetrate and affect the film, in much the same way as does visible light. The badge is worn by the operator for a week or so before being developed to ascertain accumulated dose. TLDs are also personal dosimeters but operate on a different principle. These contain a special crystal (e.g. lithium fluoride) which is excited on irradiation and remains so until heated when it emits light in amounts proportional to the degree of irradiation.

More detailed advice on sampling programmes, data interpretation and monitoring hardware can be sought in the references at the end of this chapter.

Control measures

Sealed sources constitute a potential radiation risk while unsealed sources such as radioactive chemicals pose an additional internal hazard. Details of precautions necessary for working with ionising radiations are beyond the scope of this text and cognisance of local legislation is crucial. Nevertheless, general measures applicable to minimising risk from radiation can be identified. For example, in the UK the enforcing authorities require prior notification of proposed work with ionising radiation and licensing of the premises. Classified workers must be at least 18 years of age, adequately trained, medically screened, wear personal dosimeters, and undergo periodic medical examinations and possibly biological monitoring. Appointment of a Radiation Protection Officer may be mandatory. Records must be kept of names of classified workers and their exposure details, of all isotopes entering the area and leaving, of any accidents/spillages or losses of radioactivity, of disposal of waste, of details and tests of all instruments used for monitoring, etc. Personal protective equipment may be required.

> A construction firm was to install two radioactive sources of caesium-137 as charge-level indicators in a foundry cupola. The isotopes arrived before the cupola was complete and were temporarily stored in a workman's hut among chains, wire, rope and tools. The sealed source holders indicated that each isotope contained 20 millicuries of caesium-137. The HSE had not been notified of the movement of the isotopes, the company was not licensed to store them, and the sources were not properly stored in a securely locked place. In addition no competent person had been appointed for the site and the cabinet in which the sources were held did not have a notice affixed in a prominent position indicating that radioactive sources were being kept there. The company was prosecuted and fined £2,250.

> A large company making circuit boards had a Betascope thickness gauge installed. Although it contained a low-power β-emitter it was possible for the

hands of a user to be irradiated. Furthermore, a 16-year-old youth was among the users. The company was prosecuted for failing to appoint a competent user, for not providing proper storage arrangements, for contraventions of Regulations relating to classified workers (e.g. medical examinations, monitoring) and for the fact that no person under 18 may be so appointed, for not ensuring adequate construction and maintenance of sealed sources and for not keeping a record of the source. Fines of £900 were imposed.

Sealed sources

In the UK the Sealed Sources Regulations are applicable to factories containing:
- Sealed sources at which the dose rate at the surface is greater than 10 mrem/hour.
- Machines used to produce ionising radiation.
- Machines producing adventitious radiation by operating at above 5 kV.

Radiation from sealed sources is usually less intense than from X-ray machines. Often the source is built into a piece of equipment such that radiation can be switched off. Attention to interlocks and 'fail safe' design features is important. Details of individual sources must be logged and audited at set periods to ensure satisfactory sealing. Leaking sources must be put into leakproof containers immediately and returned to the supplier for repair. In such circumstances checks must be instituted to determine the extent of contamination and, if necessary, decontamination undertaken. Since loss of sealed sources can be serious, particularly those containing hazardous isotopes such as ^{90}Sr, regular administrative checks are necessary and appropriate equipment must be available to aid in their detection and recovery.

Exposure is minimised by choice of source and by controlling duration of exposure, distance from source and shielding. Since cumulative dose is directly related to time, duration of exposure must be minimised and avoidable exposure eliminated.

Since radiation level from a point source varies inversely with the square of the distance it is important to maximise its distance from the operator; hence the reason for using tongs when handling radioisotopes. At a distance of 1 m the radiation level is reduced by a factor of almost 10.

> A qualified radiographer was in charge of work involving use of a radioactive isotope to take X-ray pictures of recently completed welding work. When the factory inspectors visited the new chemical plant the level of radiation was 20 millirems per hour. The acceptable level of radiation for the general public is 0.75 millirems per hour. Barriers which should have been 70 metres from the source were only 10 metres away. The radiographer was prosecuted under the Regulations and fined £200.

Table 8.5 Approximate HVLs for a selection of shield material and γ-emitters

Nucleide	Half-life	γ-energy (MeV)	Concrete	HVL (cm) Steel	Lead
^{137}Cs	27 years	0.66	4.82	1.63	0.65
^{60}Co	5.24 years	1.17–1.33	6.68	2.08	1.20
^{198}Au	2.7 days	0.41	4.06	–	0.33
^{192}Ir	74 days	0.13–1.06	4.32	1.27	0.60
^{226}Ra	1,622 years	0.047–2.4	6.86	2.24	1.66

Shielding the operator from the source is a valuable means of reducing exposure. The greater the mass per unit area of shield material the less radiation the operator is exposed to. α-radiation presents few problems since it is effectively absorbed by paper or dead outer layers of skin. β-radiation normally requires up to 1 cm of perspex for complete absorption; γ- and X-rays are not completely absorbed by shield material but are attenuated exponentially such that the dose rate of radiation emerging from the shield is

$$D_t = D_0 e^{-\mu t}$$

where D_0 is the dose without shielding, D_t is the dose rate emerging from shield of thickness t and μ is the linear absorption coefficient of shield material.

Values of 'half-thickness' are obtainable for different shield materials and represent the thickness required to reduce the intensity to one-half of its incident value. Table 8.5 gives the half-thickness values (sometimes referred to as half-value layer or HVL) for common materials used as shields for selected γ-ray sources.

Unsealed sources

Even small amounts of unsealed sources which represent a minor external radiation hazard can give rise to significant dose rates on contact with, or entry into, the body. Once inside, the source will continue to irradiate organs until the isotope decays or is excreted.

The three fundamental components of a programme for the safe handling of radioisotopes are:
- Minimising the amount of activity in use.
- Containing the source, usually with at least two levels of containment.
- Following agreed protocols with regard to personal protective equipment, personal hygiene, monitoring, good housekeeping, record keeping, etc.

The actual level of precaution required will be dictated by the nature and amount of isotope. This is illustrated by Table 8.6 which summarises one classification scheme for laboratories; the figures quoted for normal wet chemical operations can be adjusted for different procedures as shown in Table 8.7.

Similarly, for low levels of surface contamination a laboratory coat and gloves may be sufficient whereas a fully enclosed suit and a respirator may be necessary for higher levels of activity.

In general, cleanliness and personal hygiene are crucial. Areas need to be classified and, ideally, reserved for radioactive work and preferably restricted to one level of radioactivity. Work should be undertaken in spill trays to contain leakages. Work surfaces should be impervious (e.g. Formica, PVC, stainless steel) and walls and floors should be non-porous. Ventilation should be of high standard and may require in-duct filters to remove contamination prior to exhausting to atmosphere. Fume-

Table 8.6 Laboratory classification scheme

Isotope classification			Laboratory classification		
Class	Definition	Examples	1 Special facilities required	2 High standard as for handling toxic substances	3 Ordinary facilities for handling non-toxic substances
I	High toxicity	^{210}Pb, ^{226}Ra	>10 mCi	<10 mCi	<10 μCi
II	Medium toxicity (upper sub-group)	^{22}Na, ^{90}Sr, ^{124}I	>100 mCi	<100 mCi	<100 μCi
III	Medium toxicity (lower sub-group)	^{14}C, ^{32}P, ^{35}S	>1 Ci	<1 Ci	<1 μCi
IV	Low toxicity	^{3}H, ^{133}Xe, ^{135}Cs	>10 Ci	<10 Ci	<10 mCi

Table 8.7 Adjustment of activity limits for procedures

Procedure	Multiply permissible activity by
Storage of stock solutions in closed containers	×100
Very simple wet chemical operations	×10
Complex wet chemical operations with risk of spills	×0.1
Simple dry operations	×0.1
Dry and dusty operations	×0.01

cupboards should have a minimum air velocity of 0.5 m/second across the face of the working aperture.

Work will often need to be undertaken in enclosed systems, e.g. in glove boxes under slight negative pressure. Shielding may be required. Areas will require barriers for identification and display of appropriate notices, including the internationally recognised Trefoil symbol.

Protective clothing should be removed in a changing area equipped with wash basin, sanitary facilities, lockers for clean and contaminated clothing, warning signs and written emergency procedures.

Eating, drinking, smoking and mouth operations (e.g. pipetting) should be prohibited. Wounds must be dressed before entering the area. Samples must be labelled and display the Trefoil symbol. Apparatus must be decontaminated and precautions are required to prevent cross-contamination. Waste material must be collected for treatment or disposal.

Again local legislation must be consulted for accepted means of transport of radioactive chemicals. These are based on the deliberations of the International Atomic Energy Agency. Details vary according to the mode of transport. A recent detailed survey of all those involved with the normal transport of radioactive materials by road, rail, sea and air in the UK concluded that the annual collective dose to such workers was about 1 man-Sv.

References

(*All places of publication are London unless otherwise stated*)
1. Department of Environment, *Code of Practice for the Carriage of Radioactive Material by Road*. HMSO 1975.
2. Department of Environment, *Code of Practice for the Carriage of Radioactive Material in Transit*. HMSO 1975.
3. Department of Environment, *Code of Practice for the Carriage of Radioactive Materials through Ports*. HMSO 1975.
4. Gelder, R., Hughes, J. S., Mairs, J. H. & Shaw, K. B., *Radiation Exposure Resulting from the Normal Transport of Radioactive Materials Within the UK*. National Radiological Protection Board 1984.
5. International Commission on Radiological Protection, *Recommendations of the International Commission on Radiological Protection* (adopted 17/1/1977) ICRP Publication 26. Published by ICRP through Pergamon Press, Oxford, 1977.
6. Martin, A. & Harbison, S. A., *An Introduction to Radiation Protection*, Chapman and Hall 1976.
7. Bennellick, E. J., 'Ionising radiation', in *Industrial Safety Handbook* (2nd edn). McGraw-Hill 1977.
8. Ionising Radiations (Unsealed Radioactive Substances) Regulations 1968. HMSO.
9. Ionising Radiations (Sealed Sources) Regulations 1969. HMSO.

10. Department of Employment and Productivity, *Ionising Radiations: Precautions for Industrial Users*. Safety, Health and Welfare Booklet No. 13, New Series. HMSO 1969.
11. Department of Health and Social Security, *Code of Practice for the Protection of Persons Against Ionising Radiations Arising from Medical and Dental Use*. HMSO 1972.
12. Health and Safety Executive, *Guidance Notes for the Protection or Persons Exposed to Ionising Radiations in Research and Teaching*. HMSO 1977.
13. HM Factory Inspectorate, *Code of Practice for Site Radiography*. Kluwer-Harrap 1975.
14. National Council on Radiation, *Protection and Measurements from NCRP Report No 34 (1971)*. NCR: Washington DC.
15. *The Radiochemical Manual*. The Radiochemical Centre, Amersham.
16. International Commission on Radiological Protection, 'General principles of monitoring for radiation protection of workers', *Annals of the ICRP*, 1982, **9** (4), 1–36.
17. Atherley, G. R. C., *Occupational Health and Safety Concepts*. Applied Science 1978.
18. *Newcastle Journal*, 27 Jan. 1982.
19. *The Ionising Radiations Regulations 198-*, Consultative Document. HSC 1983.

CHAPTER 9

Safety with chemical engineering operations

Risks arising with chemicals are generally considered as functions of the following:
1. The physical properties of the materials handled, e.g. as in Chapter 3.
2. The flammability and explositivity, toxicology, radioactivity and/or chemical reactivity of the materials (Chs 4, 5, 8 and 10 respectively).

Table 9.1 Unit operations in chemical engineering

(a) *Fluid mechanics*
   Pumping
   Mixing
(b) *Heat transfer*
   Convective heat exchange
   Radiation heat transfer (e.g. in furnaces)
   Evaporation
   Condensation
   Heat transfer to boiling liquids (e.g. in boilers, vaporisers, reboilers)
(c) *Mass transfer operations*
   Distillation (either batchwise or continuous)
   Liquid–liquid extraction
   Gas absorption/desorption (stripping)
   Humidification and water cooling
   Dehumidification and air conditioning
   Drying
   Adsorption
   Leaching
   Crystallisation
   Less conventional means (e.g. dialysis, osmosis, chromatography)
(d) *Non-mass transfer operations*
   Filtration
   Sedimentation
   Centrifugation
   Gas cleaning
   Classification by screening
   Crushing and grinding

Those given in (c) and (d) are ways associated with the separation of components.

3. The scale of operation (e.g. with reference to loss of containment, potential energy release or numbers of persons likely to be exposed).
4. The condition of the surroundings, e.g. the presence of additional hazards, possible 'domino' effects, etc.

However, certain physical operations recur throughout chemical processing and manufacturing; these are referred to as chemical engineering unit operations and are listed in Table 9.1. Although the detailed plant construction may differ, the principles of design are the same irrespective of the particular application. Certain hazards may be associated with these operations and some of these will now be discussed.

In using such guidelines it is important not to overlook associated equipment hazards (e.g. electrical or mechanical as discussed in Ch. 2) or the possible consequences of deviations from normal equipment operation upon upstream or downstream processes. A single unit operation is seldom used in isolation, e.g. liquid–liquid extraction may be associated with distillation or evaporation, or crystallisation may be followed by filtration or centrifugation, so that excursions in parameters may carry through the separation or production chain. (The technique of HAZOP summarised in Ch. 12 is one means of checking for such effects at the process design stage.)

Furthermore, whatever the unit operation any inherent hazards may be increased or lessened, or the effects mitigated, by the attention given to the design considerations outlined in Chapter 13.

De-humidification – water cooling

De-humidification, i.e. air-conditioning operations which involve only the air-water system are generally considered innocuous. However, the blower and exhaust systems may constitute a fire hazard and, unless properly routed or provided with automatic shut-off, ducts may assist fire spread.[1]

Consideration should be given to possible contamination of the air due to contact with water in the system. The relatively recent recognition of the potential hazard of 'Legionnaire's disease', a respiratory tract infection caused by airborne micro-organisms (*Legionella pneumophila*) discharged from air-conditioning and shower units in hospitals, and to a lesser extent in hotels has led to guidelines in the UK for hospital systems.[2]

The recommendations are:
- All hot water must be distributed at a temperature not less than 50 °C.
- Cooling towers, evaporative condensers and humidifiers should be treated with chlorine at 5 ppm twice yearly.
- All storage and distribution of cold water shall be at a temperature near to 20 °C.

The nutritious value of the water, should also be reduced by continuous bleed-off, or regular particle dumping, i.e. eliminating as far as possible sources of organic matter such as excessive dust, grease, algae, etc.[3]

Humidifier fever (hypersensitivity to inhaled matter of microbial origin) including the development of flu-like symptoms but no infection is associated with inhalation of other living or dead microbes, e.g. from humidifiers in modern air-conditioning systems.[3] Industries which have experienced this problem are, in the main, those where organic dust is available to contaminate humidifier systems and provide growth nutrients, e.g. wood, paper and textiles.[4]

The control of microbiological contamination in building services systems requires consideration at the design stage, e.g. the selection of suitable filters to control the quantity of particulate matter able to enter the air-handling system and allowing for maintenance access. Preventive measures in operation include[4]:

- Effective and regular maintenance including inspection of plant components at specified intervals and regular, thorough cleaning of humidifier elements and/or cooling-tower packing.
- Appropriate chemical water treatment for cooling towers.
- For humidifiers, a regular schedule of draining and thorough cleaning at intervals of, e.g. 2–3 months. A final rinse with mild disinfectant solution will discourage microbial growth (e.g. commercial quality hypochlorite diluted to give approximately 1,000 ppm available chlorine in water).
- Inspection and, if necessary, cleaning and disinfection of cooling towers at least once annually.

Specific problems involving *Legionella pneumophila* necessitate the services of specialists in water treatment.

Water cooling towers for plant recirculatory water systems pose no other significant hazards, although surprisingly one or two of the large parabolic concrete natural draught type have collapsed in high winds. The addition of water-treatment chemicals may introduce some toxicity problems. In plants with a significant fire hazard the location of fin-fan coolers requires careful consideration, since if left running in a fire they can provide forced draught and accelerate fire escalation.

The water is usually circulated at low pressure and, therefore, on chemical plants water coolers and condensers often have to be provided with pressure relief on the water side for protection against a burst tube. Alternatively, the water may exit the heat exchanger discharge to atmosphere and, after collection in a tundish, flow by gravity to a cooling water sump or pond from where it is pumped to the tower; this solution is only practicable if process gas or liquid may be discharged to atmosphere near the heat exchanger without serious effects.[5]

Continuing operation of the cooling-water system is a prerequisite for safe operation of many chemical and process plants. For example in petrochemical plants cooling-water failure usually creates the largest process vapour/gas relief loads of all emergency conditions.[6] Thus consideration has to be given to inbuilt redundancy, e.g. stand-by pumps, or to emergency cooling systems and to provision for rapid plant feed reduction.

Distillation

Distillation may be performed batch-wise or continuously, which is more common for large throughputs. A typical, continuously operated distillation column and ancillaries is shown in Fig. 9.1. Operation may be at atmospheric pressure, under vacuum or under pressure; clearly the potential for ingress of air or emission of vapour, respectively, is greater in the latter cases. In any event with any flammable liquid the flashpoint is not

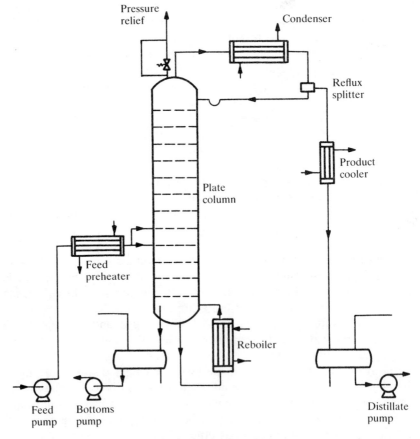

Fig. 9.1 Continuous distillation unit. Ancillaries vary with application; instruments and valves not shown

relevant since this separation depends upon the generation of vapour and generally its fractionation in either a plate or packed column through which it passes counter-current to liquid returned as reflux from an overhead condenser. Thus with flammable liquids distillation is inherently a hazardous operation; its replacement by liquid–liquid extraction may be worth considering – but only if solute and solvent recovery does not involve secondary distillation or evaporation operations. Alternatively, the strategy adopted is to minimise the inventory held up in the column and ancillaries, for example by:

- Elimination of the reflux drum.
- In a distillation column train, elimination of feed preheaters by operation of the upstream column with a partial condenser.
- Elimination of one condenser and one reboiler by combination of two distillation columns into one with a side stream.

The base of a distillation column may contain a large inventory of boiling liquid and several times this amount of liquid may be held up in the column and condenser. The inventory of boiling liquid in a continuous distillation unit may be reduced by using a thermosiphon reboiler in preference to a kettle reboiler. Such steam reboilers, or vaporisers of similar design for process vapour generation, may be prone to pin-hole tube failure; an alternative heat-transfer fluid may be sought if such contamination may give serious consequences.[7]

Consideration may need to be given to the possible effects associated with failure of a tube in the condenser, i.e. contamination of the overheads, and hence the distillate and reflux, with coolant or contamination of the coolant.

It has also been suggested that for pressure stills the selection of packing or trays should take into account the need to minimise inventory. Hence while conventional trays may have a hold-up of 40 mm to 100 mm of liquid, and packed columns 30 mm to 60 mm, per theoretical stage with film trays this reduces to 20 mm.[8] Reduced inventory in the column base can be achieved by using a narrow base, as is done with degradable bottom-products. Low inventory may also be achieved by the use of film-type evaporators.[8]

Provision is generally necessary for emergency action, e.g. isolation of the supply of heat to the reboiler, if coolant to the condenser is stopped or severely restricted. In simple terms energy is introduced into a distillation unit, either batch or continuously operated, via the still or reboiler and if this is not removed at a constant rate control will be lost.

> Hexane vapour was discharged after a loss of cooling water to the condenser of a pot still in which 6,000 litres of contaminated hexane were being distilled. A fire and explosion occurred which killed one man and injured another; the fire spread rapidly to where drums of solvent were stored. The likely source of ignition was the flame from an oil-fired steam

boiler in a nearby room. (The emergency procedure should have included shutting down the steam to the still, turning off all ignition sources and evacuating the building, and 30 minutes apparently elapsed during which this was possible.)

Columns may require vacuum relief as well as the usual pressure relief provisions.

A brewery copper vessel was used for boiling the contents and adjustment of their specific gravity. Control of specific gravity was by manual adjustment of the opening of a deadweight safety valve thus allowing steam to escape. Normal operating pressure was 2.5 psig and over-pressure protection was provided by both a safety valve set at 4 psig and the deadweight safety valve.

When steam was turned off the heater at the end of the boiling stage the pressure in the copper was allowed to fall to atmospheric. As the contents were being drained by opening a 15 cm outlet the copper collapsed.[9]

Accidental entry of water into a still where the temperatures and pressures are such that a large quantity of water immediately vaporises is dangerous (see Ch. 3, page 83) and particularly likely to damage the tray internals of, e.g. vacuum columns of crude oil stills. Injection of a cold water spray into a still full of steam, and not provided with a vacuum relief valve, has been reported; the still collapsed under the effect of external atmospheric pressure (see Ch. 3, page 88).

Relief valves may be located either on the reflux drum, giving a lower temperature release, for venting to a closed system or on the top of the column itself for relief to atmosphere.[10]

Furthermore, it is necessary to consider the effect of raising the mixture to be distilled to the operating temperature and the possible effects of overheating. Temperatures in the reboiler, on selected trays, and on the top trays of a column are commonly used as a basis for automatic control; the reliability of measurement is, therefore, very important.

A steam-heated reboiler on a column distilling an epichlorhydrin–tar mixture overheated resulting in an explosion. The cause was fouling of the thermocouple on the steam control loop requiring the operator to attempt to control the distillation on top column temperature only.[11]

Any potential hazards associated with carryover of contaminants into the column should also be considered.

Crude CMA (4-chloro-2-methylaniline) was being stripped from high boiling residue under vacuum when the still exploded. Three operators were killed and several others injured.

The crude CMA was produced by chlorination of o-toluidine hydrochloride with cuprous chloride as catalyst; this catalyst was recovered by neutralisation and filtration but some was carried with the crude CMA into the still. Subsequent investigation showed that the CMA residue at the distillation temperature would undergo oxidation in contact with air with a marked heat evolution. At the higher temperature achieved CMA would react with cuprous chloride evolving hydrogen chloride gas.

It was concluded that air had leaked into the still via a ball valve on a branch. The residue had a tendency to creep up the walls so that the thermometer was probably exposed and hence indicated an incorrect temperature; the residue also blocked the vapour line so that the hydrogen chloride gas over-pressurised the still.[12]

Consideration always has to be given to any hazards associated with the accumulation of residues in the still, in batch distillation, or on heat transfer surfaces, e.g. in the preheater or reboiler, in continuous distillation.

About 2 litres of tetrahydrofuran, which had been used in an instrument, were being recovered by distillation. This was a routine operation by an experienced technician. An explosion followed by a fire resulted in extensive damage.[13]

The cause of the explosion was an organic peroxide formed from tetrahydrofuran on exposure to air [see Ch. 10]. The original process using the tetrahydrofuran, or exposure to air after the process, is presumed to have caused removal or deactivation of the inhibitor normally present.

Care is also required during purging when residues are present, e.g. residue on packing has been ignited by purging with air.[14]

A concentration gradient is set up along the height of a fractionating column which may have a significant bearing on safety.

In a purification system involving distillation, vinylacetylene present in the feed was found to concentrate in the bottom section of the column. Therefore, continuous monitoring of the maximum vinylacetylene concentration in the column, limiting maximum operating pressure and temperature, etc., were used to ensure safe operation.[15]

It is inherent in any batch distillation that towards the end of the distillation the temperature in the still rises and the concentration of any unstable impurities in the still increases. These have been factors in a number of explosions when relatively stable substances have been distilled under vacuum.[16]

Evaporation

Evaporation refers to the concentration of a solution by vaporisation of the solvent, often water, e.g. in steam-heated vessels sometimes operated in series. The same considerations with regard to pressure and vacuum relief apply as for distillation.

Hazards are those normally associated with pressure vessels plus any which arise from concentration of the solution, e.g. with peroxides, chlorates, etc.

Absorption

In absorption components are separated, and possibly recovered from a gaseous mixture by preferential dissolution in a solvent, e.g. an aqueous solution. When the solute concentration is low as in pollution control applications it is referred to as 'scrubbing'.

Sometimes, for intermittent discharges of relatively small volumes of gas, absorption may be accomplished by simply bubbling it from a sparge pipe into liquid in a vessel. Provision is required for make-up and overflow (in particular for safety the level within the vessel should be automatically controlled or easily checked) and to prevent suck-back. A low liquid-level alarm may be necessary with toxic gases.

Most often spray columns or packed or plate columns are used for absorption with the gas flowing counter-current to the solvent.[12]

Important considerations for safe operation include:
- Monitoring of flow and composition of solvent (e.g. alkalinity of solution for scrubbing acid gases). Clearly if the absorber is relied upon to remove toxic gases from a gas stream prior to discharge to atmosphere, e.g. scrubbing chlorine with alkaline solutions or removing ammonia with water, any failure of the liquid flow should initiate emergency action.
- Monitoring of inlet gas flow and, within the limits of design, composition.
- Monitoring of outlet gas composition.
- Proper selection of materials of construction for contact with solvent–solute mixtures.
- Operation within design conditions, e.g. inlet gas flowrate, composition, temperature and pressure.
- Avoidance of mist carry over in the outlet gas stream, e.g. by the use of knitted-mesh or packed-bed demisters.

Depending on the scale and application, alarms may be required to operate in consequence of abnormal conditions of the monitored variables. Control instrumentation and operating procedures should guard against shock loading of the solute in the gas phase and fluctuations in liquid flowrate.

Liquid–liquid extraction

The range of equipment utilised in liquid–liquid extraction processes includes columns, some with provision for mechanical agitation, mixer-settlers and centrifugal contactors. Selection is based both upon the particular process requirements and tradition, the petroleum-related industries employing continuous gravity-operated columns and the minerals-extraction industries tending to use mixer-settler cascades.

Extraction processes frequently involve the use of diluents or extractants which are flammable. Therefore, since the inventory of these liquids can be significant – particularly if mixer-settlers are preferred to columns, all the precautions discussed in Chapter 4 are necessary. In addition to proper design of the solvent storage and recycle facilities, effective interface control is required on the extraction columns or liquid–liquid settlers.

Because of the phase mixing and separation operations and pumping involved, measures for the elimination of electrostatic charges are particularly important.[17] Toxic hazards are controlled as in Chapter 6 but the design of the extractor may be modified for particularly hazardous materials, e.g. pulsed columns requiring no mechanical seals are preferred for radioactive chemicals processing. Centrifugal extractors may be preferred where minimum liquid hold-up and very efficient phase separation are required.

Generally solute or solvent recovery involves distillation or evaporation so that the overall process includes any hazards from these, as discussed earlier.

Effluent streams may require treatment, e.g. using after-settlers, to ensure removal of dispersed materials; solid adsorbents such as active carbon may be used to remove soluble constituents. Consideration must be given to the effects associated with equilibration of the effluent stream with the atmosphere upon discharge (Ch. 3, page 82).

Gas adsorption

In gas adsorption a material is removed from a gas stream by retention on a specially selected porous solid, e.g. activated carbon or molecular sieves; the bulk carrier gas passes through the bed. The adsorbed material may be recovered in a regeneration cycle involving heating and purging.

When the capacity of the adsorbent is reached 'breakthrough' of solute commences; thus regeneration is required on a time cycle before this occurs. The outlet gas stream may also require monitoring. Adsorption towers may be installed in pairs, so that one is on-stream and one is either on stand-by or undergoing regeneration; a fool-proof system of valve interlocks and purge cycles is therefore essential.

When an adsorber is regenerated the contaminants collected over the extended operating period are released in minutes or hours and account must be taken of this in design and layout, e.g. in air separation plants the regenerating gas stream must be directed away from the air intake.[18] Otherwise entry of regeneration gas with the process air could cause overloading of the plant protection measures and permit unacceptably high concentrations of contaminants in the exit streams. Measures are also necessary to avoid escape of adsorbent in the process stream since combustion reactions have been sensitised by gritty adsorbent materials, e.g. the impact energy to detonate solid acetylene in liquid oxygen is reduced by a factor of 20 in apparatus containing silica gel crystals.

Crushing and grinding

Process applications

Size reduction of solids is carried out in a variety of equipment, from coarse crushers to mills.[19] The main hazards arise from entanglement with machinery (see Ch. 2), dust explosions within the machine or in the building in which it is situated, toxic hazards associated with handling fine powders (see page 192) or from rupture of high speed rotating elements.[19]

Mechanical hazards are minimised by adequate guarding and interlocks together with control of maintenance by a rigid 'permit-to-work' system (see page 975). High speed machinery should be designed with adequate safety margins to cope with foreseeable mis-operation, e.g. admission of material which is either too big, or which has mechanical characteristics outside the range of materials, for which the mill is intended. Entry of ferrous-based tramp metal can generally be prevented by magnetic separators but not all alloys of iron are magnetic.

The general rise in temperature of material as it passes through a mill can be determined; it is generally about 40 °C. Local hot spots may be very much higher in temperature and might act as a source of ignition; static generation and overheating of bearings may also be a problem. Any internal dust explosion can under certain circumstances lead to a secondary explosion. As discussed in Chapter 4 the precautions include elimination of tramp metal, earthing and bonding, adequate maintenance, provision of explosion venting and possibly suppression, good housekeeping, and in some cases operation under an inert gas blanket.

> Titanium carbide had been milled in a 0.6-mm, conical, ball mill and after removal of 95% of the solid the mill was being rotated to clean it. Ignition of a dust cloud resulted in fatal burns to the operator. The solid had apparently been milled more finely than usual and thus rendered highly reactive.[20]
> (This phenomenon is discussed in Ch. 3, page 117.)

Grinding and sieving

Similar hazards arise in grinding and in size separation, i.e. sieving. Thus any such operation which generates dust which may explode if ignited should be subjected to all practicable measures to prevent an explosion,

Table 9.2 Charge generation during powder processing

Operation	Charge per kg of powder (C)
Sieving	2×10^{-9} to 2×10^{-11}
Pouring	2×10^{-7} to 2×10^{-9}
Scroll feed transfer	2×10^{-6} to 2×10^{-8}
Grinding	2×10^{-6} to 2×10^{-7}
Micronising	2×10^{-3} to 2×10^{-7}

including enclosure of the plant, the exclusion or enclosure of ignition sources, and the removal of dust and the prevention of accumulations.[21] An inert gas blanket may be applied above particularly sensitive, combustible solids. Efficient earthing and bonding is essential to avoid static discharges.

In this respect, typical amounts of charge generated on a range of powders in a variety of processing operations are summarised in Table 9.2.[22] The quantity of charge depends more on the amount of work done on the powder than upon its chemical nature. Pneumatic transfer also generates very high levels of charge; these can reach surface densities of 5×10^{-6} to 1×10^{-5} C cm^{-2} at which spark discharges may occur between the particles and the plant.

> An operator was pouring charcoal impregnated with catalyst into a reactor from a plastic bag. The contents of the reactor were blanketed with nitrogen but the charcoal dust ignited, apparently due to a static discharge, and resulted in a flash fire.[23]

A similar example is given in Chapter 4, page 144.

Screens or magnetic separators and/or metal detectors may be used to prevent the ingress of tramp metal, nuts and bolts, etc. Special precautions are necessary with magnesium including provision for interception and removal of dust by scrubbing with water.[24] If the plant cannot be constructed to withstand the pressure generated by an internal dust explosion then means must be applied to restrict its spread, e.g. the provision of adequately sized explosion reliefs, chokes and baffles.[21,25]

The grinding of metals or metal articles using a grindstone or abrasive wheel generally requires special provisions for local exhaust ventilation[26] as does the cleaning of castings.[27] Similarly, measures are required to minimise the escape of toxic or irritant dust from the grinding of chemicals; for example all mills and screens for grinding and screening anhydrous lime, and the conveyor associated with them, have to be enclosed.[28]

The safe operation of all equipment producing dust involves consideration of the associated dust extraction and recovery plant, its efficiency (see page 328) and maintenance, and its continuous use.

Drying

Hazards may arise in drying due to the formation of a toxic or flammable dust concentration, or from the release of toxic or flammable vapours, e.g. from solvents, or due to overheating of solid. Types of drier and methods of reducing dust dispersion are discussed in Chapter 6.

With regard to fire and explosion hazards only, material to be dried can be classified (based on chemical structure and laboratory test results) into[29]:

1. Materials defined as deflagrating or detonating explosives, e.g. by Home Office tests in the UK.
2. Materials which when heated decompose exothermally with evolution of large volumes of gas, even in the absence of air.
3. Materials which undergo exothermic oxidation when heated in air.
4. Materials which do not exhibit either exothermic decomposition or exothermic oxidation when heated in air.

Precautions specific to the explosives industry are applicable to materials in (1) which are unsuitable for general purpose driers. Dependent on the amount of gas evolved and the decomposition temperature, drying of materials in (2) may also be hazardous; this applies even if drying is performed under vacuum or under inert gas. Material in type (4) presents a hazard only if it contains a flammable solvent, of if heating is by direct gas or oil firing. In general material in (3) can be dried safely by the application of heat given the correct choice of operating conditions, the rigorous exclusion of ignition sources, and the use of appropriate measures for explosion prevention and/or protection.

The potential hazards in a drier handling particulate materials may include[30]:

- Ignition of a dust cloud – resulting in a dust explosion in a confined space (see Ch. 4) or a flash fire in the open. The maximum air temperature in the drier should hence be set safely below the minimum ignition temperature from tests on a sample of the finest, driest dust likely to be encountered. Special handling precautions may be needed if the material is shown by standard tests to be particularly susceptible to spark ignition; similar tests will indicate the minimum oxygen concentration for ignition – if inerting is to be used.
- Ignition of a dust layer or deposit – because powder may be present as a layer (e.g. tray or conveyor driers) or may build up on the internal surfaces during operation (e.g. spray driers). Ignition may occur due to self-heating (see page 148); the temperature at which this may occur can be deduced, by extrapolation, from laboratory tests and hence an estimate can be made of the frequency of cleaning required.[30] In fact the initial result of ignition is generally smouldering which may not escalate to flame.[29] Evidently layers may also be ignited by sparks, burning fragments or hot surfaces.

- Ignition of bulk powder. Ignition by self-heating may also occur in those driers in which powder is present in bulk. The minimum ignition temperature is lower than for layers; it decreases with increasing dimensions. The self-heating effect accelerates slower than in layers and becomes apparent after a longer period.

  Discharge of product at too high a temperature may initiate self-heating in a bulk receiver. A product cooler may be required between drier and bulk container.

- Ignition of flammable vapour. When air drying material containing flammable solvent the drying rate and air flow must be such as to maintain the vapour-in-air concentration well below the lower flammable limit, e.g. the concentration in the exit air should be <25% of the LEL at room temperature. This precaution may make drying with inert gas more advantageous economically.

  Hybrid mixtures of combustible dust and flammable vapour present increased hazards since the explosion coefficients of the dust are considerably increased even if the vapour content is below the LEL; a hybrid explosive mixture can be produced by combination of non-explosive concentrations of dust–air and vapour–air. Low energy sparks which do not normally ignite dust–air mixtures can ignite hybrid mixtures. A number of explosions in fluid bed driers and granulators have occurred with hybrid mixtures.[31]

- Ignition of an accumulation of fuel – applicable to direct-fired driers only. Special care is required since an ignition source is always present. Purging cycles are essential on start-up and shut-down; these are normally by cycle-timers and interlocks. Flame failure devices are essential.

Mixing

Any special hazard associated with mixing, a widespread operation in all manufacturing processes, depends primarily on the phases and their inherent properties; also there are the normal mechanical hazards associated with power-driven machinery.

Liquid-liquid mixing

This is conventionally performed in either open or sealed vessels with a power-driven agitator. The agitator design depends on liquid viscosity and the process, e.g. dispersion, reaction, heat removal, dissolution, or a combination of processes.[32] A typical agitated reactor is illustrated in Fig. 9.2. It is important to ensure by an appropriate system of work reinforced by instrumentation/alarms that the agitator is, in fact, on when required during charging if layering of reactants (see page 99) or

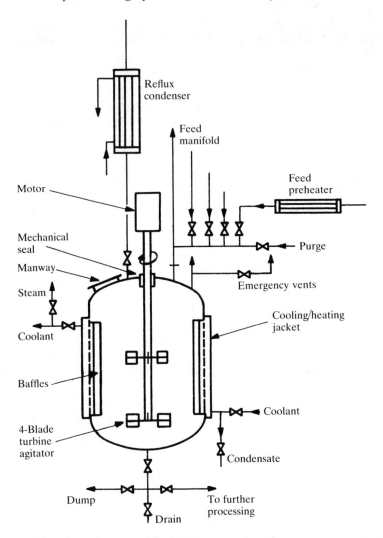

Fig. 9.2 Agitated reaction vessel (instruments not shown)

temporary crust formation can subsequently result in a hazardous reaction.

In a routine batch reactor operation chlorosulphonic acid had to be charged to a kettle followed by 98% sulphuric acid. An operator failed to switch on the agitator prior to charging the sulphuric acid; when this was noticed the agitator was started.

The chlorosulphonic acid had formed a layer over the sulphuric acid and, when mixed, hydrogen chloride gas was produced spontaneously and ejected acid out of the manway.[33]

In cases where agitation is necessary for effective heat removal via a jacket or coil (see page 93) the consequences of failure, e.g. due to a mechanical or electrical fault or power failure, must be allowed for in design. Any associated process hazards must also be considered.

> In suspension polymerisation, in which a dispersion of a monomer in water is polymerised to produce beads of, e.g. <1 mm diameter, failure of the agitator may allow coalescence of the monomer droplets leading to layering followed by exothermic bulk polymerisation in part of the reactor.

Clearly an agitator must be sized to cater for the whole range of viscosities likely to be encountered without being overloaded – a particular hazard with undersized, or magnetic, stirrers in laboratory operations.

A change in the technique used for mixing may sometimes result in more reliable and maintainable equipment, e.g.:
- Replacement of stirred vessels by in-line mixers.
- Use of a circulating pump and loop instead of a stirrer in reactors; this may be used with an external heat exchanger as in Fig. 12.15. The main advantage of this arrangement is that it can provide a much larger provision for removal of heat of reaction for processes scaled-up from a pilot plant.[34]
- Pre-mixing of difficult-to-disperse/dissolve constituents in 'carrier' liquid in a high-speed mixer prior to addition to the main process vessel.
  (This master-batching technique is also referred to for localising toxic hazards in Ch. 6.)

As with continuous reactors (Ch. 10) the use of continuous mixing may reduce the inventory of hazardous materials in process. For example glycerine and acid in nitroglycerine manufacture may be mixed in a small injector; the glycerine is drawn in through a side arm by a partial vacuum created by injection of a high velocity jet of acid. Such an arrangement is self-regulating.[8]

Gas–liquid mixing

It is sometimes possible to inject the gas via a sparger into either a vessel or column and dispense with mechanical agitation. Evidently if the liquid is flammable, use of air may produce either a flammable vapour–air mixture, or a flammable mist, or a flammable froth above the liquid interface. If mixing is performed continuously then appropriate safeguards, such as low-flow or low-pressure alarms or automatic trips, may be required on all flow lines. Appropriate precautions are necessary against static generation. If liquid is dispersed in gas, e.g. via an atomiser, then a toxic or flammable aerosol may be formed; appropriate safeguards for oil-fired burners are summarised in Table 12.28.

Solid–liquid mixing

This may be performed in agitated vessels or heavy-duty equipment, e.g. paste mixers, dough kneaders, extruders, dependent upon the phase ratio and application. If heavy-duty mixing is involved then it is essential to remove any hard extraneous matter, such as tramp metal, first. In solids dispersion/dissolution in stirred tanks the consequences of a build-up of solid on the walls, or blockage of outlet pipelines, must be considered.

Solid–solid mixing

The equipment is invariably of a heavy-duty design so that mechanical hazards predominate. If the solids are combustible then the dust explosion hazard must be minimised, e.g. by operation in an inert atmosphere, by the provision of explosion relief panels, and by the elimination of ignition sources – expecially static generation or overheated bearings. Tramp metal or 'hard' solids should be removed first by screening; magnetic separation, hand-sorting, etc.

Gas–gas mixing

Gas–gas mixing, e.g. in combustion chambers, is relatively easy to achieve effectively. For safe operation attention must be given to ensuring that the respective gas flows are established in the correct sequence and ratio, and are maintained in the correct ratio, and that they only continue so long as other process conditions are met. For example the start-up of an industrial gas-fired appliance requires establishment of the air flow *and* a pilot flame before main gas flow.

Leaching

Leaching, i.e. the extraction of a soluble solid or liquid (termed the solute) from a mixture of solids using a solvent, does not involve any special hazards other than those arising from the inherent properties of the solids and solvents involved, and from the size reduction/mechanically agitated equipment required.

> A flammable solvent was used in an extraction process. When a tank containing extracted material in solution split a seam and allowed liquid to pour over the floor of the room it was also able to spread through doorways and through floor openings to lower levels of the factory (i.e. there were no kerbs or sills to contain the spillage). In an ensuing explosion 12 people were killed.[35]

Large inventories of solid, solvent and leachate are often present, e.g. in natural oils extraction or metal ores processing.[35] Evidently selection of a non-degradable low-toxicity, low-flammability solvent is desirable.

Solute recovery may involve evaporation, or distillation, and the leached material will inevitably retain some solvent with implications for safety.

Crystallisation

Hazards in crystallisation, i.e. the recovery of a crystalline solid from a saturated or supersaturated solution, are also essentially those arising from the materials handled. For example, highly flammable solvents, such as ethers, are frequently used in laboratory purifications but not often on a production scale.

The operation is a simple one, utilising an agitated equipment often with provision for cooling and with means for easy removal of crystals.[36] Overspill of solution may inevitably result in an accumulation of solids on external equipment, surfaces and the surrounding areas with hazards dependent on the specific properties.

Filtration

Industrial filtration operations involve the recovery of solids from a suspension, or production of a purified filtrate, by passage through a porous media, e.g. a filter cloth. Equipment may vary from a pressure filter to a rotary vacuum filter but in all cases the solid is recovered as a cake. Operation may be continuous, or intermittent to allow for cake discharge.

Hazards may be inherent in the concentration of the solid, e.g. if it is toxic or combustible, or in the use of a flammable solvent. Due care must be taken to cope with the hydraulic pressures necessary and with the effects following any failure of the media, e.g. bursting of a filter cloth in a plate and frame press allowing unfiltered suspension to pass through. Adequate guards or safety cut-outs are necessary if the parts of the filter are closed by a hydraulic mechanism.

Smaller filters are also used on plants, e.g. in transfer lines, to keep hazardous materials away from sensitive parts of the system. If the material being filtered out is reactive under plant conditions then care is necessary in the design and positioning of the filter since the concentration of the material in the filter body is inherently greater than in the feed or filtrate.[18] The consequences of complete blockage of the filter must also be considered, i.e. by provision for low-flow indication/alarm and for bypassing/changeover cleaning.

Centrifugation

Centrifugation may involve the separation of solids from flammable liquids; this may result in flammable vapours accumulating in, or being released from, the equipment. Alternatively, the hazard may be due to toxic vapour release, or exposure to process materials, or to machinery in motion. Mechanical failure could result in the ejection of missiles or heat generation due to friction.

> The loaded basket of a 122 cm suspended-type centrifuge became unbalanced because of a sudden escape of cake from one side of the basket, due to a hole in the cloth. The shaft was thrown out and ruptured the outlet pipe on a second centrifuge.[37]

Production centrifuges are either batch operated, with a solid or perforated bowl from which the filter cake is removed manually, or continuous (i.e. peelers, pushers, decanters, conical bowl or clarifiers/separators). In general, because of the large number of manipulative operations, the potentials for flammable vapour formation and release and spark creation are greater with batch operation; the mechanical and material-exposure hazards are also greater.[38]

With centrifuges handling flammable solvents, e.g. in the fine chemical and pharmaceutical industries, explosion prevention is an important consideration. In addition to regular maintenance, good operating practice, elimination so far as is practicable of all ignition sources and selection of the least volatile/flammable solvent appropriate for the process, the commonest protective systems rely upon purging and blanketing with inert gas (see Ch. 4, page 132). Purging, e.g. with nitrogen, takes place with the lid closed and consumes about three times the volume of the centrifuge.[39] An interlock system permits the motor to start only after a set period of time or when the oxygen concentration is below a critical level. Blanketing may then be achieved by one of the following methods[39]:

1. Maintaining a slight positive pressure within the centrifuge and monitoring the oxygen concentration inside continuously.
2. Maintaining a slight positive pressure within the centrifuge, e.g. above 25 mm water gauge, from a non-shared nitrogen supply and relying upon this and the system design to prevent air entering. Automatic shut-down is provided if the pressure falls.
3. Maintaining a continuous flow of nitrogen through the centrifuge to dilute small air leaks, the pressure in the machine being virtually atmospheric.

The features of system (1) are illustrated in Fig. 9.3. The design criteria are:
- Continuous monitoring of oxygen level inside the centrifuge.
- Isolation of the electrical supply to the motor until the oxygen level has been reduced to < 2% by the initial nitrogen purge.
- If the oxygen level rises to 5% the nitrogen gas purge rate is doubled, audible and visible alarms are initiated, and slurry feed and wash valve to the centrifuge are closed; these valves re-open once the oxygen level has fallen to 2%.

Safety with chemical engineering operations

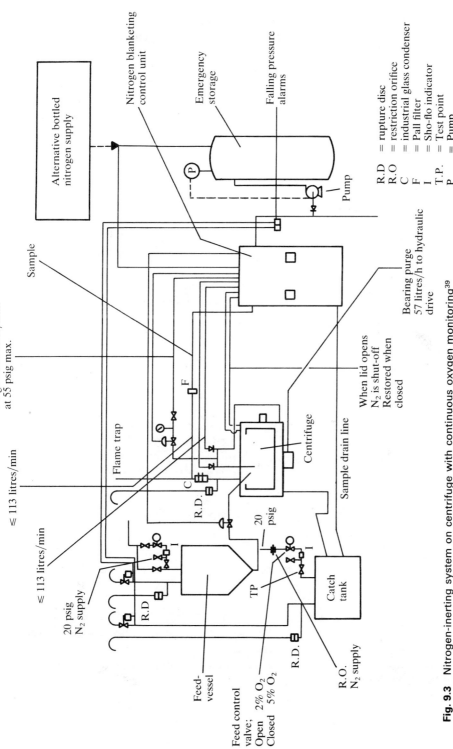

Fig. 9.3 Nitrogen-inerting system on centrifuge with continuous oxygen monitoring[39]

- If the oxygen level rises to 8% nitrogen is admitted from an emergency supply at a high flowrate and the bowl is halted by non-friction braking.
- If the oxygen analyser fails alarms are triggered.
- If the sample flow to the oxygen analyser stops, the centrifuge is shut down.
- The interlock on the centrifuge lid prevents the nitrogen purge cycle commencing until it is closed.
- An emergency supply of nitrogen at adequate pressure is maintained for each centrifuge.

Pumps and pumping

The most common pump is of the centrifugal type. Obstruction of the feed flow may result in cavitation and mechanical damage; possible causes are improperly opened valves on the suction side, vapour locking in the feed line due to temperature exceeding design conditions, partial blockage of the feed line due to fouling or blockage of a line filter.

Although they are commonly controlled by the outlet valves, problems can arise if a centrifugal pump is run 'dead-headed', i.e. against a closed delivery valve. If the delivery valve on a centrifugal pump is closed, keeping the pump full of liquid, but the pump is left running the transfer of energy will cause the temperature in the confines of the pump to rise; this can result in the production of a superheated liquid and excessive pressure. If there is no blockage in the suction line the pressure in the pump will cause a slip back to the suction line before the pump casing bursts; dry running of the pump then follows. Slip-back may not occur early enough to safeguard the pump in the case of multi-staged pumps. Dry running will generally only cause damage to the mechanical seal but this results in costly replacement.[40]

If both the inlet and outlet valves are closed a damaged pump will result, ranging from stretching of the flange bolts, to provide pressure relief, to rupture of the pump casing and destruction of the motor. Since the pressure is generated by a superheated liquid it will not be relieved by a small volume increase as with cold hydraulic testing.

In such cases the hazard is increased if the liquid involved is flammable since it will become highly flammable, and possibly prone to auto-ignition, on superheating. In any event there are numerous case histories of accidents associated with the above phenomena[40] and one was described on page 147. Protective measures may involve some combination of[40]:

- Thermal cut-out of the motor based on pump temperature.
- Provision of a pressure-relief valve set just above pump deadhead pressure.

- Provision of motorised delivery valves that open on pump start-up.
- Provision of a spill back line sink, e.g. a small line back from the delivery side to the feed storage tank.
- Provision of kick-back coolers.

If the expense of such engineered protection is not justified and the pump is small then reliance may be placed on a proper operating procedure, i.e. including a check on all valves before start-up.

Where the leakage hazard justifies it, and where it is feasible to immerse the complete pump and motor into the process fluid, a canned pump may be specified. Such a pump has no seals and, therefore, in theory, zero leakage but will be limited in pressure and flowrate. Furthermore[41]:

- The process fluid provides lubrication for the motor coil, etc., and therefore must be cleaned.
- The service factor is low and the whole line must be opened for repair; hence a flushing system may be required.

The diaphragm pump is an alternative seal-less pump but similarly can only be used for small pressures and flowrates.

Positive displacement pumps, e.g. gear or reciprocating pumps, can generate very high pressures and if pumping against a closed end may cause rupture of pipework or the pump casing. Therefore a properly adjusted and maintained pressure relief valve is required on the outlet, prior to any isolation valve.

Sometimes transfer of liquids is achieved by pressurisation with air, or in the case of flammable liquids, inert gas. The normal precautions are then necessary to prevent over-pressurisation (see Ch. 3, page 79). The receiver vessel may either be vented, in which case consideration has to be given to the effects of displaced gas/vapour, or connected back to the feed vessel by a balance line – in which case some venting will still ultimately be necessary.

In all cases the transfer vessels must be adequate to withstand the pressure, or vacuum, which can be attained.

> Two men sustained severe acid burns on the lower portions of their bodies and legs, and one also suffered a broken hip, when a 9-litre bottle of concentrated sulphuric acid exploded. They were engaged on transferring a mixture of silver sulphate and sulphuric acid from the bottle to a burette by the application of nitrogen pressure from a 75 psig supply. Tubing was butted from the nitrogen line against the nozzle of a rubber aspirator bulb previously attached to the bottle. This involved one man standing on the laboratory table. After some flow was obtained the explosion occurred.[42]

Normally manual transfer of hazardous materials from breakable bottles or carboys involves siphons, so that neither pressure nor vacuum is applied.[42]

Heat transfer operations

Process heat transfer systems cover a range of applications, e.g. heating, cooling, refrigeration, emergency cooling.

Steam is used predominantly for process heating duties but heat transfer fluids, or electrical heating, may be used as an alternative. Cooling is conventionally achieved by use of air (i.e. fin-fan) coolers, cooling water, or re-circulation of a solution, e.g. brine or glycol or by direct refrigeration. Procedures for the design of a wide range of such equipment, including heat exchangers, fired process heaters, internal coils/external jackets, reboilers/condensers are well established.[43-45]

Characteristics and applications of heat-transfer fluids are summarised in Table 9.3.[46] Clearly, for low temperature operation, the avoidance of

Table 9.3 Characteristics and applications of heat-transfer fluids

Temperature range	Fluid	Typical applications
−50 °C to +250 °C	Synthetic fluids for low temperatures (e.g. Marlotherm L)	Pharmaceuticals Fine organic chemicals
0 °C to 300 °C	Heat-transfer oils (e.g. BP Transcal 65, Mobiltherm 605, etc)	Glycol regeneration Platform heating systems Synthetic resin production Textile processing Drier heating Distillation Storage vessel heating
0 °C to 340 °C	High-temperature synthetic fluids (e.g. Santotherm 66)	Organic chemical production Polyester fibre production Alkyd resin manufacture Drying oven heating Powder drying Polystyrene manufacture
20 °C to 400 °C	Diphenyl/diphenyl oxide (e.g. ICI Thermex)	Nylon production Cumene production Dimethyl terephthalate Polyester fibre production Printing ink manufacture
150 °C to over 500 °C	Heat-transfer salts (e.g. Cassel 155)	Caprolactam production Catalytic converters Organic oxidations Sulphur distillation Caustic soda evaporation Alkyl amine converters Nuclear energy applications Waste heat recovery Melamine production Ammonoxidation reactions, e.g. acrylonitrile manufacture

freezing and of inadequate cooling capacity due to 'slush' formation or increased viscosity or crystallisation, are prime safety considerations. Consideration must also be given to conditions when the cooling/heating system is routinely shut down.

> A flow of cooling water to the tubes of a cooler containing liquid propylene was isolated during shut-down. The pressure on the propylene side was reduced and evaporation produced cooling (e.g. the temperature of liquid propylene will approach −47 °C at 1 bar). The water froze in the tubes and seven bolts in the 'floating-head' fractured. On start-up, propylene entered the cooling water system and the gas released from a pressure blow-out was ignited resulting in a serious fire.

The comparative hazards of common heating media are summarised in Table 9.4.[46] However, it follows from the discussion in earlier chapters that operation at high temperatures introduces an inherent hazard. For example contact of water, or other volatile liquid, with hot oil or molten salt may cause 'steam' explosions. Degradation and cracking of mineral oils over a period of operation may result in lowering of the flashpoint; hence there is a need for regular checking/changing of the oil.

Sometimes flammable hydrocarbons or ethers are used as heat transfer media at pressures above atmospheric. This adds to the inventory of flammables on site, indeed in some cases the inventory as 'service' fluids

Table 9.4 Some comparative hazards of steam, heat-transfer oil and high-temperature salt

Steam	Heat-transfer oil	High-temperature salt
High pressure at any reasonable temperature, e.g. 300 psig at 200 °C 1500 psig at 300 °C	Expansion tank at atmospheric pressure	Expansion tank at atmospheric pressure
High-energy release on system failure	Negligible	Negligible
Water/condensate can freeze	Diphenyl – diphenyl oxide freezes at −5 °C. Synthetic fluids can withstand temperatures down to −50 °C	Freezes at 160 °C. Salt dilution system can operate down to +65 °C
Scale forms, condensate corrodes and water treatment is required	Non-corrosive	Non-corrosive
Non-flammable	Flash points >200 °C	Non-flammable
Non degradable	Life >2 years	Life >5 years
Non toxic	Very low toxicity	Very low toxicity

exceeds that in the reaction/separation chain. It has been suggested that use of a higher boiling liquid at a lower pressure, or water albeit at high pressures, could be advantageous from safety considerations. Sometimes more reliable and maintainable heat transfer equipment can be provided by substitution, e.g.:
- Use of external heat exchangers for callandrias and/or limpet coils.
- Use of exchangers with internal hairpin tube bundles for floating head exchangers (to account for thermal expansion and contraction and shell-side cleaning requirements).

While following the general design principles summarised in Chapter 12 and the maintenance strategies mentioned in Chapter 17, safety with process heaters involves special considerations. A check-list for eliminating problems is given in Table 9.5 (after ref. 46).

One common use of heat transfer is condensation of a vapour in a condenser, the reverse of vaporisation. This is not an inherently hazardous operation since enthalpy is removed (unless failure of coolant flow occurs) but consideration should be given to any effects associated with the fractionation between components in the vapour phase.

Oxygen condenses to liquid at − 183 °C and nitrogen at − 196 °C at atmospheric pressure. Thus the constituents of air can condense on to outside surfaces of any inadequately insulated pipework, or equipment containing liquid nitrogen itself, or on to surfaces that have been cooled by immersion in it. On warming up, nitrogen evaporates first thus resulting in local oxygen enrichment.[47]

Condensation efficiency also falls off dramatically if proportions of 'non-condensibles' (e.g. inert gas) are present; clearly this has important safety implications.

Insulation is generally applied to piping or equipment at high or low temperatures to minimise heat loss/gain. However, some hot surfaces are left exposed to assist cooling. In this case there are limiting surface

Table 9.5 Check-list for process heater system safety

Radiant fired heaters	• Dead zones	For example on heat-transfer oil side of process coil due to poor flow control. Ensure constant, controlled, turbulent flow along a known and predictable flowpath.
	• Thermal degradation	Due to overheating caused by dead zones and excess flux. Design for low, evenly distributed flux.
	• Coil failure	Due to overheating or mechanical fatigue. Specify robust design, avoiding bellows and differential expansion. Allow for coil flexibility.

Table 9.5 (continued)

	• Flame impingement	Allow sufficient space in the radiation zone and adequate clearance between tubes and burner centre line.
	• Life of pressure parts	Specify higher pressure than essential, e.g. 1,000 psi to increase coil life.
	• Flow variations	May cause overheating therefore locate pump prior to heater.
	• Combustion chamber fires	May result from oil spillages. Extinguishment practicable by steam or CO_2 snuffing provided connections allowed.
Burners	• Misfiring • Poor efficiency • Lazy flame • Oil spills	May result from inadequate maintenance. Regular servicing, and replacement of parts by spares, necessary.
Pumps	• Seal failures	Careful commissioning and clean-out of system prior to full-load operation. Glandless pump specified for $> 300\ °C$. Regular maintenance. Provision of adequate instrumentation on pump and header tank, and of means to drain away leaks safely.
Vessels	• Pressure surges in expansion tanks	Sudden increases may arise from vaporisation of water from fresh oil or ingress of lower boiling process fluids. Specify expansion tanks to at least BS 5500 Class B.
	• Drainability	Provide storage/dump tank in system. This can serve as emergency drain. Avoid addition of fresh oil to hot oil in system.
Valves	• Leakage	Specify bellows-seal valves. Locate all valves with care.
	• Tampering with control valves	Avoid possibility of tampering with differential control valve in an attempt to increase outlet temp. which may result in damage to oil.
Installation	• Fire damage	Separate heater from pumps and control panel. Provide fire walls. Segregate process piping from high-temperature piping. Provide drains from pump seals and valve boxes.
	• Lagging	Use blown glass, non-adsorbent, insulation at flanged joints or where process fluid spillages may occur (see Ch. 4, page 149).
Control panels and instruments	• Intrinsic safety	Use intrinsically safe instruments on the heater and system (see Ch. 4, page 143). Purge control panel with air or nitrogen and provide interlocks to avoid operation without the purge (see page 132). Install fail-safe features which cannot be 'locked-out'.

temperatures at which protection is advisable to avoid shock or contact by operators, e.g.:

54 °C metallic surfaces } within reach from permanent floor level.
65 °C non-metallic surfaces

45 °C surfaces at higher levels within reach from portable access equipment or ladders.

If these exposed surface temperatures are exceeded they may constitute a hazard to personnel and guarding is recommended. Protection is also required for surface temperatures less than -10 °C.

Basic principles

From the above summary certain basic principles may be deduced for safety with chemical-engineering operations:

- Minimise inventory (continuous operation may be preferable to batch; high efficiency – low residence time equipment may be preferable to less expensive designs).
- Monitor the flow composition and condition (T, P, etc.) of all streams.
- Provide means to remove unacceptable contaminants from the feed materials (e.g. removal of tramp metal from solids processing stages).
- Provide for isolation (where appropriate, remotely-operated) from upstream and downstream units.
- Operate at moderate temperature and pressure (i.e. avoid high pressures or vacuum where practicable; avoid unusually high or low temperatures). Consider the consequences of over, or under, temperature and pressure.
- In chemical-engineering operations involving continuous flow consider (e.g. by a HAZOP – Ch. 12) the effect of no flow, reduced flow, increased flow, contaminated flow, reverse flow.
- Provide for safe start-up (including purging if necessary) and shut-downs (normal, stand-by or emergency).
- Provide for pressure relief and explosion suppression/venting where appropriate.
- Provide safety instrumentation, e.g. high flow/low flow alarms, high temperature/high pressure alarms, high/low level alarms, etc. (in appropriate cases linked to trips for automatic operation) *in addition to* control instrumentation.

Particular hazards may arise on start-up and shut-down. If performed wrongly these may result in fires, explosions, destructive pressure surges or release of materials into the environment. The main hazards to guard against include:

- A mixture of flammable vapour with air (or oxygen or chlorine).
- Contact of water with hot oil or molten salts or molten metal.

- Freezing of residual water or high melting point chemicals in equipment or piping.
- Over-pressurisation or drawing excessive vacuum.
- Mechanical shock, e.g. water hammer.
- Thermal shock.

References

(*All places of publication are London unless otherwise stated*)
1. Austin, G. T., 'Hazards of commercial chemical operations', in H. W. Fawcett & W. S. Wood (eds) *Safety and Accident Prevention in Chemical Operations*. Interscience 1965, p. 92.
2. Department of Health and Social Services, *Legionnaires Disease and Hospital Water Systems*, Health Notice HN(80)39. DHSS.
3. Hill, E. C., Universities Safety Association, *Safety News* (Feb. 1981), 15.
4. Ager, B. P., & Tickner, J. A., *Annals Occup. Hyg.*, 1983, **27**, No. 4, 341–58.
5. Kletz, T. A., *Chem. Engr*, No. 342 (Mar. 1979), 161–4.
6. Barnwell, J., American Instn Chemical Engineers Meeting, Atlanta, 27 Feb. 1978.
7. Lees, F. P., *Loss Prevention in the Process Industries*. Butterworths, 1980.
8. Kletz, T. A., 'Seek intrinsically safe plants', *Fire Protection Manual*, in C. H. Vervalin (ed.) *Hydrocarbon Processing Plants*, Vol. 2. Gulf, Houston, Texas, 1981, 86.
9. Anon., *Vigilance*, 1971 **2**(6), 71–2.
10. Simon, H. & Thomson, S. J., *Chem. Eng. Prog. Loss Prevention*, 1972 (6) 74.
11. Doyle, W. H., *Instrum. Technol.*, **19**(10), 38.
12. Brugger, J. E. & Wilder, I., *Journal of Hazardous Materials*, 1976, 1, 3.
13. Anon., Universities Safety Association, *Safety News*, 8 (Feb. 1977), 15.
14. Ling, K. C., 'Process industry hazards: Accidental release, assessment, containment and control', *Instn Chem. Engrs Symp. Series*, No. 47, 1976, 109–18.
15. Klaasen, P. L., *Instn Chem. Engrs Symp. Series*, No. 34, 1971, 111–24.
16. Anon., *Loss Prevention Bulletin*. Institution of Chemical Engineers, 1977 (013), 16.
17. Scuffham, J. B. & Rowden, E. A., 'Safety and environmental considerations (non-nuclear operation)', in *Handbook of Solvent Extraction*. Wiley 1983, 945–54.
18. Webster, T. J., 'Loss prevention and safety promotion in the process industries', *Proceedings of 2nd International Symposium*, Dechema 1978, 115–23.
19. Marshall, V. C., (ed.), *Comminution*. Institution of Chemical Engineers 1975.
20. Manufacturing Chemists Assoc., *Case History 618*, Vol. 2, 1966.
21. The Factories Act 1961, s. 31. HMSO.
22. Lloyd, F. C., *Fire Prevention Science and Technology*, **21**, 8–12.
23. Manufacturing Chemists Assoc., *Case History 1094*, Vol. 2, 1966.

24. The Magnesium (Grinding of Castings and other Articles) Special Regulations, 1946. HMSO.
25. Department of Employment and Productivity, 'Dust explosions in factories', *Health and Safety at Work*, Booklet 22, 1970. HMSO.
26. The Grinding of Metals (Miscellaneous Industries) Regulations, 1925. HMSO.
27. Iron and Steel Foundries Regulations 1953; Non-ferrous Metals (Melting and Founding) Regulations 1962. HMSO.
28. The Chemical Works Regulations 1922, 3. HMSO.
29. Reay, D., *Loss Prevention Bulletin*. Institution of Chemical Engineers, 1979 (025), 1.
30. I. Chem. E. Working Party on Engineering Practice, *User Guide to Fire and Explosion Hazards in the Drying of Particulate Materials*. Institution of Chemical Engineers 1977.
31. Bartknecht, W., *Chem. Eng. Prog.*, Sept. 1977, 93–105.
32. Uhl, V. W. & Gray, J. B., *Mixing-Theory and Practice*. Academic Press, New York, 1967.
33. Anon., *Loss Prevention Bulletin*. Institution of Chemical Engineers, 1977 (013), 2.
34. Kneale, M. & Foster, G. M., *Instn Chem. Engrs Symp. Series*, No. 25, 1968, 98–107.
35. Matheson, D., *Symposium on Chemical Process Hazards with Special Reference to Plant Design*. Institution of Chemical Engineers, 1960, 1–4.
36. Bamforth, A. W., *Industrial Crystallisation*. George Godwin 1965.
37. *Manufacturing Chemists Assoc., Case History 642*, Vol. 2, 1966.
38. *The Safe Operation of Centrifuges with Particular Reference to Hazardous Atmospheres*. Institution of Chemical Engineers, 1976.
39. Butler, P., *Process Engineering*, July 1974, 52–5.
40. Anon., *Loss Prevention Bulletin*. Institution of Chemical Engineers, 1979 (029), 139.
41. Jones, A. L., 'Industrial Hygiene Input to the Design Process' presented at Conference on Industrial Hygiene, Toxicology and the Design Engineer, London, Mar. 1982.
42. Russell, W. W., *Loss prevention 10*, (*Chem. Eng. Prog. Tech. Manual*). American Instn of Chemical Engineers, 1976.
43. Kern, D. Q., *Process Heat Transfer*. McGraw Hill 1950.
44. Butterworth, D., *Introduction to Heat Transfer*. Oxford University Press 1977.
45. Sinnott, R. K., 'Chemical engineering, in J. M. Coulson & J. F. Richardson *An Introduction to Design*, Vol. 6. Pergamon Press 1983.
46. 'Fluid choice takes the steam out of unsafe process heaters', *Process Eng.*,
47. Locke, B., *Loss Prevention Bulletin*. Institution of Chemical Engineers, 1979 (025), 151.